ETHYLENE DICHLORIDE:
A Potential Health Risk?

Row 1: M. Anders; K. Hooper; F. Spreafico; C. Maltoni.
Row 2: B. Van Duuren; U. Rannug; P. Infante; M. Johnson; R. Reitz.
Row 3: G. Capaldi; E. Jacobs; G. Ter Haar; B. Ames; J. Kariya.
Row 4: J. Fabricant; J. Ward, L. Gold; R. Hinderer.

5
ETHYLENE DICHLORIDE:
A Potential Health Risk?

Edited by
BRUCE AMES
University of California
PETER INFANTE
OSHA
RICHARD REITZ
Dow Chemical Company

RA1242
D45
E74
1980

COLD SPRING HARBOR LABORATORY
1980

BANBURY REPORT SERIES
Banbury Report 1: Assessing Chemical Mutagens
Banbury Report 2: Mammalian Cell Mutagenesis
Banbury Report 3: A Safe Cigarette?
Banbury Report 4: Cancer Incidence in Defined Populations

Banbury Report 5
ETHYLENE DICHLORIDE:
A Potential Health Risk?

© 1980 by Cold Spring Harbor Laboratory
All rights reserved

Printed in the United States of America

Cover design by Emily Harste

Library of Congress Cataloging in Publication Data

Main entry under title:

Ethylene dichloride.
   (Banbury report ; 5)
   Proceedings of a meeting held Nov. 14-17, 1979 at the Banbury Center, Lloyd Harbor, N.Y.
   1. Dichloroethane—Toxicology—Congresses.
I. Ames, Bruce N. II. Infante, Peter, 1941-
III. Reitz, Richard, 1940-   IV. Cold Spring Harbor,
N.Y. Laboratory of Quantitative Biology. V. Series:
Banbury Center. Banbury report ; 5. [DNLM: 1. Hydrocarbons, Chlorinated—Poisoning—Congresses.
2. Occupational diseases—Prevention and control—Congresses. W3 BA19 v. 5 1979 / QV633 E84 1979]

RA1242.D45E74   615.9'511   80-7677
ISBN 0-87969-204-9

Participants

Bruce N. Ames, Department of Biochemistry, University of California, Berkeley
Marion W. Anders, Department of Pharmacology, University of Minnesota, Minneapolis
William M. Busey, Experimental Pathology Laboratories, Inc.
Gene Capaldi, Arco Chemical Division, Atlantic Richfield Company
Jill D. Fabricant, Department of Preventive Medicine and Community Health, University of Texas Medical Branch, Galveston
Lawrence Fishbein, Department of Health, Education, and Welfare, Food and Drug Administration, National Center for Toxicological Research
Lois Swirsky Gold, Department of Biochemistry, University of California, Berkeley
Robert K. Hinderer, B. F. Goodrich Chemical Company
Kim Hooper, Department of Biochemistry, University of California, Berkeley
Peter F. Infante, Office of Carcinogen Identification and Classification, Occupational Safety and Health Administration
Emmett S. Jacobs, Petroleum Laboratory, E. I. du Pont de Nemours & Company, Inc.
Maurice N. Johnson, B. F. Goodrich Company
James Kariya, Office of Toxic Substances, Environmental Protection Agency
C. D. Kary, Shell Oil Company
Robert G. Kellam, Office of Air Quality Planning and Standards, Environmental Protection Agency
Cesare Maltoni, Institute of Oncology and Tumour Center, Bologna, Italy
Patricia B. Marlow, Office of Carcinogen Identification and Classification, Occupational Safety and Health Administration
Victor K. McElheny, Banbury Center, Cold Spring Harbor Laboratory
Jean C. Parker, Office of Environmental Criteria and Assessment, Environmental Protection Agency
Harry B. Plotnick, Division of Biomedical and Behavioral Science, National Institute for Occupational Safety and Health

Ulf Rannug, Environmental Toxiciology Unit, Wallenberg Laboratory, University of Stockholm, Sweden

K. Suryanarayana Rao, Toxicology Research Laboratory, Health and Environmental Sciences U.S.A., Dow Chemical Company

Richard H. Reitz, Toxicology Research Laboratory, Health and Environmental Sciences, Dow Chemical Company

Vincent F. Simmon, Genetic Toxicology Division, Genex Corporation

Federico Spreafico, Istituto di Ricerche Farmacologiche "Mario Negri"

Gary Ter Haar, Toxicology and Industrial Hygiene, Ethyl Corporation

Benjamin L. Van Duuren, Laboratory of Organic Chemistry and Carcinogenesis, Institute of Environmental Medicine, New York University Medical Center

Jerrold M. Ward, National Cancer Institute, National Institutes of Health

Preface

The announcement in September 1978 that ethylene dichloride, a major chemical industry intermediate, was a carcinogen for tube-fed laboratory animals in tests conducted under the bioassay program of the National Cancer Institute prompted the Banbury Center to embark on organizing a conference at which the potential carcinogenicity for humans of ethylene dichloride and related halogenated hydrocarbons could be explored.

Interest in ethylene dichloride as an urgent subject already had been quickened by Dr. Bruce Ames of the University of California at Berkeley, who had been describing the identification of this compound, of which some 10 billion pounds are manufactured each year in the United States alone, as a mutagen by means of the *Salmonella* revertance test he pioneered. Dr. Ames was centrally involved in Cold Spring Harbor Laboratory's 1976 conference on origins of human cancer, whose proceedings were published in January 1978.

In putting together this conference, Banbury Center enlisted the help of scientists in government and industry as well as academic research, in line with our goal of providing both balanced and timely reports on urgent technical problems of environmental health risk assessment. Thanks to the advice of Dr. John Burns, vice-president for research of Hoffmann-LaRoche Inc., we were able to contact Dr. Julius E. Johnson, recently retired vice-president for research of Dow Chemical Company, a leading manufacturer of ethylene dichloride. Dr. Johnson generously discussed our proposal of a meeting with colleagues at Dow Chemical, who helped identify such scientists as Dr. Richard Reitz and Dr. K. S. Rao who were close to the scientific problems and could help organize and participate in our conference.

From the U.S. government, we were able to call on the able help of Dr. Peter Infante of the Occupational Safety and Health Administration of the Labor Department, a participant not only in the Banbury Center conference on assessing chemical mutagens, published as Banbury Report 1 in April 1979, but also in the conference on origins of human cancer in 1976.

Through Dr. Ames and staff of the Chemical Manufacturers Association, we learned of the work of Dr. Cesare Maltoni and Dr. Federico Spreafico in examining the carcinogenicity of ethylene dichloride in lifetime inhalation experiments similar to those conducted on vinyl chloride. Whereas vinyl chloride has shown up as a carcinogen down to values of 10 parts per million in the Italian inhalation studies, ethylene dichloride seemed not to be producing a significant excess of tumors by this route. Thus, this conference gave promise of an exploration of the apparently conflicting results of the tube-feeding and inhalation experiments. And so, I believe, it turned out.

This was the fifth Banbury conference in a year and a half, and the third in a period of a few weeks. Throughout the organization and holding of this conference, as with previous ones, we benefited from the tireless effort of Beatrice Toliver, administrative assistant of Banbury Center, and we thank her warmly. It is a pleasure once again to express thanks and admiration for the prompt and efficient preparation of this book for publication by the Banbury editorial staff of Lynda Moran and Judith Cuddihy, editors, Kathleen Kennedy, editorial assistant, and Barbara Cowley-Durst, our free-lance editor.

Victor K. McElheny

Contents

Preface / Victor K. McElheny

SESSION 1: MUTAGENICITY AND CARCINOGENICITY OF ETHYLENE DICHLORIDE

Long-Term Carcinogenic Bioassays on Ethylene Dichloride Administered by Inhalation to Rats and Mice / Cesare Maltoni, Loretta Valgimigli, and Corrado Scarnato — 3

The Carcinogenicity of Ethylene Dichloride in Osborne-Mendel Rats and B6C3F1 Mice / Jerrold M. Ward — 35

Carcinogenic Potency: A Progress Report / Bruce N. Ames, Kim Hooper, Charles B. Sawyer, Richard Peto, William Havender, Lois S. Gold, Robert H. Harris, and Margaret Rosenfeld — 55

The Carcinogenic Potency of Ethylene Dichloride in Two Animal Bioassays: A Comparison of Inhalation and Gavage Studies / Kim Hooper, Lois Swirsky Gold, and Bruce N. Ames — 65

The Use of Different Metabolizing Systems in the Elucidation of the Mutagenic Effects of Ethylene Dichloride in *Salmonella* / Ulf Rannug — 83

Review of Nonbacterial Tests of the Genotoxic Activity of Ethylene Dichloride / Vincent F. Simmon — 97

SESSION 2: TOXICOLOGY AND OTHER TOPICS

Pharmacokinetics of Ethylene Dichloride in Rats Treated by Different Routes and Its Long-Term Inhalatory Toxicity / Federico Spreafico, Ettore Zuccato, Franca Marcucci, Marina Sironi, Silvio Paglialunga, Margherita Madonna, and Emilio Mussini — 107

Pharmacokinetics and Macromolecular Interactions of Ethylene
   Dichloride: Comparison of Oral and Inhalation Exposures /
   Richard H. Reitz, Tony R. Fox, Jeanne Y. Domoradzki, John
   F. Quast, Pat Langvardt, and Philip G. Watanabe   135

Teratogenicity and Reproduction Studies in Animals Inhaling Ethylene
   Dichloride / K. Suryanarayana Rao, Janis S. Murray, Mary M.
   Deacon, Jacqueline A. John, Linda L. Calhoun, and John T. Young   149

An Investigation of Possible Sterility and Health Effects from Exposure
   to Ethylene Dibromide / Gary Ter Haar   167

Carcinogenicity and Metabolism of Some Halogenated Olefinic and
   Aliphatic Hydrocarbons / Benjamin L. Van Duuren   189

SESSION 3: USES OF ETHYLENE DICHLORIDE: WORKER EXPOSURE

Human Exposures to Ethylene Dichloride / Lois Swirksy Gold   209

Production, Uses, and Environmental Fate of Ethylene Dichloride
   and Ethylene Dibromide / Lawrence Fishbein   227

Use and Air Quality Impact of Ethylene Dichloride and Ethylene
   Dibromide Scavengers in Leaded Gasoline / Emmett S. Jacobs   239

Medical Aspects of Ethylene Dichloride in the Workplace /
   Maurice N. Johnson   257

Human Exposure to Ethylene Dichloride: Potential for Regulation
   via EPA's Proposed Airborne Carcinogen Policy / Robert G.
   Kellam and Michael G. Dusetzina   265

SESSION 4: RELATED CHEMICALS

Dietary Disulfiram Enhancement of the Toxicity of Ethylene
   Dibromide / Harry B. Plotnick, Walter W. Weigel, Donald
   E. Richard, Kenneth L. Cheever, and Choudari Kommineni   279

Evidence for the Carcinogenicity of Selected Halogenated
   Hydrocarbons Including Ethylene Dichloride / Peter F.
   Infante and Patricia B. Marlow   287

Evidence of the Mutagenicity of Ethylene Dichloride and
   Structurally Related Compounds / Jill D. Fabricant
   and John H. Chalmers, Jr.   309

Metabolism of 1,2-Dihaloethanes / Marion W. Anders and
  John C. Livesey 331

Summary / Robert K. Hinderer 343

Appendix 347

SESSION 1: Mutagenicity and Carcinogenicity of Ethylene Dichloride

Long-Term Carcinogenic Bioassays on Ethylene Dichloride Administered by Inhalation to Rats and Mice

CESARE MALTONI, LORETTA VALGIMIGLI,
AND CORRADO SCARNATO
Institute of Oncology and Tumour Center
Bologna, Italy

Ethylene dichloride (EDC; 1,2-dichloroethane) is a colorless, oily liquid, with a chloroform-like odor detectable over a range of 6-40 ppm and with a sweet taste. It is one of the highest volume chemicals produced and used in the world with an annual production estimated to be over 15 million tons. Its major use is as an intermediate, particularly in the production of vinyl chloride but also of other chemicals, such as 1,1,1-trichloroethane, trichloroethylene, perchloroethylene, vinylidene chloride, and ethyleneamines.

EDC is a lead scavenger and therefore appears as a component of most leaded fuels. It has also been used as an extraction solvent, as a solvent for textile cleaning and metal degreasing, in certain adhesives, and as a component in fumigants for upholstery, carpets, and grain. Other miscellaneous applications include paint, varnish, and finish removers; soaps and scouring compounds; wetting and penetrating agents; organic synthesis; ore flotation; and as a dispersant for nylon, rayon, styrene-butadiene rubber and other plastics.

The following population groups may be exposed to EDC:

1. workers engaged in the production of EDC and its derivatives;
2. workers manufacturing products containing EDC;
3. residents near factories producing or using EDC;
4. people, particularly consumers, coming into contact with EDC and with goods and other materials containing EDC.

EDC easily penetrates the body by any route. At high concentrations, it is known to produce acute and subacute toxic effects on the nervous system, liver, kidney, and cardiac and respiratory systems in both animals and man. It is a strong irritant to the eyes and skin.

The current Occupational Safety and Health Administration (OSHA) standard for occupational exposure to EDC is 50 ppm (8-hr time weighted average [TWA]). In West Germany the TWA is 20 ppm. The proposal in Italy has been for 20 ppm, and recently also 10 ppm. Adverse effects on the nervous system and the liver have been found in workers exposed to 10-15 ppm.

Up to 1976 there were no known available carcinogenicity data on this compound. In that year we started a long-term experimental carcinogenicity bioassay project on EDC.

EXPERIMENTAL PROCEDURES

Our project was planned to study the effects of EDC administered by inhalation to rats and mice. The experiments were carried out in our experimental laboratories in Bentivoglio (BT), near Bologna. No other compounds were tested in the large rooms where the experiments were performed.

The exposure chambers are made basically of stainless steel and glass. Although they are large enough to contain over 500 rats, no more than 270 rats or mice were ever treated in a single chamber during the whole experiment. The EDC concentration in the chambers was controlled by continuous gas chromatography.

The fallout from the chambers was decontaminated before being dispersed into the external atmosphere to avoid general pollution and, as far as the experiments are concerned, to avoid any remote possibility of reintroducing into the exposure chambers (particularly in the control ones) any trace of the tested compound.

The EDC was supplied by Montedison. Each stock was analyzed by the Montedison Research Unit. The impurities and their maximum levels in the EDC employed are indicated in Table 1. The purity level of EDC used in our experiment was 99.82%. (See also Appendix.)

The purity of the compound in any long-term bioassay is an extremely important factor. In any case, it is important to know the composition of the samples used. The commercial EDC, whose purity is only "greater than 90%" (NCI 1978), is used for fumigation and contains various compounds that are impurities and mainly additives. The analysis of one of these types of commercial EDC has shown that the product contains up to 7% of *bis*(2-chloroethyl)ether, up to 0.5% of 2-methyl-1,3-dioxolane, and also from 0.1% to 0.2% of the following compounds: chloroform, carbon tetrachloride, 1,1-dichloroethane, dichloroethylene (*cis* and *trans*), tetrachloroethylene, chloroethanol,

Table 1
Characterization of the EDC Used in the Experiments BT 501, BT 502

Component	Percent
EDC	99.82
1,1-Dichloroethane	0.02
Carbon tetrachloride	0.02
Trichloroethylene	0.02
Perchloroethylene	0.03
Benzene	0.09

chloroprene, and trichloroacetaldehyde. It is known that *bis*(2-chloroethyl)ether produced an increased incidence of liver cell tumors in male mice of two strains following oral administration, and that when it was injected by subcutaneous route in mice it produced a low incidence of sarcomas at the injection site; a few cases of pulmonary adenomas were also observed in females of one strain treated by oral administration (IARC 1975).

Our experiments utilized Sprague-Dawley rats (12 wk old) and Swiss mice (11 wk old). The colonies of these animals have been used in our institute for many years for long-term bioassays. Whatever the bioassay, the animals have been allowed to live until spontaneous death and are housed, fed, controlled, weighed, autopsized, and histologically examined in the same way, giving us extensive information concerning their current pathology.

The animals were weaned and classified by sex when 4-5 weeks old, at which time they were numbered by ear punch and divided into groups by litter distribution. After weaning, the animals were fed, ad libitum, an adequate commercial diet, free of pesticides, antibiotics, and aflatoxins. The animals were kept in groups of ten in stainless steel wire cages, with a solid bottom of the same metal. A shallow layer of white wood shavings served as bedding. The animals were kept in a temperature-controlled laboratory at $22°C \pm 3°C$.

Four groups of 180 rats and four groups of mice of both sexes were exposed to four EDC concentrations: 250-150 ppm, 50 ppm 10 ppm, 5 ppm, respectively, 7 hours daily, 5 days a week, for 78 weeks. After several days of 250-ppm exposure, the concentration was reduced to 150 ppm because of its severe toxic effects both on rats and mice, and particularly on mice. Two groups of 180 rats and one group of 249 mice, served as controls. One of the two groups of control rats was kept in an exposure chamber under the same conditions and for the same amount of time as the exposed animals. The other group of control rats and the group of control mice were kept in a nearby room.

The plan of the experiment is given in Tables 2 and 3. After the end of the treatment period, the animals were allowed to live until spontaneous death.

During the experiment the animals were controlled every 2 weeks; they were weighed every 2 weeks during the period of treatment, and then every 8 weeks. All the detectable gross pathological changes were recorded during the control. All the animals were kept under observation until spontaneous death. The animals, when moribund, were isolated, to avoid cannibalism.

A complete autopsy was carried out on each animal. Histological examinations were performed on the brain, Zymbal glands, retrobulbar glands, interscapular brown fat, salivary glands, tongue, lungs, thymus, diaphragm, liver, pancreas, kidneys, spleen, stomach, different segments of the intestine, bladder, gonads, lymph nodes (axillary, inguinal, and mesenteric), and any other organ with pathological lesions.

The whole project was entirely carried out by the same team, ensuring highly standardized procedures.

Table 2
Plan of Experiment BT 501

Groups	Concentration	Animals[a]	
		sex	number
I	250-150 ppm[b]	M	90
		F	90
		M and F	180
II	50 ppm	M	90
		F	90
		M and F	180
III	10 ppm	M	90
		F	90
		M and F	180
IV	5 ppm	M	90
		F	90
		M and F	180
V	controls in chambers	M	90
		F	90
		M and F	180
VI	controls	M	90
		F	90
		M and F	180
V and VI	controls	M	180
		F	180
		M and F	360
Total			1080

Exposure by inhalation to EDC in air at 250-150 ppm, 50 ppm, 10 ppm, and 5 ppm, 7 hr/day, 5 days/wk, for 78 wk.
[a] Sprague-Dawley rats, 12 wk old at start.
[b] After a few weeks the dose was reduced to 150 ppm, because of the high toxicity at 250-ppm level.

RESULTS

The entire project lasted 148 weeks. All histopathological examinations are now completed.

Survival in Rats

The mortality in rats was different in the different groups, but without a direct relationship with treatment (Figs. 1 and 2). The highest survival rate was observed both in males and females of the group exposed to 5 ppm of EDC. In females the highest mortality rate was observed in the control group in the chambers and in the groups exposed to 250-150 ppm of EDC. The control

Table 3
Plan of Experiment BT 502

Groups	Concentration	Animals[a] sex	number
I	250–150 ppm[b]	M	90
		F	90
		M and F	180
II	50 ppm	M	90
		F	90
		M and F	180
III	10 ppm	M	90
		F	90
		M and F	180
IV	5 ppm	M	90
		F	90
		M and F	180
V	controls	M	115
		F	134
		M and F	249
Total			969

Exposure by inhalation to EDC in air at 250–150 ppm, 50 ppm, 10 ppm, and 5 ppm, 7 hr/day, 5 days/wk, for 78 wk.

[a] Swiss mice, 11 wk old at start.
[b] After a few weeks the dose was reduced to 150 ppm, because of the high toxicity at 250-ppm level.

group outside of the chamber had a higher survival rate than the control group in the chamber; this was particularly evident in females. The survival rates at 52 and 104 weeks of age are reported in Table 4. At these ages the overall survival rates were respectively 93.9% and 27.3%. The survival rate after 52 weeks from the start of the experiment is shown in Table 5. At this stage the overall survival rate was 83.7%.

Survival in Mice

The survival rate of mice was slightly lower in females of the group treated with 250-150 ppm of EDC, during the treatment period (see Figs. 3 and 4). Since the highest mortality rate was independent from tumor increase, it may be attributed to the general toxic effects of the compound under those conditions. The survival rates of mice at 52 and 78 weeks of age are shown in Table 6. At these ages the overall survival rates were respectively 82.4% and 45.9%. The survival rate after 52 weeks from the start of the experiment is shown in Table 7. At this stage the overall survival rate was 67.8%.

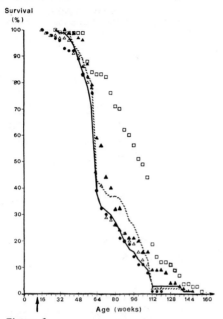

Figure 1
Survival of male Sprague-Dawley rats in experiment BT 501. (●) Group treated with 250-150 ppm EDC; (▲) 50 ppm; (△) 10 ppm; (□) 5 ppm; (———) untreated control group in chambers; (– – – –) untreated control group out of chambers.

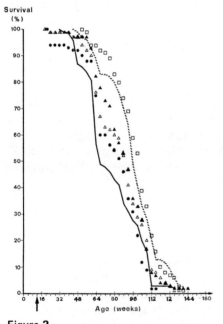

Figure 2
Survival of female Sprague-Dawley rats in experiment BT 501. (●) Group treated with 250-150 ppm EDC; (▲) 50 ppm; (△) 10 ppm; (□) 5 ppm; (———) untreated control group in chambers; (– – – –) untreated control group out of chambers.

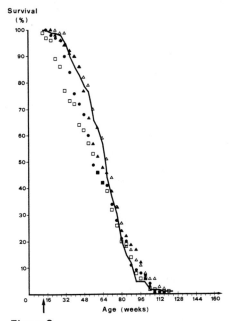

Figure 3
Survival of male Swiss mice in experiment BT 502. (●) Group treated with 250–150 ppm EDC; (▲) 50 ppm; (△) 10 ppm; (□) 5 ppm; (———) untreated control group.

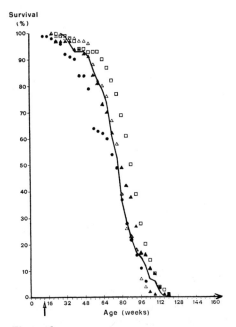

Figure 4
Survival of female Swiss mice in experiment 502. (●) Group treated with 250–150 ppm EDC; (▲) 50 ppm; (△) 10 ppm; (□) 5 ppm; (———) untreated control group.

Table 4
Experiment BT 501: Survival at 52 and 104 Weeks of Age

Groups	Concentration	sex	Animals[a] number at start	survivors at 52 weeks number	survivors at 52 weeks percent	survivors at 104 weeks number	survivors at 104 weeks percent
I	250-150 ppm[b]	M	90	79	87.8	10	11.1
		F	90	84	93.3	21	23.3
		M and F	180	163	90.6	31	17.2
II	50 ppm	M	90	87	96.7	17	18.9
		F	90	87	96.7	29	32.2
		M and F	180	174	96.7	46	25.6
III	10 ppm	M	90	81	90.0	13	14.4
		F	90	87	96.7	26	28.9
		M and F	180	168	93.3	39	21.7
IV	5 ppm	M	90	89	98.9	45	50.0
		F	90	90	100.0	48	53.3
		M and F	180	179	99.4	93	51.7
V	controls in chambers	M	90	80	88.9	12	13.3
		F	90	79	87.8	22	24.4
		M and F	180	159	88.3	34	18.9
VI	controls	M	90	83	92.2	16	17.8
		F	90	88	97.8	36	40.0
		M and F	180	171	95.0	52	28.9
	Total		1080	1014	93.9	295	27.3

Exposure by inhalation to EDC in air at 250–150 ppm, 50 ppm, 10 ppm, and 5 ppm, 7 hr/day, 5 days/wk, for 78 wk.
[a] Sprague-Dawley rats, 12 wk old at start.
[b] After a few weeks the dose was reduced to 150 ppm, because of the high toxicity at 250-ppm level.

Tumors in Rats

The distribution of the different types of tumors is shown in Table 8.

No specific types of tumors were found in treated animals. No relevant changes in the incidence of the tumors normally occurring in the strain of rats used was observed in treated groups, apart from a nondose-correlated increase of mammary tumors in some of the exposed fourth groups, particularly when compared with the group of the control animals in the chamber.

The different types of miscellaneous tumors are specified in Table 9.

The distribution of the different types of mammary tumors is given in Table 10. It appears that the increase in mammary tumors as a whole in some of the treated groups, particularly when compared to controls in the chamber, is not due to malignant tumors but to fibromas and fibroadenomas.

Table 5
Experiment BT 501: Survival after 1 Year from the Start of the Experiment

Groups	Concentration	sex	Animals[a] number at start	survivors after 52 weeks from the start number	percent
I	250–150 ppm[b]	M	90	67	74.4
		F	90	79	87.8
		M and F	180	146	81.1
II	50 ppm	M	90	70	77.8
		F	90	84	93.3
		M and F	180	154	85.6
III	10 ppm	M	90	70	77.8
		F	90	81	90.0
		M and F	180	151	83.9
IV	5 ppm	M	90	75	83.3
		F	90	85	94.4
		M and F	180	160	88.9
V	controls in chambers	M	90	64	71.1
		F	90	73	81.1
		M and F	180	137	76.1
VI	controls	M	90	72	80.0
		F	90	84	93.3
		M and F	180	156	86.0
Total			1080	904	83.7

Exposure by inhalation to EDC in air at 250–150 ppm, 50 ppm, 10 ppm and 5 ppm, 7 hr/day, 5 days/wk, for 78 wk.
[a] Sprague-Dawley rats, 12 wk old at start.
[b] After a few weeks the dose was reduced to 150 ppm, because of the high toxicity at 250-ppm level.

The increase in the incidence of mammary fibromas and fibroadenomas is statistically significant in groups exposed to 250–150 ppm, 50 ppm, and 5 ppm of EDC, when compared to controls in the chamber, but not when compared to controls outside of the chamber; the different incidence of mammary benign tumors in the two control groups is also significant (Table 11).

It is known that the onset of fibromas and fibroadenomas is age-correlated. The different incidence of these tumors observed in the various groups of our experiments is probably due mainly to the different survival rates within the groups. As a matter of fact, the greatest difference in incidence was found between the control groups in the chamber and the groups treated with 5 ppm of EDC respectively with low and high survival rates. The different survival also explains the different incidence in the two control groups.

Table 6
Experiment BT 502: Survival at 52 and 78 Weeks of Age

Groups	Concentration	sex	Animals[a] number at start	survivors at 52 weeks number	survivors at 52 weeks percent	survivors at 78 weeks number	survivors at 78 weeks percent
I	250-150 ppm[b]	M	90	56	62.2	26	28.9
		F	90	75	83.3	44	48.9
		M and F	180	131	72.8	70	38.9
II	50 ppm	M	90	68	75.6	30	33.3
		F	90	83	92.2	49	54.4
		M and F	180	151	83.9	79	43.9
III	10 ppm	M	90	74	82.2	34	37.8
		F	90	86	95.6	50	55.6
		M and F	180	160	88.9	84	46.7
IV	5 ppm	M	90	54	60.0	26	28.9
		F	90	84	93.3	68	75.6
		M and F	180	138	76.7	94	52.2
V	controls	M	115	91	79.1	42	36.6
		F	134	127	94.8	76	56.8
		M and F	249	218	87.6	118	47.4
Total			969	798	82.4	445	45.9

Exposure by inhalation to EDC in air at 250-150 ppm, 50 ppm, 10 ppm, and 5 ppm, 7 hr/day, 5 days/wk, for 78 wk.
[a] Swiss mice, 11 wk old at start.
[b] After a few weeks the dose was reduced to 150 ppm, because of the high toxicity at 250-ppm level.

Tumors in Mice

The distribution of different types of tumors is shown in Table 12. The different types of miscellaneous tumors are specified in Table 13.

No specific types of tumors, nor changes in the incidence of tumors, normally occurring in the breed of mice used, were observed in the treated animals.

DISCUSSION

Under our experimental conditions, EDC did not show carcinogenic effects. This conclusion is further supported by the following facts:

1. the high degree of purity of the compound used;
2. the strict measures adopted to avoid animal laboratory contamination;

Table 7
Experiment BT 502: Survival after 1 Year from the Start of the Experiment

Groups	Concentration	sex	Animals[a] number at start	survivors after 52 weeks from the start number	percent
I	250–150 ppm[b]	M	90	39	43.3
		F	90	58	64.4
		M and F	180	97	53.9
II	50 ppm	M	90	46	51.1
		F	90	73	81.1
		M and F	180	119	66.1
III	10 ppm	M	90	59	65.6
		F	90	72	80.0
		M and F	180	131	72.8
IV	5 ppm	M	90	42	46.7
		F	90	84	93.3
		M and F	180	126	70.0
V	controls	M	115	72	62.6
		F	134	112	83.6
		M and F	249	184	73.9
Total			969	657	67.8

Exposure by inhalation to EDC in air at 250–150 ppm, 50 ppm, 10 ppm, and 5 ppm, 7 hr/day, 5 days/wk, for 78 wk.
[a] Swiss mice, 11 wk old at start.
[b] After a few weeks the dose was reduced to 150 ppm, because of the high toxicity at 250-ppm level.

3. the range of doses tested, encompassing the highest one compatible with long-term treatment, showing subtoxic effects;
4. the knowledge of the pathology of the animals employed, on the basis of more than 10 years of experience, with more than 30,000 animals studied in a standard way;
5. the responsiveness of the animals that were used in the experiments to the cancerogenic effects of other correlated chlorinated olefins, such as vinyl chloride and vinylidene chloride (Maltoni 1977);
6. the high number of animals used and the still higher number of surviving animals at a late stage of the bioassays;
7. the systematic pathological examinations performed on all the major organs of all animals with or without pathological changes (more than 60,000 slides examined).

Table 8
Experiment BT 501: Distribution of the Different Types of Tumors

Groups	Concentration	Animals[a] sex	Animals[a] number at start	Animals[a] corrected number[c]	mammary tumors[d] total number	mammary tumors[d] percent[e]	mammary tumors[d] average latency time (wk)[f]	Zymbal gland carcinomas total number	Zymbal gland carcinomas percent[e]	Zymbal gland carcinomas average latency time (wk)[g]	leukemias total number	leukemias percent[e]	leukemias average latency time (wk)[g]	nephroblastomas total number	nephroblastomas percent[e]	nephroblastomas average latency time (wk)[g]
I	250–150 ppm[b]	M	90	89	11	12.3	92.9	0	—	—	0	—	—	0	—	—
		F	90	90	52	57.7	78.6	2	2.2	78.0	0	—	—	0	—	—
		M and F	180	179	63	35.2	81.1	2	1.1	78.0	0	—	—	0	—	—
II	50 ppm	M	90	90	10	11.1	88.8	0	—	—	1	1.1	80.0	0	—	—
		F	90	90	58	64.4	78.5	1	1.1	86.0	2	2.2	74.5	0	—	—
		M and F	180	180	68	37.8	80.0	1	0.6	86.0	3	1.7	76.3	0	—	—
III	10 ppm	M	90	89	5	5.6	60.8	0	—	—	4	4.6	74.0	0	—	—
		F	90	90	43	47.8	79.4	0	—	—	0	—	—	0	—	—
		M and F	180	179	48	26.8	77.5	0	—	—	4	2.2	74.0	0	—	—
IV	5 ppm	M	90	90	11	12.2	110.2	2	2.2	47.0	2	2.2	133.5	1	1.1	84.0
		F	90	90	65	72.2	83.2	1	1.1	47.0	6	6.7	83.5	0	—	—
		M and F	180	180	76	42.2	87.1	2	1.1	47.0	8	4.4	96.0	1	0.7	84.0
V	controls in chambers	M	90	90	8	8.9	85.5	1	1.1	78.0	1	1.1	37.0	0	—	—
		F	90	90	38	42.2	83.3	0	—	—	3	3.3	64.3	0	—	—
		M and F	180	180	46	25.5	83.6	1	0.6	78.0	4	2.2	59.5	0	—	—
VI	controls	M	90	90	5	5.5	92.0	1	1.1	74.0	0	—	—	0	—	—
		F	90	90	52	57.8	85.5	0	—	—	3	3.3	79.3	0	—	—
		M and F	180	180	57	31.7	86.1	1	0.6	74.0	3	1.7	79.3	0	—	—
Total			1080	1078												

Table 8 – Continued.

			Animals with tumors										
			angiosarcomas						angiomas and fibroangiomas				
			liver		other sites				liver		other sites		
Groups	Concentration	Sex	total number	percent[e]	total number	percent[e]	average latency time (wk)[g]		total number	percent[e]	average latency time (wk)[g]	total number	percent[e]	average latency time (wk)[g]
I	250-150 ppm[b]	M	0	–	0	–	–		0	–	–	0	–	–
		F	0	–	0	–	–		0	–	–	0	–	–
		M and F	0	–	0	–	–		0	–	–	0	–	–
II	50 ppm	M	0	–	0	–	–		0	–	–	0	–	–
		F	0	–	0	–	–		0	–	–	0	–	–
		M and F	0	–	0	–	–		0	–	–	0	–	–
III	10 ppm	M	0	–	0	–	–		0	–	–	0	–	–
		F	0	–	0	–	–		0	–	–	0	–	–
		M and F	0	–	0	–	–		0	–	–	0	–	–
IV	5 ppm	M	0	–	0	–	–		0	–	–	1	1.1	125.0
		F	0	–	0	–	–		0	–	–	1	1.1	131.0
		M and F	0	–	0	–	–		0	–	–	2	1.1	128.0
V	controls in chambers	M	0	–	0	–	–		0	–	–	0	–	–
		F	0	–	0	–	–		0	–	–	0	–	–
		M and F	0	–	0	–	–		0	–	–	0	–	–
VI	controls	M	0	–	0	–	–		0	–	–	1	1.1	130.0
		F	0	–	0	–	–		0	–	–	0	–	–
		M and F	0	–	0	–	–		0	–	–	1	0.7	130.0

Table 8 – Continued.

			Animals with tumors										
			hepatomas			forestomach epithelial tumors			skin carcinomas			subcutaneous sarcomas		
Groups	Concentration	Sex	total number	percent[e]	average latency time (wk)[g]	total number	percent[e]	average latency time (wk)[g]	total number	percent[e]	average latency time (wk)[g]	total number	percent[e]	average latency time (wk)[g]
I	250-150 ppm[b]	M	0	—	—	0	—	—	0	—	—	0	—	—
		F	0	—	—	0	—	—	0	—	—	0	—	—
		M and F	0	—	—	0	—	—	0	—	—	0	—	—
II	50 ppm	M	0	—	—	1	1.1	60.0	1	1.1	56.0	0	—	—
		F	0	—	—	0	—	—	0	—	—	0	—	—
		M and F	0	—	—	1	0.6	60.0	1	0.6	56.0	0	—	—
III	10 ppm	M	0	—	—	0	—	—	1	1.1	94.0	0	—	—
		F	0	—	—	1	1.1	78.0	0	—	—	0	—	—
		M and F	0	—	—	1	0.6	78.0	1	0.6	94.0	0	—	—
IV	5 ppm	M	0	—	—	1	1.1	101.0	0	—	—	1	1.1	30.0
		F	0	—	—	0	—	—	0	—	—	0	—	—
		M and F	0	—	—	1	0.6	101.0	0	—	—	1	0.6	30.0
V	controls in chambers	M	0	—	—	2	2.2	78.5	0	—	—	0	—	—
		F	0	—	—	1	1.1	63.0	0	—	—	1	1.1	78.0
		M and F	0	—	—	3	1.7	73.3	0	—	—	1	0.6	78.0
VI	controls	M	0	—	—	2	2.2	110.0	0	—	—	0	—	—
		F	0	—	—	1	1.1	102.0	0	—	—	0	—	—
		M and F	0	—	—	3	1.7	107.3	0	—	—	0	—	—

Table 8 – Continued.

			Animals with tumors										
			encephalic tumors					others			total[h]		
			neuroblastomas			other sites							
Groups	Concentration	Sex	total number	percent[e]	average latency time (wk)[g]	total number	percent[e]	total number	benign number	malignant number	number	percent[e]	number of different tumors/tumor-bearing animals
I	250-150 ppm[b]	M	0	—	—	0	—	9	7	2	15	16.9	1.2
		F	0	—	—	0	—	11	5	6	54	60.0	2.0
		M and F	0	—	—	0	—	20	12	8	69	38.5	1.9
II	50 ppm	M	0	—	—	1	1.1	14	14	0	20	22.2	1.6
		F	0	—	—	1	1.1	6	3	3	56	62.2	1.7
		M and F	0	—	—	2	1.1	20	17	3	76	42.2	1.7
III	10 ppm	M	0	—	—	0	—	4	2	2	13	14.6	1.1
		F	0	—	—	2	2.2	4	3	1	43	47.8	1.6
		M and F	0	—	—	2	1.1	8	5	3	56	31.3	1.5
IV	5 ppm	M	0	—	—	0	—	16	15	1	30	33.3	1.4
		F	0	—	—	0	—	9	5	4	65	72.2	1.8
		M and F	0	—	—	0	—	25	20	5	95	52.8	1.7
V	controls in chambers	M	0	—	—	1	1.1	8	7	1	17	18.9	1.3
		F	0	—	—	1	1.1	4	2	2	38	42.2	1.9
		M and F	0	—	—	2	1.1	12	9	3	55	30.6	1.7
VI	controls	M	0	—	—	0	—	7	7	0	14	15.6	1.1
		F	0	—	—	3	3.3	8	5	3	56	62.2	1.7
		M and F	0	—	—	3	1.7	15	12	3	70	38.9	1.6

Exposure by inhalation to EDC in air at 250–150 ppm, 50 ppm, 10 ppm, and 5 ppm, 7 hr/day, 5 days wk, for 78 wk. Results after 148 wk (end of experiment).
[a] Sprague-Dawley rats, 12 wk old at start.
[b] After a few weeks the dose was reduced to 150 ppm, because of the high toxicity at 250-ppm level.
[c] Animals alive after 12 wk, when the first tumor (a mammary carcinoma) was observed.
[d] Two or more tumors of the same or different types (fibroadenomas, carcinomas, sarcomas, carcinosarcomas) may be present in the same animal.
[e] The percentages refer to the corrected number.
[f] Average age at the onset of the first mammary tumor per animal detected at the periodic control or at autopsy.
[g] Average time from the start of the experiment to the detection (at the periodic controls or at autopsy).
[h] Several animals with 2 or more tumors.

Table 9
Experiment BT 501: Distribution of the Different Types of Miscellaneous ("Other") Tumors

Groups	Sex	Animals bearing other tumors				
		benign		malignant		
		number	distribution of histotypes	number	distribution of histotypes	Total
I	M	7	dermatofibroma (4) Zymbal gland adenoma (1) pheochromocytoma (2)	2	peritoneal fibrosarcoma (1) peritoneal mesothelioma (1)	9
	F	5	Zymbal gland adenoma (1) luteoma (1) adenoma of the uterus (1) polypus of the uterus (2)	6	adenocarcinoma of stomach (1) adrenal gland cortical adenocarcinoma (1) retroperitoneal leiomyosarcoma (1) granulosa cells carcinoma of ovary (1) adenocarcinoma of the uterus (1) squamous carcinoma of the uterus (1)	11
II	M	14	dermatofibroma (6) subcutaneous fibroma (1) skin cystic acanthoma (2) adrenal gland cortical adenoma (1) pheochromocytoma (3) Leydig cell tumor (1)	0		14
	F	3	skin cystic acanthoma (1) Zymbal gland adenoma (1) polypus of the colon (1)	3	arrenoblastoma (1) fibrosarcoma of the uterus (1) peritoneal mesothelioma (1)	6

	Sex					
III	M	2	dermatofibroma (1) adrenal gland cortical adenoma (1)	2	adrenal gland cortical adenocarcinoma (1) pleural mesothelioma (1)	4
	F	3	Zymbal gland adenoma (1) pheochromocytoma (1) polypus of the uterus (1)	1	adenocarcinoma of the uterus (1)	4
IV	M	15	dermatofibroma (6) cholangioma (1) adrenal gland cortical adenoma (1) pheochromocytoma (2) Leydig cell tumor (4) ganglioneuroma (1)	1	neurilemoma (1)	16
	F	5	skin acanthoma (1) Zymbal gland adenoma (1) liver histiocytosis (1) adrenal gland cortical adenoma (2)	4	intestine fibrosarcoma (1) granulosa cells carcinoma of the ovary (1) adenocarcinoma of the uterus (2)	9
V	M	7	dermatofibroma (2) skin cystic acanthoma (1) subcutaneous lipoma (1) pheochromocytoma (3)	1	skin fibrosarcoma (1)	8
	F	2	skin cystic acanthoma (1) fibroma of the uterus (1)	2	squamous carcinoma of the uterus (1) kidneys angiopericytosarcoma (1)	4

Table 9 — Continued.

| Groups | Sex | Animals bearing other tumors ||||
| | | benign || malignant ||
		number	distribution of histotypes	number	distribution of histotypes	Total
VI	M	7	skin acanthoma (1) skin cystic acanthoma (1) Zymbal gland adenoma (2) subcutaneous lipoma (1) pheochromocytoma (1) benign neurilemoma (1)	0		7
	F	5	adrenal gland cortical adenoma (3) pheochromocytoma (2)	3	liver fibrosarcoma (1) adenocarcinoma of the uterus (1) neurilemoma (1)	8

Exposure by inhalation to EDC in air at 250–150 ppm, 50 ppm, 10 ppm, and 5 ppm, 7 hr/day, 5 days/wk, for 78 wk. Results after 148 wk (end of experiment).

Table 10
Experiment BT 501: Distribution of the Different Types of Mammary Tumors

Groups	Concentration	sex	Animals[a] number at start	Animals[a] corrected number[c]	Mammary tumors total number	Mammary tumors percent[e]	Mammary tumors average latency time (wk)[f]	Mammary tumors number of tumors/tumor-bearing animals
I	250–150 ppm[b]	M	90	89	11	12.3	92.9	1.1
		F	90	90	52	57.7	78.6	1.9
		M and F	180	179	63	35.2	81.1	1.7
II	50 ppm	M	90	90	10	11.1	88.8	1.4
		F	90	90	58	64.4	78.5	1.5
		M and F	180	180	68	37.8	80.0	1.5
III	10 ppm	M	90	89	5	5.6	60.8	1.0
		F	90	90	43	47.8	79.4	1.4
		M and F	180	179	48	26.8	77.5	1.4
IV	5 ppm	M	90	90	11	12.2	110.2	1.4
		F	90	90	65	72.2	83.2	1.6
		M and F	180	180	76	42.2	87.1	1.5
V	controls in chambers	M	90	90	8	8.9	85.5	1.0
		F	90	90	38	42.2	83.3	1.8
		M and F	180	180	46	25.5	83.6	1.6
VI	controls	M	90	90	5	5.5	92.0	1.0
		F	90	90	52	57.8	85.5	1.5
		M and F	180	180	57	31.7	86.1	1.5
Total			1080	1078				

Table 10 – Continued.

Groups	Concentration	Sex	total number	fibromas and fibroadenomas			Mammary tumors[d] Histological evaluation									
							carcinomas			sarcomas			carcinosarcomas		
				percent[g]	number	percent[h]	average latency time (wk)[f]	number	percent[h]	average latency time (wk)[f]	number	percent[h]	average latency time (wk)[f]	number	percent[h]	average latency time (wk)[f]
I	250-150 ppm[b]	M	9	81.8	7	77.8	99.1	1	11.1	92.0	1	11.1	80.0	0	–	–
		F	50	96.1	47	94.0	81.3	8	16.0	71.5	3	6.0	99.3	0	–	–
		M and F	59	93.6	54	91.5	83.6	9	15.2	73.7	4	6.8	85.5	0	–	–
II	50 ppm	M	9	90.0	7	77.8	100.3	1	11.1	60.0	1	11.1	64.0	0	–	–
		F	53	91.4	49	92.4	80.7	9	17.0	73.8	1	1.9	88.0	0	–	–
		M and F	62	91.2	56	90.3	83.1	10	16.1	72.4	2	3.2	76.0	0	–	–
III	10 ppm	M	5	100.0	3	60.0	72.0	0	–	–	2	40.0	44.0	0	–	–
		F	38	88.4	33	86.8	80.9	8	21.0	88.5	2	4.6	44.0	0	–	–
		M and F	43	89.6	36	83.7	80.1	8	18.6	88.5	2	9.1	134.0	0	–	–
IV	5 ppm	M	11	100.0	11	100.0	110.2	0	–	–	1	9.1	134.0	0	–	–
		F	60	92.3	56	93.3	86.4	10	16.7	87.4	2	3.3	88.0	0	–	–
		M and F	71	93.4	67	94.4	90.3	10	14.1	87.4	3	4.2	103.3	0	–	–
V	controls in chambers	M	8	100.0	7	87.5	85.1	0	–	–	1	14.3	88.0	0	–	–
		F	35	92.1	27	77.1	85.5	15	46.8	80.1	1	2.8	124.0	0	–	–
		M and F	43	93.5	34	79.1	85.4	15	38.5	80.1	2	4.6	106.0	0	–	–
VI	controls	M	5	100.0	3	60.0	105.7	0	–	–	2	40.0	72.0	0	–	–
		F	49	94.2	47	95.9	88.1	8	16.3	80.4	0	–	–	2	4.1	73.0
		M and F	54	94.7	50	92.6	89.3	8	14.8	80.4	2	3.7	72.0	2	3.7	73.0

Exposure by inhalation to EDC in air at 250-150 ppm, 150 ppm, 50 ppm, 10 ppm, and 5 ppm, 7 hr/day, 5 days/wk, for 78 wk. Results after 148 wk (end of experiment).

[a] Sprague-Dawley rats, 12 wk old at start.
[b] After a few weeks the dose was reduced to 150 ppm, because of the high toxicity at 250-ppm level.
[c] Animals alive after 12 wk, when the first tumor (a mammary carcinoma) was observed.
[d] Two or more tumors of the same or different types (fibroadenomas, carcinomas, sarcomas, carcinosarcomas) may be present in the same animal.
[e] The percentages refer to the corrected number.
[f] Average age at the onset of first mammary tumor per animal, detected at the periodic control or at autopsy.
[g] The percentages refer to total number of animals bearing mammary tumors.
[h] The percentages refer to total number of animals bearing mammary tumors, histologically examined.

Table 11
Experiment BT 501: Chi-Square Values of Female Rats Bearing Benign Mammary Tumors (Fibromas and Fibroadenomas)

Groups	χ^2
Controls V vs controls VI	8.28[a]
Controls V vs 250-150 ppm	8.28[a]
Controls V vs 50 ppm	10.04[a]
Controls V vs 10 ppm	0.63
Controls V vs 5 ppm	17.53[b]
Controls VI vs 250-150 ppm	0.00
Controls VI vs 50 ppm	0.02
Controls VI vs 10 ppm	3.80[c]
Controls VI vs 5 ppm	1.45
Trend (controls V)	0.27

Exposure by inhalation to EDC in air at 250-150 ppm, 50 ppm, 10 ppm, and 5 ppm, 7 hr/day, 5 days/wk, for 78 wk.
[a] $p < 0.01$
[b] $p < 0.001$.
[c] Controls VI higher than 10 ppm. The latter is almost significantly lower than controls (χ^2 value at 95%, with one degree of freedom = 3.84).

In another project performed in the United States, in which EDC was administered by gavage to Osborne-Mendel rats and to B6C3F1 mice, tumors of various types were observed following the treatment (NCI 1978). The difference in the results between the two projects could be due to the different route employed and the different animals used. In our opinion, however, other factors may be involved, such as:

1. The type of EDC tested. The degree of purity of the compound used in the gavage study has been controversial (NCI 1978; Hooper et al., this volume).
2. The animal laboratory pollution. The gavage study was carried out in a room where other compounds, encompassing cancerogenic ones, were tested (NCI 1978; Hooper et al., this volume).
3. The size of both treated and control animal groups.
4. The performance of the experiment with particular reference to the probability of mix-up of chemicals during treatments, and to the professionality of the team carrying on treatment, control of animals, and autopsies.
5. The extension of histopathology, and possible difference in pathological interpretation.

We do believe that the time has come not to accept conflicting results as established data any longer, but rather to clear up and assess all the factors that form the basis of such discrepancies. In our own view, this will be an important

Table 12
Experiment BT 502: Distribution of the Different Types of Tumors

Groups	Concentration	Animals[a] sex	Animals[a] number at start	Animals[a] corrected number[c]	Animals with tumors — mammary tumors total number	Animals with tumors — mammary tumors percent[d]	Animals with tumors — mammary tumors average latency time (wk)[e]	Animals with tumors — pulmonary adenomas total number	Animals with tumors — pulmonary adenomas percent[d]	Animals with tumors — pulmonary adenomas average latency time (wk)[e]	Animals with tumors — leukemias total number	Animals with tumors — leukemias percent[d]	Animals with tumors — leukemias average latency time (wk)[e]
I	250-150 ppm[b]	M	90	81	0	—	—	0	—	—	1	1.2	75.0
		F	90	84	5	6.0	69.0	3	3.6	85.7	1	1.2	63.0
		M and F	180	165	5	3.0	69.0	3	1.8	85.7	2	1.2	69.0
II	50 ppm	M	90	87	0	—	—	3	3.4	75.3	6	6.9	69.8
		F	90	87	3	3.4	76.0	2	2.3	74.5	4	4.6	59.2
		M and F	180	174	3	1.7	76.0	5	2.9	75.0	10	5.7	65.6
III	10 ppm	M	90	89	1	1.1	32.0	4	4.5	85.7	6	6.7	54.5
		F	90	88	6	6.8	73.7	2	2.3	53.5	5	5.7	69.6
		M and F	180	177	7	4.0	67.7	6	3.4	75.0	11	6.2	61.4
IV	5 ppm	M	90	69	0	—	—	1	1.4	44.0	2	2.9	79.0
		F	90	89	5	5.6	82.6	4	4.5	85.0	6	6.7	69.8
		M and F	180	158	5	3.2	82.6	5	3.2	76.8	8	5.1	72.1
V	controls	M	115	111	0	—	—	4	3.6	78.5	6	5.4	65.8
		F	134	133	7	5.3	77.4	4	3.0	51.2	15	11.3	61.3
		M and F	249	244	7	2.9	77.4	8	3.3	64.9	21	8.6	62.6
Total			969	918									

Table 12 – Continued.

			Animals with tumors										
			nephroblastomas			kidneys adenocarcinomas			liver		angiosarcomas	other sites	
Groups	Concentration	Sex	total number	percent[d]	average latency time (wk)[e]	total number	percent[d]	average latency time (wk)[e]	total number	percent[d]	total number	percent[d]	average latency time (wk)[e]
I	250-150 ppm[b]	M	0	—	—	0	—	—	0	—	1	1.2	40.0
		F	0	—	—	0	—	—	0	—	0	—	—
		M and F	0	—	—	0	—	—	0	—	1	0.6	40.0
II	50 ppm	M	0	—	—	0	—	—	0	—	0	—	—
		F	0	—	—	0	—	—	0	—	0	—	—
		M and F	0	—	—	0	—	—	0	—	0	—	—
III	10 ppm	M	0	—	—	1	1.1	62.0	0	—	0	—	—
		F	0	—	—	0	—	—	0	—	0	—	—
		M and F	0	—	—	1	0.6	62.0	0	—	0	—	—
IV	5 ppm	M	0	—	—	0	—	—	0	—	0	—	—
		F	0	—	—	0	—	—	0	—	0	—	—
		M and F	0	—	—	0	—	—	0	—	0	—	—
V	controls	M	0	—	—	0	—	—	0	—	1	0.9	69.0
		F	0	—	—	0	—	—	0	—	0	—	—
		M and F	0	—	—	0	—	—	0	—	1	0.4	69.0

Table 12 – *Continued.*

			Animals with tumors									
				angiomas and fibroangiomas					hepatomas		forestomach epithelial tumors	
				liver		other sites						
Groups	Concentration	Sex	total number	percent[d]	average latency time (wk)[e]	total number	percent[d]	average latency time (wk)[e]	total number	percent[d]	average latency time (wk)[e]	total number	percent[d]	average latency time (wk)[e]
I	250-150 ppm[b]	M	0	–	–	0	–	–	0	–	–	0	–	–
		F	0	–	–	0	–	–	0	–	–	0	–	–
		M and F	0	–	–	0	–	–	0	–	–	0	–	–
II	50 ppm	M	0	–	–	1	1.1	76.0	0	–	–	0	–	–
		F	1	1.1	83.0	1	1.1	83.0	0	–	–	0	–	–
		M and F	1	0.6	83.0	2	1.1	79.5	0	–	–	0	–	–
III	10 ppm	M	0	–	–	0	–	–	0	–	–	0	–	–
		F	0	–	–	1	1.1	93.0	0	–	–	0	–	–
		M and F	0	–	–	1	0.6	93.0	0	–	–	0	–	–
IV	5 ppm	M	0	–	–	0	–	–	0	–	–	0	–	–
		F	0	–	–	0	–	–	0	–	–	0	–	–
		M and F	0	–	–	0	–	–	0	–	–	0	–	–
V	controls	M	1	0.9	69.0	1	0.9	76.0	4	3.5	82.0	0	–	–
		F	0	–	–	0	–	–	0	–	–	0	–	–
		M and F	1	0.4	69.0	1	0.4	76.0	4	1.6	82.0	0	–	–

Table 12 — Continued.

			Animals with tumors											
			skin epithelial tumors			subcutaneous sarcomas			others			total[f]		number of different tumors/tumor-bearing animals
Groups	Concentration	Sex	total number	percent[d]	average latency time (wk)[e]	total number	percent[d]	average latency time (wk)[e]	total number	benign number	malignant number	number	percent[d]	
I	250-150 ppm[b]	M	0	—	—	0	—	—	0	0	0	2	2.4	1.0
		F	0	—	—	0	—	—	2	2	0	11	12.9	1.0
		M and F	0	—	—	0	—	—	2	2	0	13	7.7	1.0
II	50 ppm	M	0	—	—	0	—	—	1	0	1	9	10.6	1.2
		F	1	1.1	103.0	0	—	—	4	3	1	15	17.2	1.1
		M and F	1	0.6	103.0	0	—	—	5	3	2	24	14.0	1.2
III	10 ppm	M	0	—	—	1	1.1	77.0	1	1	0	12	13.5	1.1
		F	0	—	—	1	1.1	25.0	2	1	1	17	19.1	1.0
		M and F	0	—	—	2	1.1	51.0	3	2	1	29	16.3	1.1
IV	5 ppm	M	0	—	—	0	—	—	1	0	1	4	5.7	1.0
		F	1	1.1	79.0	0	—	—	4	3	1	19	21.3	1.0
		M and F	1	0.6	79.0	0	—	—	5	3	2	23	14.5	1.0
V	controls	M	1	0.9	109.0	1	0.9	106.0	1	1	0	14	12.4	1.4
		F	1	0.7	77.0	0	—	—	4	2	2	27	20.1	1.1
		M and F	2	0.8	93.0	1	0.4	106.0	5	3	2	41	16.6	1.2

Exposure by inhalation to EDC in air at 250-150 ppm, 50 ppm, 10 ppm, and 5 ppm, 7 hr/day, 5 days/wk, for 78 wk. Results after 119 weeks (end of experiment).
[a] Swiss mice, 11 wk old at start.
[b] After a few weeks the dose was reduced to 150 ppm, because of the high toxicity at 250-ppm level.
[c] Animals alive after 24 wk, when the first tumor (a leukemia) was observed.
[d] The percentages refer to the corrected number.
[e] Average time from the start of the experiment to the detection at the periodic control or at autopsy.
[f] Several animals with two or more tumors.

Table 13
Experiment BT 502: Distribution of the Different Types of Miscellaneous ("Other") Tumors

		Animals bearing other tumors				
		benign		malignant		
Group	Sex	number	distribution of histotypes	number	distribution of histotypes	Total
I	M	0	—	0	—	0
	F	2	polypus of the uterus (2)	0	—	2
II	M	0	—	1	adrenal gland cortical carcinoma (1)	1
	F	3	fibroma of the uterus (1) tumor of the granulosa (1) fibroma of the ovary (1)	1	Harderian gland adenocarcinoma (1)	4
III	M	1	Leydig cell tumor (1)	0	—	1
	F	1	polypus of the uterus (1)	1	adenocarcinoma of the ovary (1)	2
IV	M	0	—	1	oligodendroglioma of the brain (1)	1
	F	3	polypus of the uterus (3)	1	adenocarcinoma of the uterus (1)	4
V	M	1	papilloma of the bladder (1)	0	—	1
	F	2	Harderian gland adenoma (1) leiomyoma of the ovary (1)	2	intestinal fibrosarcoma (1) leiomyosarcoma of the uterus (1)	4

Exposure by inhalation to EDC in air at 250-150, 50, 10, 5 ppm, 7 hr/day, 5 days/wk, for 78 wk. Results after 119 wk (end of experiment).

tool not only to improve bioassays and the quality of information they supply, but also to provide contributions to our knowledge on the scientific methodology in biomedical research and on the basic mechanisms of carcinogenesis.

REFERENCES

IARC (International Agency for Research on Cancer). 1975. *Monographs on the evaluation of carcinogenic risk of chemical to man,* vol. 9. *Some aziridines, N-, S- & O-mustards and selenium.* International Agency for Research on Cancer, Lyon.

Maltoni, C. 1977. Recent findings on the carcinogenicity of chlorinated olefins. *Environ. Health Perspect.* **21**:1.

NCI (National Cancer Institute). 1978. *Bioassay of 1,2-dichloroethane for possible carcinogenicity.* NCI carcinogenesis technical report series no. 55, DHEW publication number (NIH) 78-1361. Government Printing Office, Washington, D.C.

COMMENTS

HOOPER: We will try to determine the purity of EDC used in the NCI gavage bioassay (see further information on this in Hooper et al., this volume).

REITZ: This is an absolutely vital question to resolve, and Dr. Hooper and I have agreed that he's going to send a sample of the material used in the NCI bioassay to Dow Chemical. We will have it analyzed within our laboratory. Although it is unfortunate, we cannot resolve this question at the conference today. I think we can arrange to have that analysis as part of the proceedings. (Editor's note: see Appendix.)

HOOPER: I'd like to interject that we inquired or Dr. Hugh Farber of Dow Chemical if there were any records of purity of the same batch of EDC that was sent to NCI. He tried to oblige, but there is a company policy not to retain records past 5 years. Since the material was sent out in about 1972, these records are no longer available.

MALTONI: The commercial EDC with a 90% purity, to be used for fumigation, whose analysis is reported in the "experimental procedures" of our presentation, is the one marketed in Europe by Dow Chemical in 1972.

HOOPER: I noticed in your mortality table for high-dose rats that survival was good through 52 weeks of age, but thereafter mortality began to increase markedly, reaching about 30-40% at 64 weeks. Additionally, mortality at 64 weeks seemed to be dose-related: That is, mortality seemed higher at higher doses. Is this true? If it is, then as you go to higher doses, fewer animals survive past 64 weeks, and consequently fewer are at risk of getting tumors during the latter half of the experiment during which you would expect most tumors to appear (because cancer increases at t^3 or t^4). If the number of tumors, say of mammary fibroadenomas in female SDA rats, is about the same in each of the higher-dose groups, as I believe is true, it is possible that an analysis of the life-table data would indicate a significant dose-related effect because of the fewer animals at risk in the higher-dose groups.

MALTONI: At 64 weeks of age the survival of female rats was as follows: I group (250-150 ppm) 87.8%, II group (50 ppm) 93.3%, III group (10 ppm) 90.0%, IV group (5 ppm) 94.4%, V group (controls in chamber) 81.1% and VI group (controls out of chamber) 93.3%.

PLOTNICK: Do you rotate the cages within the chamber each day?

MALTONI: Cages have been rotated every 2 weeks. The homogeneity of the distribution of the compound in the chambers has been controlled by continuous monitoring by gas chromatography in five different positions. To summarize the results of our bioassays, no relevant differences between the incidence of tumors in the various groups of mice (experiment BT 502) have been observed, and as far as rats (experiment BT 501) are concerned, the only important difference is seen in mammary fibroadenomas and fibromas, particularly striking in the group treated with 5 ppm. Their incidence appears to correlate with survival rate, as it is shown by survival figures at 104 weeks from the start of the experiment.

INFANTE: Are those your denominators, your surviving animals?

MALTONI: I am referring particularly to surviving females.

INFANTE: Where are the numerators to go with that?

HOOPER: I thought that you said it was the same number of fibroadenomas all the way along and then you have fewer survivors.

MALTONI: No. As relevant differences in mortality rates were found between groups of the experiment on rats, we are now planning to statistically evaluate our results on mammary benign tumors, using a statistical method which allows for such variations, to strengthen our interpretation and conclusions.

HOOPER: But right here you have a high number of survivors and it's a low dose. As you increase your dose, it looks to some extent dose-related.

MALTONI: The survival in the low-dose group was higher than in other treated groups and in the control groups. The increase in fibroadenomas was not paralleled by an increase of mammary carcinomas.

HOOPER: So it's different in some sense?

MALTONI: The progression of mammary fibroadenomas to carcinomas in rats is not a frequent event.

SPREAFICO: Those are very, very limited conditions. Normally, once it's benign, it's benign.

MALTONI: Mammary carcinomas in general arise grossly as such, and not ex-fibroadenomas.

WARD: Ken Chu and Bob Tarone at NCI are now evaluating the distribution of incidence of tumors of different control groups of 50. You'd expect a normal statistical variation of tumor incidence in groups of 50 animals that may be binomial or nonbinomial. In this study, the fibroadenomas were not included because comparisons with the historical controls at the lab and the pool controls were not statistically significant.

If you take fibroadenomas, there is a 30% incidence in 20,000 Osborne-Mendel rats that have lived during the same time and survival is the same. In any group of 50, what is the distribution in 10 groups of 50, 100 groups of 50? You could have a binomial distribution—you might have one group out here with 50% incidence and one group out here with 5% incidence, and maybe even one with zero incidence.

In the analysis that I've seen so far (it's not finished) it's surprising that for many tumors you have this nice binomial distribution, but with a few other tumors you don't. There are a lot of factors that have to be taken into account, but until it's published and Dr. Maltoni publishes his work, to show that there is a normal variation, we won't have the data at hand.

If you did a life-table analysis comparing these two groups—controls and lower dose—it's based on age because this had more survival and the tumors were late in appearing. So there's a difference in incidence. This result may be a normal variation that in this case can be explained on survival. In the other cases let's assume the survival's the same. We have to consider that.

TER HAAR: I have one other comment about all of this. It seems to me that one of the problems is that we've got altogether too much toxicity in these studies. The animals ought not to be dying from toxicity if we're trying to make some sense out of this.

WARD: Right. That's another factor.

TER HAAR: We're going to start getting statistics that will take care of this toxicity. Instead of going to more reasonable levels where we can understand what's happening, we're going to go up to higher-than-maximum tolerated doses.

INFANTE: Well, the fact is you should have a high mortality in the controls as well, right? I mean you're talking 60% over the lifetime?

MALTONI: After 2 years from the start of the experiment we still had more than 27% of rats surviving.

TER HAAR: I think the mortality in Maltoni's study was fine. It was the NCI study that has the high mortality. We ought not to see life-shortened toxicity in these studies.

HOOPER: There may be some interface between the toxicity. It may be that you need to go to a large number of animals, and that's the solution to a smaller dosage.

MALTONI: Again I have to say that I am more than surprised not to have observed in our animals any of the specific tumors described in the report of NCI bioassays, particularly when one thinks of the larger number of animals we have used, and the number of animals still alive at a later stage of the experiment.

REITZ: I'd like to point out that the guidelines for carcinogen bioassay in rodents suggest very strongly that the maximum tolerated dose (MTD) should be one that has no life-shortening effects other than production of the tumors. The reason that that was put into the NCI policy was that you get all sorts of confounding factors if you exceed MTD. These confounding factors will not be eliminated just by using life-table analysis.

WARD: The survival in the low-dose female mice was similar to that of the controls and they did have significant increases of a few types of tumors, so that would be more relevant. At least the mortality would be similar.

RETIZ: That's a much more significant observation than this one with early mortality.

TER HAAR: One of the things you gain in Kim's [Hooper] approach is if they are dying from tumors and you do a life-time mortality, you get a lot better estimate of risk. The risk is higher.

HOOPER: It's a more accurate assessment of what is going on that experiment. That's basically it.

TER HAAR: If the life shortening is due to the tumor and you're going to do risk analysis, you've got a lot better risk analysis if you do it this way.

References

NCI (National Cancer Institute). 1978. *Bioassay of 1,2-dichloroethane for possible carcinogenicity*. Technical report series number 55. DHEW publication number (NIH) 78-1361. Government Printing Office, Washington, D.C.

The Carcinogenicity of Ethylene Dichloride in Osborne-Mendel Rats and B6C3F1 Mice

JERROLD M. WARD
National Cancer Institute
National Toxicology Program
National Institutes of Health
Bethesda, Maryland 20205

The causes of the majority of human cancers are unknown at this time. To better understand and identify potential human carcinogens and to identify structure-activity relationships of chemicals, scientists at the National Cancer Institute (NCI) initiated a series of animal experiments in nongovernment contract laboratories during the early 1970s. Halogenated compounds were screened for their potential carcinogenicity by gavage studies in Osborne-Mendel rats and B6C3F1 mice at one laboratory. This paper describes the animal experiments conducted to determine if ethylene dichloride (EDC; 1,2,-dichloroethane) is carcinogenic in rodents and compares these results to similar experiments on structurally related halogenated chemicals.

MATERIALS AND METHODS

The complete materials and methods are given in the NCI Technical Report (NCI 1978a). In summary, technical grade EDC was obtained from Dow Chemical Company, Midland, Michigan. Gas-liquid chromatography revealed a purity of 98-99% EDC with four to ten minor peaks.

Solutions of EDC were prepared in corn oil, and Osborne-Mendel rats and B6C3F1 mice were gavaged five times weekly. After preliminary toxicity studies, doses for chronic studies were estimated and the animals received these dosages as in Tables 1 and 2. Animals dead or sacrificed received a complete necropsy and slides were prepared from many tissues and lesions (NCI 1978a).

RESULTS

Conduct of the Study

Because of the inadequate preliminary toxicity studies, the initial doses administered to test animals were found to be inappropriate. Signs of toxicity in these animals early in the course of the study led to changes of the dosages several times (Tables 1 and 2). Weight depression was observed in most groups exposed

Table 1
Design Summary for Osborne-Mendel Rats EDC Gavage Experiment

	Initial number of animals	EDC dosage[a]	Observation period (weeks) treated	Observation period (weeks) untreated	Time-weighted average dosage[b]
Male					
Untreated control	20	–	–	106	–
Vehicle control	20	0	78	32	0
Low dose	50	50	7		47
		75	10		
		50	18		
		50[c]	34	9	
		0		32	
High dose[d]	50	100	7		95
		150	10		
		100	18		
		100[c]	34	9	
		0		23	
Female					
Untreated control	20	–	–	106	–
Vehicle control	20	0	78	32	0
Low dose	50	50	7		47
		75	10		
		50	18		
		50[c]	34	9	
		0		32	
High dose[d]	50	100	7		95
		150	10		
		100	18		
		100[c]	34	9	
		0		15	

Data from NCI (1978a).

[a] Dosage, given in mg/kg body weight, was administered by gavage 5 consecutive days per week.

[b] Time-weighted average dosage (over a 78-week period) = Σ (dosage $\times$ weeks received) $\div$ 78 weeks.

[c] These dosages were cyclically administered with a pattern of 1 dosage-free week followed by 4 weeks (5 days per week) of dosage at the level indicated.

[d] All animals in this group died before the bioassay was terminated.

to EDC. By 50 weeks, the weight depression averaged 12% in the high-dose rats and over 20% in the high-dose female mice. Mortality was early and severe in most dosed animals, especially those given the highest dosages. Mean survival was approximately 55, 55, and 67 weeks on test for high-dose male rats (Fig. 1), female rats, and female mice, respectively, all values considerably below that of

Table 2
Design Summary for B6C3F1 Mice EDC Gavage Experiment

	Initial number of animals	EDC dosage[a]	Observation period (weeks)		Time-weighted average dosage[b]
			treated	untreated	
Male					
Untreated control	20	–	–	90	–
Vehicle control	20	0	78	12	0
Low dose	50	75	8		97
		100	70		
		0		12	
High dose	50	150	8		195
		200	70		
		0		13	
Female					
Untreated control	20	–	–	91	–
Vehicle control	20	0	78	90	0
Low dose	50	125	8		149
		200	3		
		150	67		
		0		13	
High dose	50	250	8		299
		400	3		
		300	67		
		0		13	

Data from NCI (1978a).

[a] Dosage, given in mg/kg body weight, was administered by gavage 5 consecutive days per week.

[b] Time-weighted average dosage = Σ (dosage × weeks received) ÷ Σ (weeks receiving chemical).

the low-dose and control group. Mortality was not dose-related in male mice. The early deaths were usually not due to cancer; rather, the toxic effects of EDC appeared to be responsible for these deaths. Animals dying early had a variety of lesions, including bronchopneumonia and endocardial thrombosis (rats), that may have been associated with death. The pneumonia may have been a result of viral, bacterial, or mycoplasmal infection, and the exposure to the chemical may have increased the susceptibility of the animals to develop severe pulmonary lesions, which would lead to death.

Appearance of Induced Tumors in Rats and Mice Exposed to EDC by Gavage

A variety of tumors were seen in both control animals and animals receiving EDC. Except for those described in Table 3, most were thought to be

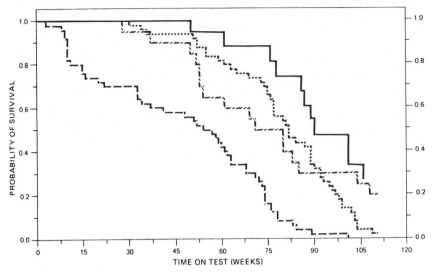

Figure 1
Survival curves for male rats exposed to EDC and controls. (———) Untreated control; (– – – – –) vehicle control; (·····) low dose; (– – – –) high dose. (Data from NCI 1978a).

spontaneous and not the result of animal exposure to EDC. With use of the Cochran-Armitage and Fisher exact tests, there was a statistically significant increased incidence of the following rat tumors: males—subcutaneous fibromas, gastric squamous cell carcinomas, and hemangiosarcomas and females—mammary adenocarcinomas. In mice the following tumors were significantly increased: alveolar-bronchiolar adenomas in males and females, and mammary adenocarcinomas and endometrial tumors in females.

Life-table methods (Gart et al. 1979) revealed dose-related effects for several tumor types (Fig. 2). However, these effects may be associated with death owing to toxicity rather than to cancer. For male rats, stomach cancers usually did not eventuate in death, since the majority of lesions were small, whereas hemangiosarcomas frequently resulted in death as a result of hemorrhage. Mammary carcinomas in female rats also appeared to cause death. Lung tumors in mice were usually not the cause of death, but stomach, uterine, and mammary tumors frequently contributed to early mortality.

It must be noted that the statistics (NCI 1978a) were not age-adjusted for early mortality owing to toxicity, so that the induced tumors actually occurred in incidences higher than those in the statistical tables and Table 3. Thus, in Table 3 hemangiosarcomas in female rats, and stomach and uterine carcinomas in female mice may well be induced tumors. The histopathologic review of these lesions supports this view, as does comparison with incidence in historical controls (Table 3).

Table 3
Incidence of Specific Tumor Types in Osborne-Mendel Rats and B6C3F1 Mice Exposed to EDC

Species	Sex	Tissue	Tumor type	Matched vehicle control	Pooled vehicle control	Doses low	Doses high[b]	NCI[a] vehicle control (%)
Rat	male	subcutaneous	fibroma	0/20	0/60	5/50	6/50 (12)	(1.6)
		stomach	squamous cell carcinoma	0/20	0/60	3/50	9/50 (18)	(0)
		blood vessels	hemangiosarcoma	0/20	1/60	9/50	7/50 (14)	(2.4)
	female	mammary gland	adenocarcinoma	0/20	1/59	1/50	18/50 (36)	(2.0)
		blood vessels	hemangiosarcoma	0/20	0/59	4/50	4/50 (8)	(0)
Mouse	male	lung	alveolar-bronchiolar adenoma	0/19	0/59	1/47	15/48 (31)	(3.9)
	female	lung	alveolar-bronchiolar adenoma	1/20	2/60	7/50	15/48 (31)	(3.3)
		stomach	squamous cell carcinoma	1/20	1/60	2/50	5/48 (10)	(0.5)
		mammary gland	adenocarcinoma	0/20	0/60	9/50	7/48 (15)	(0)
			adenocarcinoma	0/20	1/60	3/49	4/47 (9)	(1.1)
		uterus	endometrial stromal polyp or sarcoma	0/20	0/60	5/49	5/47 (11)	(0.5)

[a] 250 male and 250 female Osborne-Mendel rats which received corn oil by gavage; 180 male and 180 female B6C3F1 mice which received corn oil by gavage.
[b] Percentage is given in parentheses.

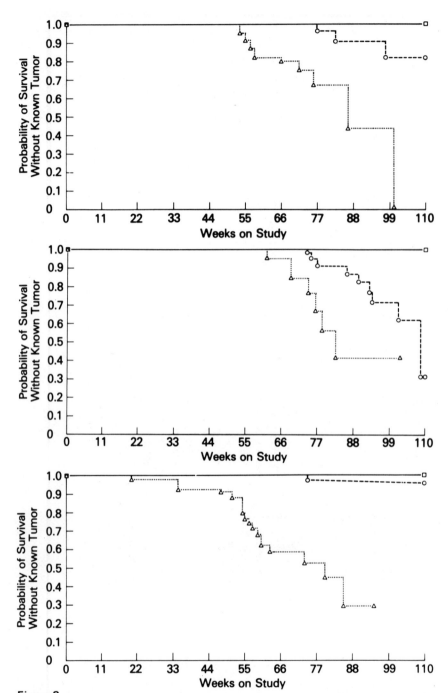

Figure 2
Life tables for rats receiving EDC. (*Top*) Male rats–stomach squamous cell carcinoma; (*middle*) male rats–hemangiosarcoma; (*bottom*) female rats–mammary adenocarcinoma. (□) Control; (○) low dose; (△) high dose.

In Figure 3 it is shown that the vast majority of high-dose male rats that developed the induced tumors (stomach, subcutis, blood vessels) died after week 60, when the majority of high-dose rats had died without induced tumors (Fig. 1). If a larger population was at risk after week 60, additional induced tumors may have appeared.

Pathology of Induced Hyperplastic and Neoplastic Lesions

A summary of the induced tumors have already been reported (NCI 1978a). Other features of these lesions will be emphasized here. Hyperplastic lesions of the forestomach in several rats and mice included diffuse and focal acanthosis, hyperkeratosis, and basal cell hyperplasia (Figs. 4, 5). A few animals had multiple hyperplastic or neoplastic lesions. Papillomas were seen in a few rats and mice with carcinomas (Fig. 6). The 12 squamous cell carcinomas in male rats invaded to the submucosa, muscularis, serosa (Fig. 7), or peritoneal cavity. Five of 12 mouse gastric carcinomas in males or females invaded the serosa or peritoneal cavity (Figs. 8, 9). Carcinomas of the forestomach were rare in untreated or vehicle control rats or mice (Table 3). Two high-dose male rats had sarcomas that originated in the stomach and then metastasized to the peritoneal cavity. These unusual tumors may have been induced by the chemical.

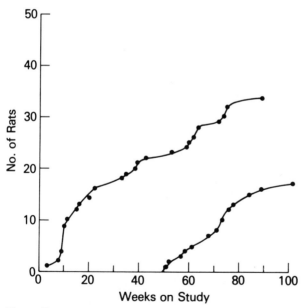

Figure 3
Mortality of high-dose male rats receiving EDC and dying without or with induced tumors (stomach, blood vessels, subcutis). (*Top plot*) without induced tumors; (*bottom plot*) with induced tumors.

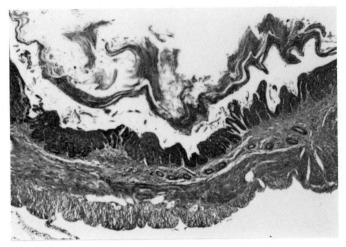

Figure 4
Diffuse acanthosis and hyperkeratosis in the forestomach of a high-dose male rat. Magnification, 40×.

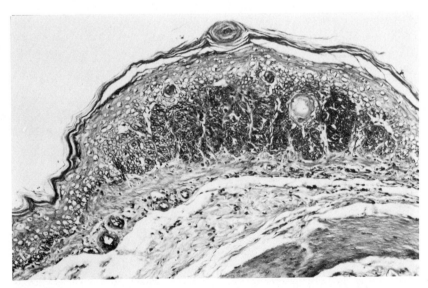

Figure 5
Focal basal cell hyperplasia of the forestomach in a low-dose male rat. Magnification, 130×.

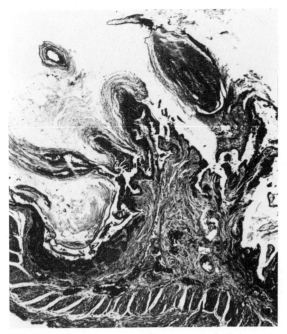

Figure 6
Forestomach papilloma in a high-dose male rat. Magnification, 40×.

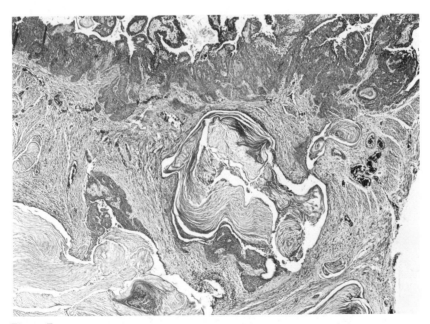

Figure 7
Gastric squamous cell carcinoma invading wall of stomach in high-dose male rat. Magnification, 40×.

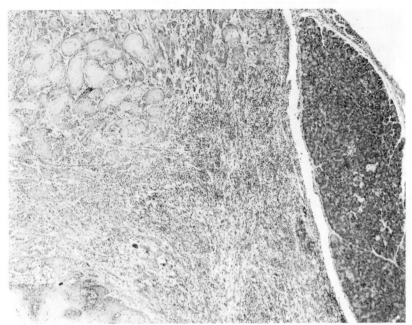

Figure 8
Gastric squamous cell carcinoma metastatic to pancreas in low-dose female mouse. Magnification, 55×.

Figure 9
Gastric squamous cell carcinoma in low-dose female mouse. Note formation of keratin pearls. Magnification, 130×.

Vascular tumors of the rat spleen and other tissues were frequently composed of large vascular spaces and cords of neoplastic cells (Fig. 10). Hemorrhage of the lesions led to death.

Mammary tumors in rats were tubular and papillary adenocarcinomas and were multiple in only a few rats but metastasized to the lungs in two rats (Fig. 11). The majority of mouse mammary tumors were adenosquamous tumors (Fig. 12), which spread to the lungs of two mice (Fig. 13). The usual spontaneous mammary tumors are type-C adenocarcinomas. Mouse uterine sarcomas and carcinomas were highly malignant, invading all layers of the uterus; three invaded peritoneal cavity (Fig. 14). These reproductive tumors are rare in control animals (Table 3).

Metastatic Tumors in Rats and Mice Exposed to EDC

Nine rats and seven mice receiving EDC developed metastatic tumors of organs in which significantly increased incidences of tumors were found (Table 4). Metastatic tumors of these types are rare in these rats and mice (Ward et al. 1979). Several are illustrated in Figures 8, 11, and 13.

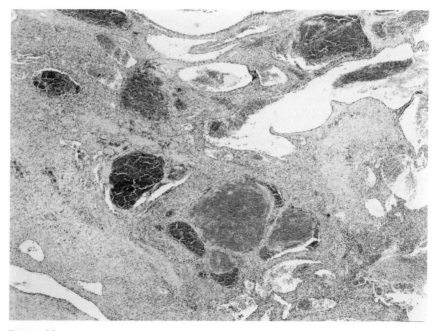

Figure 10
Hemangiosarcoma of spleen in high-dose male rat. Note large vascular spaces and thick cellular trabeculae. Magnification, 40×.

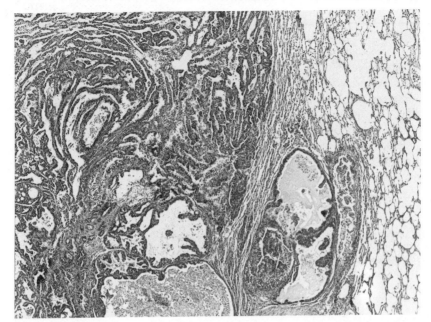

Figure 11
Mammary adenocarcinoma metastatic to lung in high-dose female rat. Magnification, 40×.

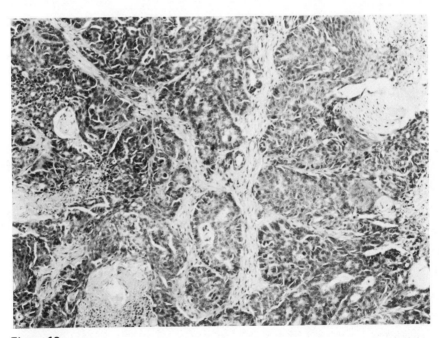

Figure 12
Mammary adenosquamous carcinoma in low-dose female mouse. Note glandular and squamous differentiation. Magnification, 130×.

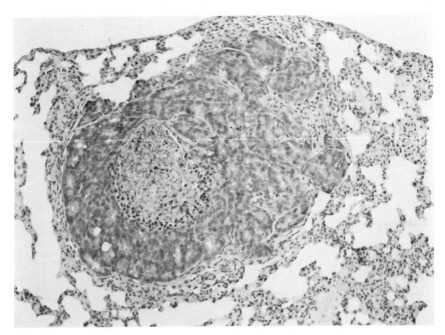

Figure 13
Metastatic mammary adenosquamous carcinoma in lung of low-dose female mouse. Magnification, 130×.

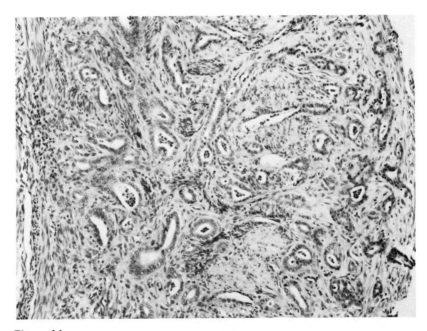

Figure 14
Uterine adenocarcinoma invading serosa of high-dose female mouse. Magnification, 130×.

Table 4
Origin of Metastatic Tumors in Rats and Mice Exposed to EDC

Primary site	Site of metastases	low dose	high dose
Rats			
Males			
stomach			
squamous cell carcinoma	peritoneum		1
leiomyosarcoma	peritoneum		1
hemangiosarcoma	peritoneum	1	
spleen			
hemangiosarcoma	adrenal		1
multiple organs			
hemangiosarcoma	peritoneum		3
Females mammary gland			
adenocarcinoma	lung		2
Mice			
Females			
stomach			
squamous cell carcinoma	peritoneum	2	
mammary gland			
adenocarcinoma	lung	2	
uterus			
endometrial stromal			
sarcoma	peritoneum		2
adenocarcinoma	peritoneum		1

DISCUSSION

The carcinogenicity of EDC was demonstrated in rats and mice in the NCI bioassay. This chemical was also shown to be carcinogenic by Van Duuren et al. (1979), as lung tumors were induced in mice after repeated skin application of EDC. Tumors induced by EDC after gavage exposure were similar to those found in animals receiving other halogenated hydrocarbons by gavage—ethylene dibromide (EDB; 1,2-dibromoethane) (NCI 1978c), dibromochloropropane (DBCP; 1,2-dibromo-3-chloropropane) (NCI 1978b), and allyl chloride (NCI 1978d). Stomach tumors were seen in these animals, although the highest incidences were seen in animals given DBCP and EDB. Mammary tumors were induced by DBCP and EDC. The finding of lung tumors in mice in our study was similar to the results of Van Duuren et al. (1979). Additional comparative aspects of the carcinogenicity of halogenated compounds is discussed by Infante and Marlow (this volume) and Weisburger (1977).

The carcinogenicity of EDC in laboratory animals and the mutagenicity of EDC in vitro suggests that EDC may pose a potential health risk to humans.

The degree and importance of that risk is a subject requiring further research and evaluation.

ACKNOWLEDGMENTS

The bioassay was performed under Public Health Service Contract NO 1-CP-43350 to Tracor Jitco, Inc. at Hazleton Laboratories America, Inc. The personnel responsible for various phases of the bioassay are noted in the NCI 1978a reference. The additional aid of Dr. Jim Joiner, Robert Nye, Ann Johnson, and Barbara Coolidge was appreciated.

REFERENCES

Gart, J. J., K. C. Chu, and R. E. Tarone. 1979. Statistical issues in interpretation of chronic bioassay tests for carcinogenicity. *J. Natl. Cancer Inst.* **62**:957.

NCI (National Cancer Institute). 1978a. *Bioassay of 1,2-dichloroethane for possible carcinogenicity.* NCI Carcinogenesis technical report series No. 55, DHEW Publication No. (NIH) 78-1361, Government Printing Office, Washington, D.C.

———. 1978b. *Bioassay of dibromochloropropane for possible carcinogenicity.* NCI Carcinogenesis technical report series No. 28, DHEW Publication No. (NIH) 78-828, Government Printing Office, Washington, D.C.

———. 1978c. *Bioassay of dibromoethane for possible carcinogenicity.* NCI Carcinogenesis technical report series No. 86, DHEW Publication No. (NIH) 78-1336, Government Printing Office, Washington, D.C.

———. 1978d. *Bioassay of allyl chloride for possible carcinogenicity.* NCI Carcinogenesis technical report series No. 73, DHEW Publication No. (NIH) 78-1323, Government Printing Office, Washington, D.C.

Van Duuren, B. L., B. M. Goldschmidt, G. Loewengart, A. C. Smith, S. Melchionne, I. Seidman, and D. Roth. 1979. Carcinogenicity of halogenated olefinic and aliphatic hydrocarbons. *J. Natl. Cancer Inst.* **63**:1433.

Ward, J. M., D. G. Goodman, R. A. Squire, K. C. Chu, and M. S. Linhart. 1979. Neoplastic and nonneoplastic lesions in aging (C57BL/6N X C3H/HeN) F1 (B6C3F1) mice. *J. Natl. Cancer Inst.* **63**:849.

Weisburger, E. K. 1977. Carcinogenicity studies on halogenated hydrocarbons. *Environ. Health Perspect.* **21**:7.

COMMENTS

PLOTNICK: Seventeen other chemicals were being employed in gavage studies done at Hazleton in the same room.

WARD: Yes, 17 with the mice and 4 with the rat.

PLOTNICK: What were these chemicals? Was EDB used?

WARD: All the chemicals used were halogenated compounds. A few of them induce the types of tumors seen in the EDC study, but the majority of them do not. For example, DBCP induces mammary carcinomas at a low incidence level in the rats, but at a much higher incidence in this EDC study. DBCP induced vascular tumors at a fairly low incidence level, so that the EDC-treated animals would, theoretically, have to receive the dose of the DBCP animals to develop this incidence of vascular tumors.

PLOTNICK: What about interactions, interference with enzymatic detoxification, and so on?

WARD: There are a lot of possibilities.

PLOTNICK: We've published much information, and there is much in the literature, indicating that with oral doses of volatile materials, depending on vapor pressure and rate of biotransformation, a percentage of the administered oral dose appears unchanged in the expired air. For example, with EDB administered to guinea pigs it is 10–14%.

I won't challenge your results, but you have an awful lot of faith in a study that you considered not to be flawed, and which I consider to be so utterly flawed that you can't rely upon it.

WARD: You can argue that point. If you consider the other studies and their results, EDC is carcinogenic and mutagenic. You can make this point for any given study, and we've had this argument about many different studies. For example, half of the chemicals tested on animals in the same room were not carcinogenic, and yet the animals didn't have the same response as EDC animals. You can make that same argument: Why didn't they have the same kinds of tumors?

HOOPER: Yes, it is possible, but it seems unlikely (see Hooper et al., this volume) that those negative or weakly carcinogenic chemicals may act synergistically, even at low levels, to induce enzymes which activate DBCP or EDC to reactive metabolites that would not be produced from exposures to EDC alone, giving a positive result that EDC alone would not have given.

PLOTNICK: There is no question of that.

TER HAAR: One of the things we are seeing here is a great need for detail and care in conducting studies and also the need to dig into the background of some of our old studies. What we have heard here today would certainly support that the given situation with this given material, whatever it is, certainly indicates that it causes a carcinogenic response. But in the future we will be looking more at the question of promotion and induction and the potential for other materials interacting. This question really deals with the risk of the material and will require a great deal of detail and care in designing studies. In the case of EDC, is the question of risk with respect to EDC itself or is it other factors and components? This discussion really tells us that we need to be taking a closer look at exactly how we're dosing animals because it's not the simplified situation that we thought it was in the past. And this makes it very difficult to interpret what the real conclusions of your studies are.

WARD: You're exactly right. In the last couple of years, we have been using one animal room for one chemical only in testing. Also, now there is detailed chemical analysis, not only at the beginning of a study but in intervals throughout the study by two different laboratories. So it's much better than previously.

MALTONI: I am concerned about the chamber rooms and also about contamination. In addition, I am concerned that the control group, which has already been shown to be small, may be the same control group for all these different experiments.

These experiments were run all together, and you have spoken about 60 pooled controls, which means that the experience of the laboratory itself was not very great with that fact alone and that the number of gavage is very little. These same 20 animals probably were the control group for a number of experiments. In addition it will never be known if the same people who conducted the experiments in the third group were the same people observing the control group. Do you understand what I mean? Because probably this EDC experiment could have been grouped with one of the ten other types of experiments being run.

Also I would like very much not to see slides of the same sort of tumor. I would very much like to see the tumor produced by the compound itself, not analogous slides. This is particularly important for squamous cell tumors of lung and stomach, because you know many people take the adenomatous hyperplasia of the epithelium in the area between glandular stomach and forestomach and call it squamous cell carcinoma. A lot of so-called experienced pathologists have made this mistake.

WARD: You're correct.

MALTONI: And so we must see the pathology reports. In 1972, our pathological skill was much less sophisticated than it is today. But the possibility of misinterpretation among blood lakes angioma and angiosarcoma of the spleen, as you know, is something that also deserves a lot of attention.

I would also like to know if the only animal systematically and pathologically observed is the animal with some gross pathology. It is a very important question.

WARD: The original protocol was limited in histopathology. However before the study was finished, the protocol was expanded and 30 or more tissues were taken from each animal, including all the routine tissues as well as any grossly visible lesion. The original protocols did, however, require complete gross necropsies of all animals in the study.

MALTONI: How many animals were done this way?

WARD: All of them.

MALTONI: Is all the pathology in the data?

WARD: Yes, but not for the acute studies. For the 2-year study, the slides were reviewed from all the animals.

MALTONI: No, I mean of this type of study. All of the animals have been systematically, currently, pathologically reviewed?

WARD: Yes.

MALTONI: Because after 1975, it was rather unusual in many studies.

WARD: In 1972 all the tissues were put in formalin, and only a few were sectioned. Before the study was finished the protocol was changed so that 30 or more tissues were taken from each animal. All the tissues were always preserved in formalin, although they may not have prepared on slides. Only a few autolytic or cannabalized animals may have been saved.

MALTONI: And finally, a new question. Do you make a distinction, as we do, between fibromas in males, subcutaneous hemangiomas that are mammary fibromas (fibroadenomas with some epithelial components), and dermatofibromas (which are some sort of collagenic tissue, nodulated or plastic) that I consider as a benign tumor? It is a matter of concern over interpretation to compare our results.

WARD: Yes, you're right. These were all in the mammary area, and they were large masses.

MARLOW: Jerry [Ward], you didn't mention anything about time to tumor. In the bioassay I notice that most of the tumors started after 60 weeks, with the exception of the female mammary tumors in the rats, which started at 20 weeks.

WARD: Well, there was only one animal at 20 weeks with a mammary tumor.

HOOPER: We calculated carcinogenic potency of EDC in female Osborne-Mendel rats using life-table data for mammary gland adenocarcinomas and fibroadenomas, and the results were significant.

KARY: Yes, this is an area to which more discussion should be directed. In this experiment we have seen a number of changes of dose, and so I would like to raise a general type of question. What effect does this type of parameter have on the overall pharmacokinetics of how EDC is going to be handled in these particular animal experiments with regard to metabolism?

WARD: The experiment to perform to answer this question would be a gavage study where you have a peak in the blood levels, a change in doses, and then compare this experiment to one where there is a uniform exposure, by inhalation, feed, etc., and see if it makes a difference in the outcome of the carcinogenesis experiment.

KARY: The question here, Jerry, that we shouldn't lose sight of, is: Does changing doses like this have an effect on the experiment?

WARD: You're right, but no one has ever shown that it may have an effect on the incidence of induced tumors or whether the chemical is a carcinogen under certain conditions only. Until someone does that experiment, we'll always speculate.

Carcinogenic Potency: A Progress Report

BRUCE N. AMES, KIM HOOPER, CHARLES B. SAWYER,
ALAN D. FRIEDMAN, RICHARD PETO,* WILLIAM HAVENDER,
LOIS S. GOLD, THOMAS HAGGIN, ROBERT H. HARRIS, AND
MARGARET ROSENFELD
Department of Biochemistry
University of California
Berkeley, California 94720

*Radcliffe Infirmary
University of Oxford
Oxford OX2 6HE, England

Chemicals in the environment, both natural and man-made, are now recognized as increasing the risk of human cancer. Current regulatory policy and scientific research have focused mainly on determining whether a chemical is capable of inducing cancer, but little systematic effort has been directed at quantifying how much hazard a given substance poses. It has become increasingly evident, however, that quantifying the intrinsic carcinogenic power of different chemicals is crucial to developing a sensible policy response to these hazards. This realization has come about because the relative differences in carcinogenic potency that may even now be inferred from animal experiments are enormous. For example, a daily dose of saccharin that would give cancer to 50% of exposed rats would be more than 10^6-fold larger (mg per kg body weight basis) than the dose of aflatoxin that would yield the same incidence of tumors. Regulatory decisions aimed at avoiding the largest number of cancer deaths should take into account the extent and level of human exposure to chemicals and the vast differences in intrinsic carcinogenicity.

To improve risk assessments, risk-benefit judgments, and regulatory policy, an index number of carcinogenic strength (or potency) for each chemical is desirable. Ideally, we would like to have quantitative information on the capacity of various chemicals to cause cancer in man, but with rare exceptions this is not available. An alternative source of information is animal bioassays. In our laboratory, we have been engaged for several years in creating a comprehensive data source incorporating the animal bioassays reported in the world's literature that are suitable for determining a potency value. This paper details our progress in attaining the following project objectives:

1. The calculation of a quantitative measure of potency.
2. A thorough search of the world's literature to identify tests that are sufficiently complete to allow potency estimates to be made and, from this raw data, development of a computerized data base that enables convenient storage and manipulation of the information.
3. Analysis of the sources of variation in the data.

DEVELOPMENT OF A POTENCY INDEX

Attempts to quantify estimates of potency began as early as 1930 (Twort and Twort 1930, 1933; Iball 1939; Bryan and Shimkin 1943; Druckrey 1967; Meselson and Russell 1977). The most generally satisfactory of these indices was devised by Meselson and Russell (1977), where the potency, k, was given by

$$k = \frac{-\ln(1-I)}{D \times t^{n+1}} \qquad (1)$$

and where I is the cumulative single-risk incidence observed at time t (expressed as the fraction of a normal lifetime), D is the administered dose rate (expressed as mg per kg body weight per day), and n was taken as 3 (selected as the best estimate of the dependence of tumor appearance upon duration of exposure).

This index has two limitations. First, it does not take into account the incidence of spontaneous tumors in control animals, which in practice can vary widely and has a substantial effect on the magnitude of the calculated potency. Second, it does not take into account the progressively smaller number of animals at risk because they have died from causes other than tumors during the course of the experiment. (The effect of failing to account for this mortality is to underestimate the true potency.) Richard Peto (Oxford University) together with Charles Sawyer and Alan Friedman of our group have developed the theory of calculating carcinogenic potency from animal bioassays and converted this to computer programs. Our index, the TD_{50} (tumorigenic dose$_{50}$), is the daily dose rate required to decrease by half the probability of an animal remaining tumor-free at the end of a standard lifetime (taken as 104 weeks for rats and mice). The calculation of this index takes into account whatever spontaneous tumor incidence occurs in control animals and, where life-table data is available, corrects for intercurrent mortality. It has the additional merit that the dose rate to be estimated (that which gives cancer to half of otherwise tumor-free animals) is usually not far from an actual dose used in an experiment that yields statistically positive results; therefore, only a small extrapolation from experimental observation is necessary. This means that the choice of a particular dose-response function (ranging from Mantel-Bryan to linear) will not greatly affect the estimate of TD_{50}.

Computer programs have been developed to estimate a TD_{50} together with its confidence limits; this program also analyzes the shape of the dose-response relation and the probability that the TD_{50} is significant. It is now a routine task to determine a TD_{50} from any suitable set of data. Separate programs have been developed to calculate life-table TD_{50} values (where complete data are available giving the time of death and tumor occurrence for each animal) and summary TD_{50} values (where tumor data have been reported only as a summary for a group of animals).

We have also developed a measure of the sensitivity of negative bioassays, which can differ enormously. The sensitivity depends on the dose levels used and on the experimental design. A negative test is described as excluding TD_{50} values below a certain limit rather than simply as "negative." Some experiments have such small numbers of test animals and use such low doses that they could not have detected any but the most potent carcinogens. The comparison of research designs and dose levels will sometimes make it possible to reconcile positive and negative results with the same compound: if two such tests examined different regions of the TD_{50} range, they need not be contradictory.

THE CREATION OF A DATA BASE

Whereas there is a paucity of data on human carcinogens, there is an abundance of research reports of animal bioassays on hundreds of chemicals. We have conducted a painstaking search of the world's literature to collect and evaluate all tests that would be suitable for the calculation of a TD_{50}. In addition to exhaustively scanning the major cancer journals and *Current Contents,* we have consulted several major bibliographies of cancer tests, including the monographs on chemical carcinogens prepared by the International Agency for Research on Cancer (1972-1979) and the PHS survey of carcinogens (Shubik and Hartwell 1948-1973). In addition, we have obtained all the bioassays carried out by the National Cancer Institute (NCI) that have been released.

Most of these tests have utilized quite diverse and unsystematized protocols; this makes direct comparisons difficult. We have tried to cope with this problem by selecting only those tests in which:

1. exposure occurred chronically over at least one-half the animal's normal lifespan,
2. the route of exposure was by diet, gavage, water, or inhalation (i.e., analogous to the major human exposures),
3. the whole body was exposed rather than only a specific site, as with subcutaneous injection or skin painting, and
4. there was a control group.

We now have analyzed over 1500 experiments that meet these criteria, chiefly rat and mouse-feeding studies, which have sufficient data for calculations. We define an experiment as the control group and the various dose groups for one chemical in one species, strain, and sex from one research report.

Most of the papers we have collected report only the cumulative number of tumor-bearing animals seen over the course of the experiment and the number at risk at the start. TD_{50} values calculated from such summary data are more subject to bias than TD_{50} values calculated from life-table data. Fortunately, the NCI has recently completed tests on nearly 200 chemicals using comparable protocols, and they have supplied us with the full life table from each experiment. Here it has been possible to estimate unbiased TD_{50} values using the Peto-Cox theory.

This vast sum of data has been stored in a computerized data base, which facilitates rapid retrieval and manipulation of information. Thus, it is possible to make comparisons in potency between sexes, strains, and species, as well as to carry out mathematical analyses.

This phase of the project is nearly complete and currently includes information on over 600 chemicals. Many of these include mulitple tests that use different strains and species. The output includes the estimated TD_{50}, confidence limits, tumor type and site, and information on the dose-response relationship. We have partially completed our error check on the data base.

ANALYSIS OF THE DATA

A goal of this analysis is to see how well results in one species of animal predict those in another and to examine the reasons for aberrations. Questions of interest that can readily be approached with the computerized data base include:

1. How similar are the TD_{50} values calculated from independent tests on the same compound?
2. How well do males and females compare within a strain in a single bioassay?
3. How well do strains within a species correlate with each other?
4. How well do rats and mice compare in overall sensitivity as well as in the preferred target organ? On the basis of preliminary results, we anticipate that interspecies extrapolations, at least among rodents, will usually agree within a factor of ten. Seen against the possible variation of potency of some 10^7-fold, such results will be clearly useful. Chemicals that deviate far from the usual correlations between sexes and species can alert us to special circumstances—unusual pharmacokinetics, or a peculiar metabolic route—that could be investigated. This would result in a better understanding of the conditions under which useful predictions can be made between laboratory animals and humans.
5. How well do rodents compare with other species? An area of particular theoretical interest will be the comparison of long-lived and short-lived species; the very fact that long-lived animals are long-lived, that is, they do not succumb to cancer after only 2 years, as do rats, points to the existence of mechanisms of cellular and tissue control that are quantitatively, and perhaps qualitatively, different from those of short-lived mammals. This quantitative difference is, in fact, much larger than the simple ratio of lifetimes would suggest because of the fourth- to fifth-order dependence of cancer incidence upon age within a species (Peto 1977, 1979). This comparison bears directly on the problem of extrapolating risk estimates from rodents to man. Therefore, a major and important focus of this project will be to analyze the potency of chemicals that have been tested in both rodents and monkeys. We are collaborating with R. Adamson (NCI), who is currently conducting extensive cancer tests with monkeys. He will give us life-table data on the seven compounds that he finds positive in monkeys and data on 19 test compounds that have either failed to induce

tumors or have not been under test a sufficient length of time. These TD_{50} values will provide us with an important set of reference points for making rodent-primate risk extrapolations. We also intend to explore the available human data on carcinogenic potency following up previous work (Meselson et al. 1975).
6. One would like to know which environmental chemicals pose the greatest hazard to humans, though this is a very difficult and complicated subject. In the meantime, we will examine the TD_{50} values of the chemicals that have already aroused public concern because of widespread human exposure (such as DDT, dioxin, benzene, saccharin, benzo[a]pyrene, vinyl chloride, ethylene dibromide, and ethylene dichloride).
7. How do TD_{50} values compare when the same chemical is administered by different routes (e.g., inhalation versus diet) or different dosing schedules?

ACKNOWLEDGMENTS

This work was supported by Department of Energy contract DE-AM-03-76SF00034 PA156, by a California Policy Seminar grant to B. N. A., and by National Institute of Environmental Health Sciences Center grant ES-01896. We are indebted to Jade Goldstein and Elizabeth Higgins for help with the data base and to Ken Chu and Jerrold Ward of NCI for much help with the NCI bioassays.

REFERENCES

Bryan, W. R. and M. B. Shimkin. 1943. Quantitative analysis of dose response data obtained with three carcinogenic hydrocarbons in strain C3H male mice. *J. Natl. Cancer Inst.* 3:503.

Druckrey, H. 1967. Quantitative aspects in chemical carcinogenesis. *UICC Monogr. Ser.* 7:60.

Iball, J. 1939. The relative potency of carcinogenic compounds. *Am. J. Cancer* 35:188.

International Agency for Research on Cancer. 1972--1979. The evaluation of carcinogenic risk of chemicals to man. *IARC Monogr.* 1-19.

Meselson, M. and K. Russell. 1977. Comparisons of carcinogenic and mutagenic potency. *Cold Spring Harbor Conf. Cell Proliferation* 4:1473.

Meselson, M. S., B. N. Ames, P. M. Dolinger, V. H. Freed, J. Kolojeski, R. L. Metcalf, M. A. Schneiderman, E. K. Weisburger, C. F. Wurster, and R. Hansberry. 1975. Health hazards of chemical pesticides. In *Pest control: An assessment of present and alternative technologies, Vol. 1. Contemporary pest control practices and prospects: The report of the executive committee*, pp. 4 and 54. National Research Council, National Academy of Sciences, Washington, D.C.

Peto, R. 1977. Epidemiology, multistage models and short-term mutagenicity tests. *Cold Spring Harbor Conf. Cell Proliferation* 4:1403.

Peto, R. 1979. Detection of risk of cancer to man. In *Proc. R. Soc. Lond. B Biol. Sci.* **205**:111.

Shubik, P. and J. L. Hartwell, ed. 1948-1973. *Survey of compounds which have been tested for carcinogenic activity*. Public Health Service Publication number 149. Government Printing Office, Washington, D.C.

Twort, C. C. and J. M. Twort. 1930. The relative potency of carcinogenic tars and oils. *J. Hyg.* **29**:373.

———. 1933. Suggested methods for the standardisation of the carcinogenic activity of different agents for the skin of mice. *Am. J. Cancer* **17**:293.

COMMENTS

KARY: Dr. Ames, where do you differentiate between routes of exposure?

AMES: We put down the route of exposure so that it is in the computer print-out. For both an inhalation study and with a feeding study we calculate the dose. Whether inhalation is going to be exactly the same as feeding, we don't know. We can ask the computer to compare all the inhalations and feedings. We'll start to do that kind of thing soon. Getting this computer print-out has been an enormous amount of work, and there are still many errors that we have to weed out. As a result, we really haven't had time to sit down and try and analyze it in any great detail.

MALTONI: I want to compliment Dr. Ames on this tremendous amount of work that he is doing. I think everybody is aware of how much we need a quantitative approach for this. If a substance is a carcinogen, we must know if it is more or less carcinogenic than a possible subacute compound.

Our laboratory is working on this problem in a different way from Dr. Ames. We are not working so much on world-wide data, but rather working in our own system, always using the same type of animals with some 60 different compounds and keeping them entirely and strictly homogeneous and serialized to the following extent. We would like to be able to record all the subtle types of differences within the doses, the routes, the schedule of treatment, the role of species, of sex, age, etc., for a series of compounds that may have a similar type of structure. We also would like to assess what effect molecular structure will have—what will be the bearer of the slightest changes and to find out, really, what is the practical value in selecting one compound over another.

An important element in these assessments is the weight of the lab variability and the experimental situation, which may affect your type of quantitative monitoring. An aspect really bound to the monitoring you have studied is the biological model that is engaged. Time of survival of this animal, when it has been examined, and where it has been kept, observed, and interpreted may bring enough information so that when you go to a difference from a picogram down to a gram, data are coming up pretty well, but when you are working in a range of the milligram you will have extremely wide sources of data.

AMES: I know. There are 2000 labs using our *Salmonella* test, and I know there are some people messing it up completely, even though we write the directions down very precisely. Some people do mess up experiments, and animal cancer tests are more complicated than *Salmonella*, so all we can do here is put down what people report.

But I am surprised, so far, on how much labs agree, given that we're not very concerned about a factor of ten in all of this. That will all come out in the analysis. It may be that for discrepancies some expert will have to look at that paper and say "garbage." We don't know enough now to say that any paper is garbage. Experts will have to start analyzing, using our data base as a guide, and say, "Here's a discrepancy; what do we make of this paper or that one?"

MALTONI: You know that just 2 years ago we published the results of tests on benzene—claimed to be a negative compound for some 30 years. All of the previous experiments in their entirety were too little, too inadequate, too short, too impure. But just by performing a very small experiment on 200 animals kept in a very controlled way, you can pick up very quickly that benzene is a carcinogen (Maltoni and Scarnato 1977). And so, to pick up the false negative is quite a good job.

AMES: I'd like to make one more point. One of the reasons we got into this is that Meselson and Russell (1977) had published in the *Origins of Human Cancer* an analysis of our data on mutagenic potency in our test versus carcinogenic potency (which they calculated) for about a dozen chemicals. That was brave. Clearly, ground up rat liver is only some crude first approximation of a rat. Can ground up rat liver plus *Salmonella* tell you anything about potency in an animal?

Meselson found a rough correspondence between mutagenic and carcinogenic potencies for most, but not all, of those chemicals. There was a million-fold spread in the potency range for both.

We became interested in the mutagenic and carcinogenic potency. We soon found that carcinogenic potency needed our full attention, and we have been so busy thinking about carcinogenic potency we haven't really thought about mutagenic potency. But I think out of this will come calibration points for calibrating all of the short-term tests. Scientists are finding new mutagens by the bucketful—there are so many mutagens out there. Cancer tests are never going to catch up; only a few hundred cancer tests are being done in the world in a year. We need some way of determining if a powerful mutagen is more likely to be a greater human hazard than a weak mutagen as determined by a battery of short-term tests. We don't know, but at least I think we will have the calibration points. (Joyce McCann will look into all of this.) One can use human liver in some of the short-term tests, and compare human liver with rat liver. If there's a species difference between rats and mice, one can put in mouse liver and rat liver and perhaps explain a difference and then try human autopsy liver. We hope that with these calibration points we can see how good or bad short-term tests are quantitatively.

References

Maltoni, C. and C. Scarnato. 1977. Le prime prove sperimentali dell'-azione cancerogena del benzene. *Gliospedali della Vita* **4**:111.

Meselson, M. and K. Russell. 1977. Comparisons of carcinogenic and mutagenic potency. *Cold Spring Harbor Conf. Cell Proliferation* **4**: 1473.

The Carcinogenic Potency of Ethylene Dichloride in Two Animal Bioassays: A Comparison of Inhalation and Gavage Studies

KIM HOOPER, LOIS SWIRSKY GOLD, AND BRUCE N. AMES
Department of Biochemistry
University of California
Berkeley, California 94720

As part of our project on carcinogenic potency (Ames et al., this volume), we have analyzed the two long-term cancer bioassays of ethylene dichloride (EDC; 1,2-dichloroethane) that are presently available—the gavage study by the National Cancer Institute (NCI 1978) and the inhalation study by Maltoni (this volume). The NCI bioassay indicated significant increases in a variety of tumors in rats and mice of both sexes, whereas the preliminary conclusion from the Maltoni study was that there were no significant carcinogenic effects in either rats or mice. Thus, the results of cancer tests on EDC apparently were different when the dose was administered by inhalation rather than by gavage. We discuss seven factors that alone or in combination might account for this apparent discrepancy, including:

1. impurities in the test chemicals;
2. contaminants in air from other chemicals tested in the same room in the gavage bioassay;
3. contaminants in diet and effect of vehicle (corn oil) in gavage study;
4. dose levels and patterns of dosing;
5. routes of exposure;
6. strain differences in test animals;
7. statistical considerations of the effect of intercurrent mortality: potency analysis of NCI gavage bioassay, and preliminary potency analysis of Maltoni's inhalation bioassay.

From our evaluation of these factors, we conclude that the difference in experimental results may be due in part to several of these factors, and that an adequate comparison requires life-table data from the inhalation study to assess the effects of intercurrent mortality on tumor incidence. The information that would permit such a life-table analysis was not available at the time of writing of this paper. A future communication will contain an analysis of this data.

PURITIES OF TEST CHEMICALS

The EDC preparations used in the gavage and inhalation bioassays were of equivalent high purity, and it seems unlikely that impurities played an important role in the outcomes. This was not clear when the meeting started. Although the NCI report stated, not very precisely, that the sample tested in the bioassay was a technical grade chemical with a purity "greater than 90%," the NCI Chemical Repository indicated a purity of greater than 99.9% (NCI Chemical Repository, Midwest Research Institute [Tracor Jitco subcontract #74-24-106002], IIT Research Institute, 3441 South Federal Street, Chicago, Illinois 60616. EDC data sheet, June 8, 1978, from James Keith [October 29, 1979]).

We obtained a sample, and reanalysis of this material by R. Reitz of Dow Chemical and H. Plotnick of the National Institute for Occupational Safety and Health (NIOSH) indicated a purity by gas chromatography of between 98.5% and 99.8% for the 7-year-old stock samples (see Appendix). A minor impurity of chloroform of 0.02% was found, as well as 14 other trace contaminants.

(The analysis of purity of EDC by Dow Chemical Corporation was an average of 4 determinations and gave 99.3 ± 0.23% purity, with impurities of $CHCl_3$ (2200 ppm), acetone (350 ppm), and seven halogenated compounds at less than 100 ppm (Reitz, this volume). Dibromochloropropane [DBCP; 1,2-dibromo-3-chloropropane] was present at 40 ppm, possibly indicating that test samples of EDC and DBCP were stored in the same room. The analysis of purity of EDC performed by NIOSH was a peak-area analysis indicating greater than 98% purity, with $CHCl_3$ as the major contaminant and 12 other minor contaminants [Plotnick, this volume].)

Maltoni reported a purity of 99.8% for the EDC used in his inhalation bioassay, which is virtually identical to the purity of the NCI chemical. Although no chloroform was found in the Maltoni sample, minor impurities of trichloroethylene (TCE; 1,1,2-trichloroethane), perchloroethylene (PCE; tetrachloroethylene), carbon tetrachloride, and benzene were present.

CONTAMINANTS IN AIR IN THE GAVAGE BIOASSAY

In the NCI bioassay, several compounds were tested simultaneously in the same room as EDC. Some of these chemicals were carcinogens. We estimate that the exposure to each of these contaminants would likely be less than 0.002% of the highest dose administered in each test and therefore would be unlikely to contribute appreciably to the observed tumor incidence.

(In the NCI bioassay with rats, four other chemicals [1,1-dichloroethane, DBCP, TCE, and carbon disulfide] were under test in the same room with EDC. A rough calculation indicates that exposures to these contaminants should be quite low. The significant exposure to these contaminants should be by inhalation of volatiles in the room atmosphere, which arise largely from the exhaled breath of dosed rats [there should be no volatilization from feed because all were administered by gavage]. There were 200 rats [50 animals of each sex for each of two dosed groups] in each bioassay. The air changed 300 times/day [12-15 times/hr] [NCI 1978]. Rats exhale 10% of the gavage dose [Yllner 1971; Sopikof and Gorshunova 1979], and rats breathe in about 0.2 m^3/day. We estimate that the average daily exposure by inhalation to each contaminant would be about 0.002% of the highest gavage dose administered in each test. Mice in this study were housed with mice used in the gavage bioassays of 17 other chemicals. However, ventilation conditions were similar, and thus it seems unlikely that these contaminants contributed appreciably to the observed tumor incidences.)

Synergism is hard to evaluate between EDC and the low levels of these impurities, but appears improbable. It is also difficult to estimate the probability that any mix-up of chemicals occurred during the administration of dose.

CONTAMINANTS IN DIET—EFFECT OF VEHICLE IN GAVAGE STUDY

The Maltoni and NCI studies differ in the lab chow, and only the NCI study used a corn oil vehicle. Although vehicle controls were run in the gavage study, we cannot exclude the possibility that corn oil or various unanalyzed factors in the chow (such as antioxidants, aflatoxin, pesticides, or nitrosamines) could have had promoting effects on tumor frequency. We calculate the amount of corn oil as about 1 ml/(day · rat) at the NCI high dose (100 mg/kg EDC was dissolved in about a 6% solution in corn oil). The NCI lab chow was Wayne Lab-Blox® (Allied Mills); Maltoni used a different chow.

DOSE LEVELS AND PATTERNS OF DOSING

A possible explanation for the discrepancy between the results of the two bioassays is that the nonpositive inhalation study used doses that were much lower than those used in the NCI gavage study. Calculations show, however, that the two highest dose levels in the inhalation study were entirely comparable on a mg/(kg · day) basis to those yielding a strongly positive result in the NCI study (Table 1).

(We calculate daily doses [mg/kg body weight] extrapolated over the lifetime of the animals in the experiment, as determined by the age of the last survivor. For all dose routes, we calculate a daily administered dose, without adjusting for any differences in efficiency of absorption or retention by the various routes.

The calculated doses in the gavage study and in the two highest dose groups in the inhalation study were quite similar, each with two roughly equivalent dose levels within the dose range where significant increases in tumor incidences were observed in the gavage bioassay. Although the inhalation study by Maltoni was of more thorough design, employing four dose levels and a larger number of animals per dose group, two of the inhalation dose levels were very

Table 1
Number of Animals and Average Daily Lifetime Doses in EDC Experiments

	mg/(kg · day)	
	gavage (NCI)[a]	inhalation (Maltoni)[b]
Rats		1.6
		3.2
	24.0	16.0
	48.0	48.0
Mice	60.0[c]	5.6
	120.0[c]	11.0
	92.5[d]	56.0
	185.0[d]	171.0

Data for males and females unless otherwise indicated.
[a] 50 Animals/dosed group; 20 animals/matched vehicle controls; 60 animals/pooled vehicle controls; dosed for 5 days/wk for 78 wk; daily dose calculated for 90-wk experiment period for mice and 110 wk for rats; B6C3F1 mice and Osborne-Mendel rats of both sexes.
[b] 90 Animals/dosed group; 45 animals/in-chamber control group; 45 animals/out-of-chamber control group; dosed 7 hr/day, 5 days/wk for 78 wk; observed for life; Swiss mice and Sprague-Dawley rats of both sexes.
[c] Male.
[d] Female.

much lower [one-tenth to one-fifth] than the NCI gavage low dose [see Table 1]. The gavage low dose produced significant increases of tumors at some sites, but it is unlikely that significant responses would be produced at inhalation dose levels an order of magnitude lower.

The doses shown in Figure 1 are calculated from the following data. We assume that rats weigh 0.5 kg, eat 5% of their weight per day, and breathe 0.14 liter/min; that mice weigh 0.03 kg, eat 10% of their weight per day, and breathe 0.03 liter/min. In the NCI gavage study, the animals are exposed 5 days each week for 78 weeks, and the lifetimes are taken as 90 weeks for mice and 110 weeks for rats. In rats, for the last 34 weeks of the exposure period, a cycle of 4 weeks of dose alternating with one dose-free week was maintained.)

In both tests the exposure time was the same: doses were administered 5 days per week for 78 weeks. In the NCI bioassay there were two minor adjustments in dose levels for each dose group in rats and female mice, and one change in male mice. The dose we report is a time-weighted arithmetic average of these dose levels (see Table 1). The pattern of dosing was also changed in the NCI bioassay in rats. During the final half (34 weeks) of the exposure period, the dosing was interrupted for 1 week every fifth week. In the inhalation study, there was one reduction in dose level in the highest dose groups of both rats and mice.

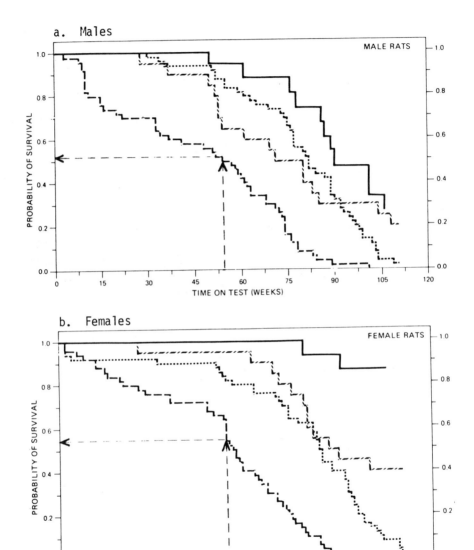

Figure 1
Survival comparisons of EDC chronic study rats (by gavage; NCI); (*a*) males, (*b*) females. We have not reproduced here the survival data for mice; premature mortality among mice was less than for rats. (———) Untreated control; (/ – / – / –) vehicle control; (······) low dose; (– – – –) high dose. (Reprinted from NCI 1978.)

ROUTES OF EXPOSURE

Would the different routes of exposure used in the two bioassays give different effective whole-body doses of EDC, i.e., would doses administered by inhalation be less effective than those given orally or by gavage? Published evidence available at the date of this meeting indicates that the effective doses of EDC delivered to tissues would be similar whether the chemical was administered by gavage or by inhalation, except for a transient high concentration in the liver in the gavage exposure due to the first-pass effect.

In uptake studies of vinyl chloride in Sprague-Dawley rats (Table 2), it was calculated that similar doses administered orally (20 ppm in drinking water for 24 hours; 0.9 mg/kg) or by inhalation (2 ppm for 24 hours; 1.7 mg/kg) would produce the same concentration-time dependence of the chemical in blood (Withey 1976). In a separate experiment, after intravenous administration of EDC and three related compounds (methylene chloride, chloroform, and TCE) to Wistar rats, the concentration in all tissues except adipose tissue was very similar to that found in blood (Withey and Collins 1980).

Gut flora might produce carcinogenic metabolites from doses of EDC administered by gavage that would not be produced (or produced in lesser amounts) in the lung from doses of EDC administered by inhalation. From existing data we cannot evaluate the likelihood of this occurring. However, the doses of EDC required to produce similar toxic effects by the two routes are quite similar. The acute toxic doses from a single gavage or a short inhalation exposure are roughly equivalent using rats of the same strain (Table 3). In addition, daily chronic exposures at equivalent dose levels over a longer period of time by the two routes also produced comparable toxic effects (8-week exposure period with Osborne-Mendel and Wistar rats; 52–64 week exposures with Osborne-Mendel and Sprague-Dawley rats; see Tables 4 and 5).

Thorough studies by Reitz and Spreafico (both, this volume), which directly address the question of effective doses by different routes, conclude that the blood concentration, metabolism, and tissue distribution of EDC are approximately the same for a given administered dose by gavage or inhalation.

Table 2

Doses by Inhalation and Gavage Give the Same Concentration of Vinyl Chloride in Blood

Route	Dose	AUC[a] (min-µg/ml)
Gavage[b]	0.9	19.2
Inhalation[c]	1.7	19.5

[a] Area under curve of plot of blood concentration of vinyl chloride versus time.
[b] Withey and Collins (1976).
[c] Withey (1976).

Table 3
Acute Toxicity of EDC: Wistar Rats, Single Exposure

	$LD_{50}{}^a$ (mg/kg · day)
Gavage[b]	680
Inhalation[c]	
1000 ppm, 7 hr	519
3000 ppm, 2.7 hr	611
12,000 ppm, 0.5 hr	444

[a] Median lethal dose.
[b] McCollister et al. (1956).
[c] Spencer et al. (1951).

Table 4
8-Week EDC Exposure: Rats

	Mortality	
	males (%)	females (%)
Gavage dose[a]	60	20
134 mg/(kg · day)	(3/5)	(1/5)
Inhalation dose[b]	100	100
148 mg/(kg · day)	(15/15)	(15/15)

[a] Osborne-Mendel rats, NCI bioassay (1978).
[b] Wistar rats, Spencer et al. (1951).

Table 5
52–64-Week EDC Exposure: Rats

	Mortality	
	males (%)	females (%)
Gavage dose[a]	48	46
37 mg/(kg · day)	(24/50)	(23/50)
Inhalation dose[b]	67	40
53.6 mg/(kg · day)	(60/90)	(36/90)

[a] Osborne-Mendel rats, NCI bioassay (1978); mortality at 52 wk of experiment.
[b] Sprague-Dawley rats, Maltoni (this volume); expected mortality at 64 wk of experiment.

STRAINS OF TEST ANIMALS

Different strains of test animals were used in the two cancer tests. The NCI used Osborne-Mendel rats and B6C3F1 mice. Maltoni used Sprague-Dawley rats and Swiss mice. Both strains of mice are responsive to tumor induction by chlorinated compounds. The B6C3F1 mouse is responsive to aldrin; chlordane; heptachlor; DDE (1,1-dichloro-2,2-*bis*(*p*-chlorophenyl)ethylene); kepone; chloroform; carbon tetrachloride; hexachloroethane; 1,1,2,2-tetrachloroethane; PCE; TCE; and 1,1,2-trichlorethane by oral route (R. A. Griesemer and C. Cueto, in prep.). The Swiss mouse is responsive to vinyl chloride (VC) and vinylidene chloride (VDC; 1,1-dichloroethene) by inhalation. In general, the rat is less responsive to chlorinated chemicals than the mouse. The Osborne-Mendel rat is responsive to chloroform, chlordane, and kepone by oral route but not to the other chlorinated compounds listed above. The Sprague-Dawley rat is responsive to VC by inhalation but not to VDC. It is not responsive to TCE by gavage (Maltoni 1980).

In the inhalation study using Sprague-Dawley rats, three dose groups had significantly higher incidence of mammary tumors than controls (Fisher exact $p < 0.05$). The majority of these tumors were fibromas and fibroadenomas. Because of the high incidence of these tumors in the control group (36/90 = 40%) and their variability in historical controls, intrepretation of this finding is difficult. In the NCI gavage study in Osborne-Mendel rats, the incidence of mammary fibroadenomas and adenocarcinomas in the control group was 0/20, and a significantly positive result was found in this tissue.

There was widespread murine pneumonia among rats in the NCI gavage study that was not present in the inhalation study. This may have had some effect on the bioassay for rats.

EFFECT OF INTERCURRENT MORTALITY ON NUMBER OF ANIMALS AT RISK

In the gavage study there was a significant degree of premature mortality among the test animals (as shown in Fig. 1) and this is also apparently the case in the inhalation study. This could significantly alter the determination of carcinogenicity. Because most tumors arise very late in the natural lifetimes of test animals, animals that die early will not have lived long enough to manifest tumors that otherwise might have developed. Premature mortality tends to be most prevalent in the highest dose groups. In a bioassay in which significant early mortality occurred in the high-dose group, the calculated proportion of tumor-bearing animals, based on the initial number of animals in the experiment, would be too low and thus would lead to an underestimate of the carcinogenic effect of the chemical.

Where significant mortality has occurred during any experiment, as in the case of the gavage study and apparently in the case of the inhalation study of EDC, life-table analysis will correct for its effects and therefore should be used in the proper statistical interpretation of the experimental results (Pike and Roe 1963; Peto 1979). Life-table (actuarial) analysis relates the occurrence of a tumor to the number of animals still alive at the time the tumor appears, rather than to the number that started the experiment.

Most published animal cancer tests report only the final percentage of tumor-bearing animals in each dose group. This data is insufficient for an actuarial analysis.

Potency Analysis of NCI Gavage Bioassay

NCI concluded that EDC was a carcinogen in its bioassay and that the chemical significantly increased tumor incidence in both sexes of Osborne-Mendel rats and B6C3F1 mice at the following sites: squamous cell carcinomas of the forestomach, hemangiosarcomas, and subcutaneous fibromas in male rats; mammary adenocarcinomas in female rats; mammary adenocarcinomas and endometrial tumors in female mice; and lung alveolar-bronchiolar adenomas in mice of both sexes (see Ward, this volume). Moreover, there were a number of nonstomach metastatic tumors produced in both species (hemangiosarcomas in male rats, lung adenocarcinomas in both sexes of mice, and endometrial adenocarcinomas in female mice). EDC is clearly a carcinogen in these strains of test animals under the conditions of this experiment.

Our potency project (Ames et al., this volume) has analyzed the NCI data and the preliminary Maltoni data to estimate the carcinogenic potency of EDC. As described by Ames et al., we calculate a TD_{50}, the daily dose in mg/(kg · day) which if administered over a standard lifetime would decrease by half the probability that an animal remains tumor-free.

The estimate of TD_{50} is subject to both life-table and summary analysis. Depending upon the data available, two types of estimates of carcinogenic potency can be made, one more accurate than the other. Summary TD_{50} values are calculated when only tumor incidences are reported for each dose group. Life-table TD_{50} values are calculated when there is information on the time of death of each animal and the type of tumor, if any, present at death (Ames et al., this volume).

The potency analysis of the NCI gavage study is shown in Figure 2 where we have plotted the TD_{50} values for the combined tumor incidences of those tissues found significant by NCI, as mentioned above (see Fig. 2 legend for details of plot). In this plot of the summary data in Figure 2 (enclosed by single-dot confidence limits), the TD_{50} values of EDC in Osborne-Mendel rats lie to the left of the TD_{50} values plotted for B6C3F1 mice, indicating that the chemical is more potent in rats than in mice. Thus, EDC is more potent in Osborne-Mendel rats by gavage than in B6C3F1 mice, and more potent in female rats than in male rats.

In the life-table analysis of the same experiment (enclosed by double-dot confidence limits in Fig. 2), it is evident that in all cases the life-table TD_{50} values are to the left of the corresponding summary TD_{50} values, at three- to tenfold more potent values—a shift to lower doses in mg/(kg · day) from 180 to 53 in female mice; from 256 to 90 in male mice; from 55 to 5 in female rats; and from 74 to 11 in male rats. These dramatic increases in the calculated potency are due to significant early mortality which occurred in the experiments. The mortality in the high-dose group in rats was nearly 50% at 52 weeks and 90% at 78 weeks; this compares to the mortality in the control group of 5% and 40%, respectively (see Fig. 1). Thus, only one-half of the high-dose animals remained at risk for the crucial second half of their 2-year life-span; it is during this time that we would expect most tumors to appear because of the higher-order dependence of cancer incidence upon age. Using the life-table data and averaging the fairly similar results from males and females, our estimates of potency of EDC by gavage are about 70 mg/(kg · day) in B6C3F1 mice and about 7 mg/(kg · day) in Osborne-Mendel rats.

In addition to influencing the TD_{50}, life-table analysis can influence the determination of whether the data is compatible with a linear model. This effect can be seen in Figure 2. The life-table analysis of a number of the NCI experiments indicates that more tumors were observed in the high-dose group than would be expected on the basis of a linear model.

Potency Analysis of the Inhalation Bioassay (Maltoni)

The preliminary data from the inhalation study only allow us to perform a summary analysis. A comparison of the inhalation study and the gavage study in terms of their calculated potencies is presented in Figure 3. TD_{50} values from the inhalation study are plotted for those sites that had at least some response, although that response was judged nonsignificant. For comparison we plot life-table and summary TD_{50} values from the NCI data for similar single tissue sites and pathology. None of the summary TD_{50} values from the inhalation study were significant for tissues that were significant in the NCI gavage study. This agrees with Maltoni's conclusion that there is no significant dose-related increase in tumor incidence for any tissues in rats or mice.

We have plotted the symbol (· >) for the inhalation results, which indicates that based on this test we are 97.5% confident that a dose level which might be calculated to induce tumors in 50% of otherwise tumor-free animals would have to be greater than the value plotted, in this case, greater than 86.0 mg/(kg · day). This dose value is greater than the TD_{50} values calculated for the same tissues in the NCI study (i.e., less potent). Thus, even with the potency analysis, the results of the gavage and inhalation studies cannot be reconciled. However, the analysis here is based on preliminary summary data only. If there were a high rate of early mortality in the high-dose group, then it is possible that a life-table analysis might shift the location of this lower confidence limit downward and might also result in significant tumor incidences in some groups of animals. Early mortality may be expected here, based on results from the NCI gavage bioassay and other toxicity studies (Tables 3, 4, 5).

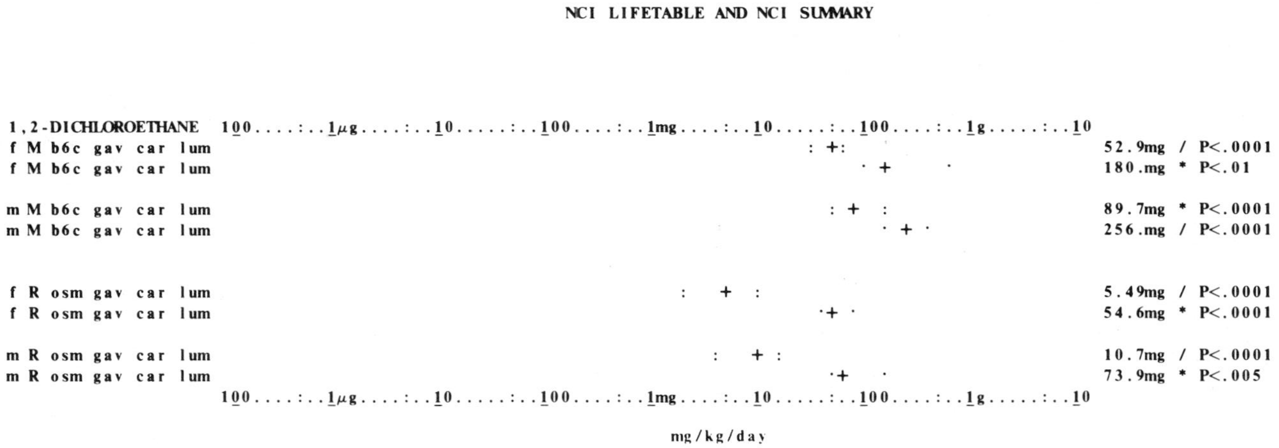

NCI LIFETABLE AND NCI SUMMARY

```
1,2-DICHLOROETHANE  100......1μg......10......100......1mg......10......100......1g......10
f M b6c gav car lum                                              :+:                          52.9mg  / P<.0001
f M b6c gav car lum                                                    +      .               180.mg  * P<.01

m M b6c gav car lum                                                         :+:               89.7mg  * P<.0001
m M b6c gav car lum                                                         ∴+∴               256.mg  / P<.0001

f R osm gav car lum                            ∴∴+∴                                            5.49mg  / P<.0001
f R osm gav car lum                                  ∴+∴                                       54.6mg  * P<.0001

m R osm gav car lum                                         ∴+∴                                10.7mg  / P<.0001
m R osm gav car lum                                              ∴+∴                           73.9mg  * P<.005
                    100......1μg......10......100......1mg......10......100......1g......10
                                                    mg/kg/day
```

Figure 2

In this figure each experiment is contained in a single line, with information presented in the following sequence. (*Left*): Sex, species, and strain of test animal; route of exposure; and tumor pathology (tissue and tumor type). (gav) Gavage; (inh) inhalation; (f) female; (m) male; (M) mice; (R) rats; (b6c) B6C3F1; (swi) Swiss; (osm) Osborne-Mendel; (sda) Sprague-Dawley; (car lum) TD₅₀ is calculated for the combined tumor incidences of all sites which were considered statistically and biologically significant by the NCI for that test group. (*Middle*): The TD₅₀, together with its 95% confidence interval, is plotted to the right on a logarithmic scale of dose, ranging from 100 mg/(kg · day) to 10 gm/(kg · day). (+) TD₅₀ when $p < 0.02$; (·) TD₅₀ values calculated from summary data (single-dot confidence limits); (:) life-table TD₅₀ values (double-dot confidence limits). (*Right*): Numerical value for the TD₅₀; how compatible the data are with a linear model; probability that the dose-response is different from zero. (*) The data are compatible with a linear model; (/) the high-dose group had more tumors than would be predicted by a linear model.

NCI LIFETABLE, NCI SUMMARY, AND MALTONI SUMMARY

1,2-DICHLOROETHANE   100......1μg......10......100......1mg......10......100......1g......10

```
f M b6c gav mam acn                                           : +                    133.mg  P<.0001
f M b6c gav mam acn                                           . ±                    238.mg  P<.02
f M swi inh mam mx1                                               (·>)               >1.21gm P<.7

m M b6c gav lun a/a                                         : +                      89.7mg  P<.0001
m M b6c gav lun a/a                                         . . +                    256.mg  P<.0001
m M swi inh lun ade                                                        (·>)      4.05gm  P=1.0

f R osm gav mam mx2                              : +                                 5.49mg  P<.0001
f R osm gav mam mx2                              . +                                 54.6mg  P<.0001
f R sda inh mam mx3                                    (·>)                          >86.0mg P<.6

m R osm gav sct fib                                    : +                           43.2mg  P<.002
m R osm gav sct fib                                          . ±                     >137.mg P<.07
m R sda inh ski fib                                                  (·>)            >297.mg P<.5
```

100......1μg......10......100......1mg......10......100......1g......10

mg/kg/day

Figure 3

The summary and life-table TD$_{50}$ values from the gavage study are compared with summary TD$_{50}$ values from the inhalation bioassay for simiar tissues. The inhalation data are enclosed in parentheses as they must be considered preliminary until the final data are published by Maltoni. Data are presented as in Fig. 2. Tumor pathology (*left*): (gav) gavage; (inh) inhalation; (mam) mammary gland; (acn) adenocarcinoma; (lun) lung; (car) carcinoma; (ski) skin; (fib) fibroma; (sct) subcutaneous; (a/a) alveolar-bronchiolar adenoma; (ade) adenoma; (mxl) mixture of chiefly carcinomas with carcinosarcomas and possibly fibroma-fibroadenomas; (mx2) mixture of fibroadenomas and adenocarcinomas; (mx3) mixture of fibroadenomas, fibromas, and carcinomas. (*Middle*): (±) The probability is between 0.01 and 0.10 that the dose-response is different frcm zero—when this probability is greater than 0.025, the upper 95% confidence limit on the TD$_{50}$ is infinite and hence is not plotted; (• >) the experiment did not give a significant carcinogenic response, but if the chemical is a carcinogen, we have 97.5% confidence that the TD$_{50}$ will be greater than the value of the lower confidence limit which is plotted here and listed to the right.

(From the summary incidence data of mammary tumors in female rats [in-chamber control, 36/90; dose 1, 64/90*; dose 2, 43/90; dose 3, 55/90*; dose 4, 52/90*, where * = Fisher exact $p \leqslant 0.05$], if significant early mortality occurs in higher dose groups, the denominator in these incidences [the number of animals at lifetime risk] effectively becomes smaller, and it is possible that the nonsignificant dose response might become significant. Similar increases in significance might occur at other sites if life-table data were used in the analysis.)

CONCLUSION

The most likely explanations for the apparently discrepant results are:

1. the strains of test animals differ in responsiveness;
2. the route of exposure does make a difference concerning the carcinogenic action of EDC;
3. an artifact has been introduced by intercurrent mortality that would be corrected by life-table analysis.

We cannot choose among these possibilities on the basis of the information at hand.

ACKNOWLEDGMENTS

We wish to acknowledge Charles B. Sawyer for performing the potency analysis presented here as part of the carcinogenic potency project and also Thomas Haggin, William Havender, and Cesare Maltoni for their help. This work was supported by Department of Energy contract EY-76-S-03-0034 PA156 and a California Policy Seminar grant to B. N. A. and by National Institute of Environmental Health Sciences Center grant ES-01896. (Part of this work was performed while K. H. was employed by the Hazard Alert System, State of California Department of Health Services, Berkeley, California.)

REFERENCES

Maltoni, C. 1980. Results of long-term carcinogenicity bioassays of trichloroethylene: Experiments by oral administration of Sprague-Dawley rats. *Medicina del Lavoro* (Milan). (In press)

McCollister, D. D., R. L. Hollingsworth, F. Oyen, and V. K. Rowe. 1956. Comparative inhalation toxicity of fumigant mixtures. *A.M.A. Archives of Industrial Health* **13**:1.

National Cancer Institute. 1978. *Bioassay of 1,2-dichloroethane for possible carcinogenicity*. NCI Carcinogenesis technical report series No. 55, DHEW publication number (NIH) 78-1361, Government Printing Office, Washington, D.C.

Peto, R. 1979. Guidelines on the analysis of tumor rates and death rates in experimental animals. *Br. J. Cancer* **29**:101.

Pike, M. C. and F. J. C. Roe. 1963. An actuarial method of analysis of an experiment in two-stage carcinogenesis. *Br. J. Cancer* **17**:605.

Sopikof, N. F. and A. I. Gorshunova. 1979. Investigation of the intake, distribution and excretion of ethylene dichloride in rats. *Gig. Tr. Prof. Zabol.* **1SS4**:36.

Spencer, H. C., V. K. Rowe, E. M. Adams, D. D. McCollister, and D. D. Irish. 1951. Vapor toxicity of ethylene dichloride determined by experiments on laboratory animals. *A.M.A. Archives of Industrial Hygiene and Occupational Medicine* **4**:482.

Withey, J. R. 1976. Pharmacodynamics and uptake of vinyl chloride monomer administered by various routes to rats. *J. Toxicol. Environ. Health* **1**:381.

Withey, J. R. and B. T. Collins. 1976. A statistical assessment of the quantitative uptake of vinyl chloride monomer from aqueous solution. *J. Toxicol. Environ. Health* **2**:311.

――――. 1980. Chlorinated aliphatic hydrocarbons used in the foods industry: The comparative pharmacokinetics of methylene chloride, 1,2-dichloroethane, chloroform and trichloroethylene after I.V. administration in the rat. *J. Environ. Pathol. and Toxicol.* **3** (in press).

Yllner, S. 1971. Metabolism of 1,2-dichloroethane-^{14}C in the mouse. *Acta Pharmacol. Toxicol.* **30**:257.

COMMENTS

WARD: Kim [Hooper], I note that the numbers from the NCI Technical Reports (NCI 1978) for dose calculation are 47 and 95.

HOOPER: Yes. The dose levels (mg/kg · day) given by NCI are calculated for a 78-week exposure period, using a 5-day week. We calculated a daily dose using a 7-day week for an experiment time, which is taken as 90 weeks in mice and 110 weeks in rats (see Hooper et al., this volume). That's why our doses are lower.

AMES: When you calculate an inhalation, what percentage of absorption do you use?

HOOPER: We assume that 100% of the dose is effectively retained. We do not make any adjustments for efficiency of absorption, or attempt to calculate an effective dose. We simply assume at this point that everything given to the animal, by any dose route, stays in the animal.

WARD: Have you ever approximated TD_{50} values in humans to the known human carcinogens to see how they applied to the animal TD_{50} values?

HOOPER: Yes, we are trying to do that using, for example, cigarette smoking data or benzo[a]pyrene exposure. And there is agreement, that is, we don't see huge discrepancies in those two cases. But I think we want to just hold off on that. That's another area we could work on—the differences between rats, mice, and people, and how we make that jump in extrapolation.

I think Dr. Ames' point was that we wanted to take monkey data as an approximation and see how that compares. For example, if we have a compound that's been tested in rats, mice, hamsters, and monkeys and the potency in all those four species appears to be the same, then I think we can state that we have a higher probability, or more confidence, that the human point would be somewhere near that. If there's a great spread— a thousandfold difference between all of them—then I am not sure what we would do. But that's the approach we're taking at present.

WARD: The discrepancy between the interpretation of your statistics and Dr. Maltoni's is based on mammary tumors in the rat—the fibroadenomas. I wonder if Dr. Maltoni can add any data on the incidence of fibroadenomas in other Sprague-Dawley controls in his laboratory. Are they compatible with most of his other controls, or are they groups of controls where you might have 50% incidence under the same conditions? Besides the life-table methods, these data might give you more weight of the evidence one way or the other.

HOOPER: Right. So that the one control used might be abnormally low.

WARD: Well, he had two controls, 25-30%, I believe.

MALTONI: We have an average of 50% of our female rats with spontaneous mammary tumors; fibroadenomas in males range from 7-11%. We will soon publish all spontaneous pathology in these 6000 animals and give data on the year of birth, because it may change from 5 to 6 years; their weights; the survival at this time; those neoplastic pathologies found out at this time; and those neoplastic pathologies found out at invervals of 4 weeks. With this data I think that one can get a sort of meter to co-equate our data within the spontaneous group.

Since we also are discussing scientific methodology here, a good new model for bioassays would be to give data on incidence of the tumor in a particular situation, in treated animals as well as in the proper controls, plus the incidence of historical control (within the laboratory) with the range of the population.

I would also like to make an observation. It is surprising that in the NCI project for EDC you have a greater sensitivity for rats than for mice, because in our laboratory studies mice always have been far more sensitive to things, in line with their enzymatic profile.

HOOPER: In the NCI study, the rats seem more sensitive than mice. They cannot tolerate such a high dose as mice (about half that of mice), and their mortality curve is much steeper, even at these lower doses. Part of this may be due to the significant amount of murine pneumonia present in the rat colony and virtually absent from mice.

Just to add to Dr. Maltoni's comments, we feel we cannot stress too much the importance of people reporting life-table data for their cancer bioassays so that the effects of intercurrent mortality can be corrected for.

REITZ: Although epidemiological studies, as you pointed out, Kim, don't have the statistical power that carefully controlled animal studies do, there are cases where you can take human data and use it to put perspective on risk analysis. One of these is the ethylene dibromide (EDB; 1,2-dibromo-ethane) case.

I think, if you recall, that the threshold limit value (TLV) for EDB allowed workers to be exposed to many times higher concentrations than your calculated TD_{50}. But there is data in the literature by Drs. Ott and Gehring, which indicate that the actual incidence of cancer in these workers is much less than predicted, in fact, not distinguishable from background. So this points out the need to check our predictions against actual human data wherever possible.

INFANTE: If you're going to try to assess the risk epidemiologically, then you need to have adequate sample size, latency, and follow-up period. Now, I haven't seen the study that you're speaking of with EDB, but tomorrow I'm going to present some of the data that's been published on some of the structurally related compounds (see Infante and Marlow, this volume). In every instance, there are so many study deficiencies that these are all no-decision studies. Right now I don't think that you can be saying that there is something different between animals and humans unless you have the adequate population to follow for an adequate period to make that evaluation.

TER HAAR: There may be a chance of risk in humans compared to animals, but I think if you look at the EDB data on its surface there is a clear-cut difference in both between humans.

References

NCI. 1978. *Bioassay of 1,2-dichloroethane for possible carcinogenicity.* Technical Report Series No. 55. DHEW publication number (NIH) 78-1361. Government Printing Office, Washington, D.C.

The Use of Different Metabolizing Systems in the Elucidation of the Mutagenic Effects of Ethylene Dichloride in *Salmonella*

ULF RANNUG
Environmental Toxicology Unit
Wallenberg Laboratory
University of Stockholm
106 91 Stockholm, Sweden

Ethylene dichloride (EDC; 1,2-dichloroethane) was first shown to be mutagenic on *Drosophila melanogaster* by Rapoport (1960). Several mutagenicity studies on a variety of test organisms have been carried out since that time, but few have dealt with the metabolism of EDC in connection with its mutagenicity. One reason may be that, although EDC is activated by the metabolism, it also functions as a directly acting mutagen in microorganisms, that is, in the absence of a metabolizing system. This review is based on mutagenicity tests on *Salmonella typhimurium* in combination with different metabolizing systems carried out at our laboratory to illustrate the metabolism and mutagenic properties of EDC. To verify and explain our first findings in the *Salmonella* mutagenicity test, which indicated that the S-9 mix potentiated the mutagenic effect of EDC (Rannug and Ramel 1977), two different approaches were used. The first alternative involves metabolizing systems less complex than the S-9 mix—namely, different fractions of the S-9 or pure enzyme preparations (Rannug et al. 1978). The other alternative involved more complex or intact systems, liver perfusion and in vivo metabolism (Rannug and Beije 1979). Other aspects, including experimental conditions and strain differences in the metabolism, are discussed. All mutagenicity studies reported here have been carried out on *S. typhimurium* TA1535 if not stated otherwise (for a detailed description of the experimental procedures, see Rannug et al. 1978; Rannug and Beije 1979).

RESULTS AND DISCUSSION

Experiments with Different S-9 Mixes

The first thing that was noticed regarding the activation of EDC was its NADPH independence. This indicated that microsomal mixed function oxygenase was not responsible for the activation. From Table 1 it can be concluded that a relatively high amount of S-9 in the S-9 mix is necessary to detect the activation

Table 1
The Mutagenic Effect of EDC on *S. typhimurium* TA1535 in the Presence of Different Amounts of 9000*g* Supernatant from Phenobarbital-Induced Rat Livers (R and Sprague-Dawley Strains Mixed)

Amount of S-9/plate (μl)	Control ethanol	Number of mutants/plate ± S.E.		
		EDC (μmole/plate)		
		20	40	60
25	7.0 ± 1.22[a]	11.6 ± 1.36*[c]	16.6 ± 1.83**[d]	16.6 ± 1.12***[e]
75	10.0 ± 1.00[b]	22.2 ± 3.12*	44.0 ± 4.14***	67.6 ± 8.49***
150	10.6 ± 0.81	36.0 ± 3.86***	58.4 ± 4.18***	85.4 ± 6.50***

[a] Four plates.
[b] Three plates.
[c] $0.01 < P < 0.05$.
[d] $0.001 < P < 0.01$.
[e] $P < 0.001$.

and to distinguish it from the direct effect seen in *Salmonella*. This also indicated that other enzymes than the cytochrome P450 system were involved. The noted activation could either be a result of a nonenzymatic reaction with thiols as shown for ethylene dibromide (EDB; 1,2-dibromoethane) (Kondorosi et al. 1973) or a reaction with cysteine or rather glutathione (GSH) as shown by Nachtomi and colleagues for EDB and EDC (Nachtomi et al. 1966; Nachtomi 1970). Another metabolic pathway was, however, suggested for EDC in mice (Yllner 1971). To study these aspects in further detail, the mutagenicity of EDC was tested on TA1535 in the presence of an S-9 mix supplemented with either of the following thiols (1.6 μmoles/ml S-9): 2-mercaptoethanol, L-cysteine, *N*-acetyl-L-cysteine, and GSH. Of these thiols, only GSH gave an enhancement of the mutagenic effect (Fig. 1). From the same figure it can also be seen that there is a strain difference between S-9 fractions from Sprague-Dawley rats or R-strain rats (a Wistar strain) (to be discussed later). To study if the enhancement seen in the presence of GSH was enzymatic or nonenzymatic, experiments with denatured S-9 fractions were carried out. The results show that by denaturing the S-9 fraction and thereby inactiviating the enzymes, the mutagenic effect is reduced to the level seen in the absence of a metabolizing system. It can thus be concluded that the activation is enzymatic and GSH-dependent. Other liver fractions were used as a metabolizing system in subsequent experiments.

Fractions of S-9 and Pure Enzyme Preparations

Experiments were also carried out in which the S-9 fraction was divided into a microsomal fraction and a soluble fraction (115,000*g* supernatant), which were used as metabolizing systems. The microsomal fraction was used both in the

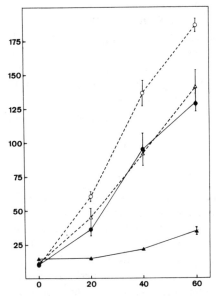

Figure 1
The mutagenic effect on *S. typhimurium* TA1535 of EDC in the presence of postmitochondrial liver fraction (S-9) from phenobarbital pretreated Sprague-Dawley or R-strain rats. (▲) Sprague-Dawley rats with addition of GSH (1.6 μmole/ml S-9 mix), (△) Sprague-Dawley rats without addition of GSH; (●) R-strain rats with addition of GSH (1.6 μmole/ml S-9 mix); (○) R-strain rats without addition of GSH.

absence and presence of an NADPH-generating system. The total activity seen with the S-9 fraction was found in the soluble fraction. Thus far the data indicated that a GSH-dependent enzyme in the soluble fraction was responsible for the activation of EDC to a more potent mutagen. A group of enzymes, having the characteristics of conjugating various compounds with GSH, are the GSH *S*-transferases. The mutagenicity of EDC was thus tested in the presence of GSH and a highly purified enzyme preparation of one of three types of GSH *S*-transferases (A, B, and C). With form B of the enzyme no unequivocal enhancement was noted. GSH *S*-transferases A and C, on the other hand, caused a time-dependent increase in the mutagenicity of EDC in the presence of GSH. The results show that EDC can act as a substrate for at least GSH *S*-transferase A and C.

Synthetic Conjugates

The first step in the mercapturic acid synthesis in vivo is a conjugation with GSH (Chasseaud 1976). The initial conjugate is then enzymatically degraded to the corresponding cysteine conjugate, which finally is acetylated to form the

N-acetylcysteine conjugate (mercapturic acid). To see if such conjugates with EDC exhibit mutagenic properties, three conjugation products were synthesized and the direct mutagenic effect was tested on *Salmonella*. Both S-(2-chloroethyl)-L-cysteine and N-acetyl-S-(2-chloroethyl)-L-cysteine gave approximately the same mutagenic effects (Fig. 2). Whereas S-(2-hydroxyethyl)-L-cysteine showed no mutagenic effects in any of the concentrations tested, neither in the presence nor in the absence of a metabolizing system (S-9 mix). These results show that the enzymatic degradation of the GSH moiety does not abolish the mutagenic properties of the conjugate, whereas a substitution of the chlorine with an hydroxyl group does.

Liver Perfusion and Treatment In Vivo

The above-mentioned results indicate that an enzymatic conjugation with GSH results in an activation of EDC. The results also imply that these reactions take place in vitro when testing EDC in the *Salmonella* mutagenicity test, at least under certain conditions. It was of great interest to know whether or not these reactions occur in a more intact system and in vivo. Most GSH conjugates are normally excreted in the bile; therefore the use of a liver perfusion system seemed appropriate for studying the mutagenicity and metabolism of EDC. In these experiments, EDC was added to the perfusate and samples from the perfusate and the produced bile were tested at different intervals. The bile revealed a high mutagenic effect 15-30 minutes after the addition of EDC to the system, whereas the perfusate revealed no mutagenic effects. A small but significant difference was also noted with the two strains of rat (Sprague-Dawley

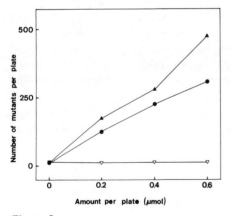

Figure 2
The mutagenic effect on *S. typhimurium* TA1535 of EDC (▽) and the two synthetic conjugates S-(2-chloroethyl)-L-cysteine (▲) and N-acetyl-S-(2-chloroethyl)-L-cysteine (●) in the absence of a metabolizing system.

and R strain). The bile produced by the liver from strain R thus showed a somewhat higher mutagenic effect as compared to the Sprague-Dawley strain after EDC treatment. As mentioned earlier a similar difference in mutagenicity was noted in the experiments with S-9 fractions from the two strains of rat. From these results it can be concluded that the intact liver metabolized EDC to a potent mutagenic intermediate excreted in bile.

Experiments with mice treated in vivo also showed that the bile is mutagenic after treatment with EDC, which indicates that the same metabolism also can take place in the in vivo situation. These results are also in accordance with other metabolism studies in vivo (Nachtomi et al. 1966; Nachtomi 1970).

SOME FURTHER CHARACTERISTICS OF THE MUTAGENICITY

The results reported earlier imply that the significant metabolic pathway for EDC, in terms of biological activity, proceeds through a conjugation with GSH. This metabolic pathway turned out to be connected with some difficulties when making quantitative comparisons between experiments. The experimental variation has to be taken into consideration when dealing with these types of experiments and attempts should be made to eliminate the source of this variation.

Thus, after storage on ice of the S-9 mix, the capacity to activate EDC decreases with time (Fig. 3). This is probably due to oxidation of GSH, since an addition of reduced GSH immediately before the experiment almost totally

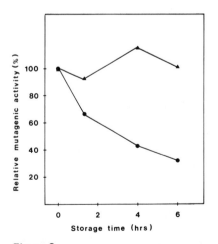

Figure 3
The mutagenic effect on *S. typhimurium* TA1535 of 2-aminoanthracene (▲) and EDC (●) in the presence of a metabolizing system (S-9) stored on ice for different lengths of time (0–6 hours) before the experiment.

restored this capacity. Such an effect of storage was not noted with 2-aminoanthracene used as a positive control in our experiments (Fig. 3). In the case of 2-aminoanthracene, the activation is carried out by the microsomal mixed function oxygenase system, and if a conjugation with GSH takes place, this will reduce the mutagenic effect. 2-Aminoanthracene produced approximately the same number of mutants in the presence of a freshly prepared S-9 mix or an S-9 mix stored up to 6 hours on ice. This shows that the balance between activating and deactivating processes is unaltered, which in turn implies a decreasing efficiency with time for the activating enzymes. The mutagenic effect of most compounds is the result of the balance between activating and deactivating processes. For a substance such as EDC, this balance can very easily change and can give not only a large variation in the test results but also false negative results.

Experimental conditions in favor of the activation have also been noted. Preincubation, for instance, gives a much higher mutagenic effect than the standard plate incorporation test (see Fig. 4). With preincubation 1 μmole/plate of EDC can readily be detected, without preincubation, approximately 20 μmole/plate is needed.

Besides the large variation in mutagenicity from experiment to experiment a strain difference between the two strains of rat (Sprague-Dawley and R) could be noted. Thus S-9 mixes from the strain R potentiated EDC to a higher

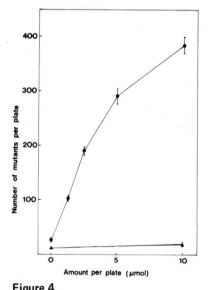

Figure 4
A comparison of the mutagenic effect on *S. typhimurium* TA1535 of EDC in the presence of S-9 mix – NADP from noninduced R-strain rats with preincubation at 37°C for 30 min (●) and without preincubation (▲).

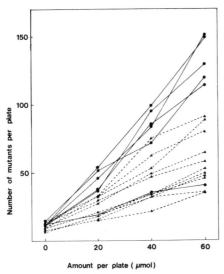

Figure 5
The variation of the mutagenic effect on *S. typhimurium* TA1535 of EDC (20, 40, and 60 μmole/plate) in the presence of metabolizing systems (S-9 mix − NADP) from phenobarbital pretreated Sprague-Dawley (▲) and R-strain (●) rats. Regression analysis between strains significant ($p < 0.001$).

extent than did S-9 mixes from the strain Sprague-Dawley (see Fig. 5). A corresponding difference could also be seen in the liver perfusion experiments.

Even higher differences in mutagenicity could be noted with S-9 mixes from the same rat strain when comparing test results over a much longer period of time (Fig. 6). The results showed that experiments carried out in 1976 gave a much higher mutagenic effect than those performed in 1978. These fluctuations in mutagenicity both within and between the two strains of rat most likely reflect the status of the rat livers in terms of GSH *S*-transferase activity, amount of available GSH, and other biological factors rather than differences in experimental procedures. However, attempts to increase the mutagenic effect to the 1976 level by adding GSH were unsuccessful in experiments carried out in 1978 and 1979.

Different types of induction have also been tested. Phenobarbital pretreatment gave no detectable increase in mutagenicity compared with untreated controls. It has been shown that *trans*-stilbene oxide specifically induces epoxide hydratase and GSH *S*-transferases (Seidegård et al. 1979). In preliminary experiments the mutagenic effect of EDC has been tested in the presence of S-9 mixes from rats pretreated with *trans*-stilbene oxide. A comparison of the two rat strains is shown in Figure 7. The induction increases the mutagenic effect with a metabolizing system from the Sprague-Dawley strain. This effect was only seen when the S-9 mix was supplemented with extra GSH. However, strain R

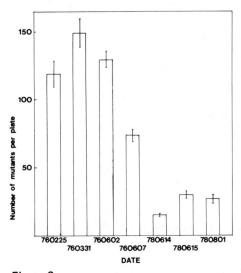

Figure 6
A comparison of the mutagenic effect on *S. typhimurium* TA1535 of EDC (60 μmole/plate) in the presence of a S-9 mix – NADP from phenobarbital pretreated rats of strain R. Identical experiments have been carried out at different times over a period of 2 years.

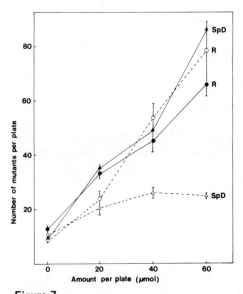

Figure 7
The mutagenic effect on *S. typhimurium* TA1535 of EDC in the presence of a S-9 mix – NADP + GSH (0.5 mg/plate) from *trans*-stilbene oxide pretreated (440 mg/kg body weight) and two strains of control rats (Sprague-Dawley and R). (▲) Induced Sprague-Dawley; (△) control Sprague-Dawley; (●) induced R strain; (○) control R strain.

has so far not given any enhancement in the mutagenic effect after *trans*-stilbene oxide pretreatment. Further studies are in progress involving both mutagenicity tests and biochemical characterization of the strain differences on the one hand and the induction pattern on the other.

CONCLUSION

The results discussed here illustrate some important features regarding metabolism and mutagenicity testing. First, the results with EDC have shown that enzymes other than the microsomal enzymes may be of importance, not only by deactivating but also by activating certain chemicals. Although the results using EDC may be unique in some respects, they nevertheless point to the fact that no enzyme or enzyme system can be referred to as either an activating or a detoxifying principle, as the same enzyme can act in both directions, depending on the circumstances. Second, the importance of the balance between activating and detoxifying processes must be emphasized. In the case of EDC this balance can easily be changed in vitro, as can be seen from the test results under different conditions. Also, in the in vivo situation this balance is influenced by different factors, resulting in a large biological variation, strain, and species differences regarding the genotoxic effects of EDC. This also gives rise to conflicting results and reduces the reproductivity and sensitivity of different investigations.

REFERENCES

Chasseaud, L. F. 1976. Conjugation with glutathione and mercapturic acid excretion. In *Glutathione: Metabolism and function* (ed. I. M. Arias and W. B. Jakoby), p. 77. Raven Press, New York.

Kondorosi, A., I. Fedorcsák, F. Solymosy, L. Ehrenberg, and S. Osterman-Golkar. 1973. Inactivation of Qβ RNA by electrophiles. *Mutat. Res.* **17**:149.

Nachtomi, E. 1970. The metabolism of ethylene dibromide in the rat. The enzymatic reaction with glutathione in vitro and in vivo. *Biochem. Pharmacol.* **19**:2853.

Nachtomi, E., E. Alumot, and A. Bondi. 1966. The metabolism of ethylene dibromide in the rat. I. Identification of detoxification products in urine. *Is. J. Chem.* **4**:239.

Rannug, U. and B. Beije. 1979. The mutagenic effect of 1,2-dichloroethane on *Salmonella typhimurium*. II. Activation by the isolated perfused rat liver. *Chem.-Biol. Interact.* **24**:265.

Rannug, U. and C. Ramel. 1977. Mutagenicity of waste products from vinyl chloride industries. *J. Toxicol. Environ. Health* **2**:1019.

Rannug, U., A. Sundvall, and C. Ramel. 1978. The mutagenic effect of 1,2-dichloroethane on *Salmonella typhimurium*. I. Activation through conjugation with glutathione in vitro. *Chem.-Biol. Interact.* **20**:1.

Rapoport, I. A. 1960. The reaction of genic proteins with 1,2-dichloroethane. *Dokl. Akad. Nauk. SSSR.* **134**:1214.

Seidegård, J., R. Morgenstern, J. W. DePierre, and L. Ernster. 1979. Transstilbene oxide: A new type of inducer of drug-metabolizing enzymes. *Biochem. Biophys. Acta* **586**:10.

Yllner, S. 1971. Metabolism of 1,2-dichloroethane-^{14}C in the mouse. *Acta Pharmacol. Toxicol.* **30**:257.

COMMENTS

HOOPER: Do you use normal bile as control? The mutagenic activity of the normal bile is subtracted as background?

RANNUG: Yes.

HOOPER: So that background is subtracted, that is, the control bile is subtracted.

RANNUG: No, I prefer to present the actual number of mutants for each sample, rather than subtract the control value. In the liver perfusion experiments the bile sample taken before the addition of the test compound serves as a control for that particular experiment. We did also run separate control perfusions where we checked the function of the liver as well as the mutagenicity of the perfusate and the bile, so we know that a perfusion without test compound with or without solvent does not produce any mutagenic effects.

WARD: Did you actually compare the levels of enzymes with the different strains to show that there were different levels, and that may be one of the reasons for the differences?

RANNUG: That's what we are doing now. I have only some preliminary results indicating that both the GSH level and overall GSH S-transferase activity are approximately the same for the two strains of rat. We don't know yet if the distribution of the different types of GSH S-transferases is different for the two strains.

HOOPER: The purity of the compound was the same?

RANNUG: Yes. We tested samples from the same batch, we tested different batches, and we also tested EDC from different suppliers but we still got a low effect. The most recent experiments, however, show a small increase in the mutagenic effect compared to the lowest effect (Fig. 3).

PLOTNICK: What happens when you add just GSH and S-9, but no EDC?

RANNUG: Nothing in terms of mutagenicity.

PLOTNICK: But you didn't discuss that specifically.

RANNUG: No, but that is an obvious control and is included in all experiments where we compare the mutagenic effect of EDC in the presence of S-9 with and without GSH.

PLOTNICK: How would you postulate a mechanism for the effect of a GSH conjugate. You know that at least the β-hydroxyethyl was without activity. How do you postulate that? That's considered to be a detoxification reaction.

RANNUG: Although conjugation with GSH normally is considered to be a detoxication pathway, and in most cases it also results in a detoxication, it doesn't in the case of EDC. That is because the conjugate formed, S-(2-chloroethyl)-GSH, is a half sulfur mustard and, as such, a strong electrophilic agent. It is well known that electrophilic compounds are mutagenic. The S-(2-hydroxyethyl)-GSH does not have these properties.

ANDERS: Your data suggest very strongly that the episulfonium ion that would be formed would be the actual electrophile. That would be an extremely reactive compound capable of alkylation, although our most recent data suggest that sulfenyl halides may be formed from related compounds. We have no idea as to their toxicity.

AMES: I think it might help just to draw a picture on the board.

$R\text{-}S \longrightarrow CH_2CH_2Cl \rightarrow R\text{-}S \oplus]$

You have the sulfhydryl group on GSH and you just have a monofunctional alkylating agent, and this sulfhydryl group detoxifies it. The EDC conjugate makes the sulfur half-mustard, which is a known alkylating agent.

RANNUG: In this connection it could be added that the first reactive conjugate can also react with another molecule of GSH. This double conjugate —S, S'-ethylene-bis(GSH)—has been identified in rat livers, at least after treatment with the analog ethylene dibromide (EDB; 1,2-dibromoethane) (Nachtomi 1970). If this second step occurs it also means a complete detoxification of EDC. So, I think the level of available reduced GSH in vivo may be very important since it can contribute to an activation as well as a complete detoxication of both EDB and EDC.

ANDERS: In recent work from Holland, the mutagenicity of cis- and trans-1,2-dichlorocyclohexene was examined and it was shown that the cis isomer is the mutagenic one, which is consistent with episulfonium ion formation. If it were simply the half-mustard doing the reacting it wouldn't matter.

HINDERER: This sort of mechanism has been proposed for the compound β-napthylamine as a situation where conjugation might provide the answer to why you're seeing bladder tumors appear, that is, you have a conjugate formed that is *trans*-carried to the bladder where it's transposed into a radical state. That's theoretical, but it's one way of explaining why you get bladder tumor formation when you don't have activation, or a great deal of activation capability, at the bladder.

ANDERS: The very recent studies of the episulfonium ion reported from Russia indicate that the positive charge is carried more on the carbons than on the sulfur, which is quite consistent with the metabolic products that have been described.

KARY: Is the composition of bile obtained from the liver perfusion preparation similar to the composition of bile from the intact animal?

RANNUG: We haven't done any chemical analyses of the bile produced during the perfusion and compared it to the bile produced in vivo, but several tests of the liver are carried out during or after the perfusions to detect liver damage or impairments of its function. These tests haven't indicated any differences, which of course doesn't rule out that minor changes may occur.

KARY: And at what temperature do you maintain the liver?

RANNUG: 37°C.

SPREAFICO: Did you do any with mice?

RANNUG: We haven't done any perfusion experiments with mice, only treatment in vivo. In these experiments the gall bladder was taken out and the bile was tested in the plate incorporation test. Samples were taken 30 and 60 minutes after a single i.p. injection and both samples gave a significant mutagenic effect. The 30-minutes sample gave the highest number of mutants.

KARY: In relation to the bile question, do you have any idea of what percentage of EDC metabolism would go through bile?

RANNUG: No, I don't know that because we haven't done any direct comparison. Since the conjugate formed is very reactive, even a small amount could account for the mutagenic effects seen in the bile during liver perfusion.

Review of Nonbacterial Tests of the Genotoxic Activity of Ethylene Dichloride

VINCENT F. SIMMON
Genetic Toxicology Division
Genex Corporation
Rockville, Maryland 20852

The mutagenic potential of ethylene dichloride (EDC; 1,2-dichloroethane) was first reported by Rapoport in 1960. In this publication, Rapoport discusses a series of experiments with *Drosophila melanogaster* that date from 1947.

In the earliest experiments, eggs and larvae (18–40 hr postoviposition) were exposed to EDC on the surface of nutrient medium in a small test tube. The exposure period was estimated to be less than 1 hour, but no effort was made to contain the vapors of the volatile liquid. Sex-linked recessive lethals were determined in the second filial generation. Of 1405 chromosomes from the treated population, 37 sex-linked lethal mutations were observed (2.6%). The unexposed control population contained two mutations out of 931 chromosomes tested (0.21%).

When larvae of males were exposed to 53 mg of EDC vapors in a 20-liter container for 48 hours, 33 sex-linked lethal mutations were observed among 721 chromosomes (4.58%). No control values were reported. The mutation frequency was increased by exposing adult males to 105 mg of EDC vapors in a 20-liter container for 24 hours. In this experiment 20 mutants out of 194 chromosomes (10.3%) were observed.

To increase survival at higher doses, the exposure was fragmented; 20-minute exposures were separated by 30 minutes of no EDC. Using this procedure, researchers exposed adult male flies to 35 mg of EDC vapors in a 1.5-liter container for a total of 6 hours. In two separate experiments, mutation frequencies of 22.2% and 20.8% were observed (20 mutants out of 90 chromosomes and 22 mutants out of 106 chromosomes, respectively).

Shakarnis (1969) reported that EDC induced recessive sex-linked lethal mutations and chromosome nondisjunction in 3-day-old Canton-S female *D. melanogaster*. The flies were exposed to EDC vapors (approximately 700 mg in 1.5 liters) for 4 or 8 hours. Immediately after exposure, individual females were mated singly with Muller-5 males or groups of five females were mated with five males carrying a marked, inverted X chromosome. The average number of progeny per female was 53% of control after the 4-hour exposure and 8% of control after the 8-hour exposure. The F_1 offspring were examined for exceptional individuals; exceptional females were mated for genetic analysis and exceptional males were tested for sterility. The F_2 generation was analyzed for recessive sex-linked lethal mutations. The 4-hour exposure resulted in an increase in XXY females but not in XO males. After 8 hours of exposure, the ratio of exceptional females and that of exceptional males was increased significantly (15 out of 8452 and 7 out of 7780, respectively).

An exposure-time-related effect also was observed in sex-linked recessive lethal mutations in the F_2 generation. After 4 hours of exposure, 67 lethal mutations were observed among the 2031 chromosomes analyzed (3.30%); after 8-hours of exposure, 281 lethal mutations were observed among 4750 chromosomes (5.92%).

Ehrenberg et al. (1974) reported on the effects of EDC on barley kernels. The seeds were treated by exposure to 30.3 mM EDC for 24 hours at 20°C in sealed containers. The EDC exposure increased sterility (reduction of fertile spikelets possibly the result of recessive lethal mutations) and increased number of chloroplast mutants.

A third *Drosophila* experiment by Nylander and coworkers (Nylander et al. 1978) also indicated that EDC was mutagenic. In these assays, newly hatched larvae were exposed at 25°C to medium containing 0.1% and 0.5% EDC. Males carrying either an unstable or a stable sex-linked eye color marker were mated to attached X females. The results indicated a dose-related increase in mutants, with the unstable strain yielding a higher mutation frequency. The percentage of mutants was 0.045% (2/441) for the control, 4.2% (263/6260) for the 0.1% treatment, and 7.21% (44/610) for the 0.5% treatment with the stable strain. The percentage of mutants in the unstable strain was 0.075% (4/5363) for the control, 9.48% (274/2889) for the 0.1% treatment, and 24.88% (50/201) for the 0.5% treatment.

The most recent investigation for the mutagenic activity of EDC in *Drosophila* has confirmed previous results.

King and colleagues (1979) fed 50 mM EDC to 1- to 2-day-old Berlin-K males in 5% sucrose for 3 days. Individual males were then mated with three virgin Basc females (Muller 5) for 3 days. This 3-day mating scheme was repeated twice. This permitted an assessment of the effects of EDC on mature sperm, spermatids, and spermatocytes. The average recessive lethal mutations in the untreated controls for the three broods was 47 lethals among 22,048 X chromosomes (0.21%). Recessive lethal mutations in the three broods were 6 of 1185 (0.51%) for mature sperm, 41 of 1179 (3.48%) for spermatids, and 2 of 156 (1.28%) for spermatocytes.

The effect of EDC on mitotic recombination was examined in *Saccharomyces cerevisiae* D3 (Simmon, unpubl. results; Simmon 1979). The yeasts were exposed in buffer (pH 7.4) to various concentrations of EDC for 4 hours at 37°C. At a concentration of 0.5% (47% survival), six mitotic recombinants were observed, compared with three mitotic recombinants (at 100% survival) in the untreated controls. The number of mitotic recombinants per 10^5 survivors were six (control) and 25 (0.5% EDC). Higher concentrations of EDC were toxic and no increases were observed at 0.4% or lower EDC concentrations. This weak effect may not be genetically significant.

SUMMARY

The exposure of *Drosophila*, whether larvae or adults, males or females, results in heritable mutations and nondisjunction. Point mutations have also been demonstrated in *Salmonella* and possibly in barley. If EDC is recombinogenic in *S. cerevisiae*, it is only weakly so.

The assays in *Drosophila*, *Saccharomyces*, and bacteria (Rannug, this volume) have been used as predictive assays for chemical carcinogens. Thus, the positive response in several of these assays suggests that EDC may be a carcinogen.

Furthermore, the experiments using *Drosophila* clearly indicate the mutagenic activity of EDC to gonadal DNA. Thus, it would seem prudent to be concerned about the possible mutagenic activity of EDC to humans. The overall burden of mutation to individuals, or to society as a whole, is not known. However, nearly 2000 human defects that are genetically determined have been catalogued (Childs et al. 1972) and it has been estimated that genetic defects appear in 55 to 105 of every 1000 live births (Neel 1979). It seems reasonable to expect that exposure to EDC could increase this burden.

REFERENCES

Childs, B., S. M. Miller, and A. G. Bearn. 1972. In *Mutagenic effects of environmental contaminants* (ed. H. E. Sutton and M. I. Harris), p. 3. Academic Press, New York.

Ehrenberg, L., S. Osterman-Golkar, D. Singh, and U. Lundqvist. 1974. On the reaction kenetics and mutagenic activity of methylating and β-halogenating gasoline additives. *Rad. Botany* **14**:185.

King, M. T., H. Beikirch, K. Eckhardt, E. Gocke, and D. Wild. 1979. Mutagenicity studies with X-ray-contrast media, analgesics, antipyretics, antirheumatics and some other pharmaceutical drugs in bacterial, drosophila and mammalian test systems. *Mutat. Res.* **66**:33.

Neel, J. V. 1979. Mutation and disease in humans. In *Banbury report 1: Assessing chemical mutagens: The risk to humans* (ed. V. K. McElheny and S. Abrahamson). Cold Spring Harbor Laboratory, Cold Spring Harbor, New York.

Nylander, P., H. Olofsson, B. Rasmuson, and H. Svahlin. 1978. Mutagenic effects of petrol in *Drosophila melanogaster* I. Effects of benzene and 1,2-dichloroethane. *Mut. Res.* **57**:163.

Rapoport, I. A. 1960. Reaction of genic proteins with 1,2-dichloroethane. *Dokl. Akad. Nuak S. S. R.* **134**:1214.

Shakarnis, V. F. 1969. Induction of X-chromosome non-disjunction and recessive sex-linked lethal mutations in females of *Drosophila melanogaster*. *Genetika* 5:89.

Simmon, V. F. 1979. In vitro assays for recombinogenic activity of chemical carcinogens and related compounds in *S. cerevisiae* D3. *J. Natl. Cancer Inst.* 62:901.

COMMENTS

FABRICANT: What dose levels did you use in your work with the mitotic recombination?

SIMMON: Those experiments were taken up through 0.5% for a 4-hour exposure of a resting cell in buffer to EDC. In concentrations greater than 0.5%, the survival in the population was less than 10%. At about 0.5%, the surviving population was in the order of 50%, and, there was an insignificant random increase at that particular level, but it couldn't be considered positive by many repeated tests.

MARTONI: Do you know if there is any work ongoing on the genetic effects on man?

TER HAAR: I've seen studies with other chemicals, but not with EDC.

KARIYA: The group here may be interested to know that a chemical company submitted the results of a dominant lethal study to the Occupational Safety and Health Administration (OSHA) in response to OSHA's request (December 5, 1978) for information on EDC. The results are rather ambiguous: "Statistical analysis of the data comparing numbers of implants and deaths per litter in control versus test groups indicate that 1,2-dichloroethane be tested again." The study has not been published, as far as I know.

VOICE: Are you aware of any cell transformation studies on DNA repair?

SIMMON: I'm not aware of any DNA repair now.

BUSEY: Jerry [Ward], were there any nontumor testicular effects in the NCI study?

WARD: With ethylene dibromide (EDB; 1,2-dibromoethane) and dibromochloropropane (DBCP; 1,2-dibromo-3-chloropropane), we had no testicular effects in the two.

REITZ: Isn't it true that in the bioassays some time intervenes between the termination of dosing and actual examination of the animals?

WARD: When most of these animals died they were still on dosing. Most of them had died before 78 weeks. And the testicular atrophy is more common in the control Osborne-Mendel rats, than in the F344 rats, so there is no relation to treatment.

PLOTNICK: There was a study of DBCP by the Russians in which, I think, 5 mg/kg per day was administered orally. It shows the same thing.

HOOPER: Are plans underway to begin studies of possible genetic effects in workers exposed to EDC?

SIMMON: Well, obviously, the only place you could do that would be in the industrial setting, where the exposure is "natural." You couldn't have an inhalation study set up at NCI. So I imagine it would have to be done with the cooperation of industry.

KARY: I know of no plans for an in vivo cytogenetic study of workers. Also there may be an underlying factor here to consider, that is, the difficulty of trying to sustain that kind of study, because in the real-life situation of the workplace the average worker is exposed potentially to a number of compounds, and we don't have what we would call pure, clean compound-related exposures. So you are dealing with a situation of mixed compounds.

HOOPER: Viral infections (influenza, colds, etc.) in workers can affect the results of in vivo cytogenetic studies.

FABRICANT: With respect to the problem of viral seasonal effects in cytogenetics, I don't think that was actually a problem because you'd be having the controls at the same time. Also, it's very minor in terms of any kind of real significant differences of consistency, in terms of seasonal changes. There have only been a couple of studies, and I don't think that's all that damaging.

SIMMON: One could also look for excretion of EDC and related products, in a noninvasive test. This has been done with some other compounds, for example epichlorohydrin, that have been found in human urine after a very large massive exposure in amounts sufficient to prove positive in the *Salmonella* test. Using rather sensitive tests, such as gas chromatography and mass spectrometry, it may be possible to detect levels as low as parts per million in human blood or urine.

SESSION 2:
Toxicology and other Topics

Pharmacokinetics of Ethylene Dichloride in Rats Treated by Different Routes and Its Long-Term Inhalatory Toxicity

FEDERICO SPREAFICO, ETTORE ZUCCATO,
FRANCA MARCUCCI, MARINA SIRONI,
SILVIO PAGLIALUNGA, MARGHERITA MADONNA,
AND EMILIO MUSSINI
Istituto di Ricerche Farmacologiche "Mario Negri"
20157 Milan, Italy

This report deals with two aspects of the biointeraction of ethylene dichloride (EDC; 1,2-dichloroethane) because a better understanding of its bioreactivity and toxic potential will contribute to a better appraisal of the risk-benefit ratio for this chemical, which has such widespread use, and, therefore, its socioeconomic relevance. In this paper, data is presented on the effect of long-term exposure of rats to inhalatory EDC on a series of standard clinical chemistry parameters to contribute to the assessment of the general toxicity of this compound. In addition, after describing the analytical procedure developed specifically to detect and quantify this chemical in the organism, we discuss results obtained in studies on the kinetics and distribution of EDC when administered by the oral, intravenous (i.v.), or inhalatory routes in the same species. The importance of information on the metabolism and distribution of a chemical for the rational conduct of carcinogenicity-chronic toxicity studies does not need to be belabored.

MATERIALS AND METHODS

EDC (99.82% pure) supplied by Montedison (Milan, Italy) was used for the clinical chemistry studies. According to the supplier, it contained as impurities 1,1-ethylene dichloride (0.02%), CCl_4 (0.02%), benzene (0.09%), trichloroethylene (0.02%), and perchloroethylene (0.03%). For the pharmacokinetics investigations, EDC was supplied by Dow Chemical Company.

Clinical Chemistry Investigations

For this study, Sprague-Dawley rats were provided, randomized, maintained, and treated with EDC at the Central Tumour and Oncology Institute of the University of Bologna, Italy, under the supervision of C. Maltoni. Details of the treatment conditions used can be found in the paper by Maltoni et al. (this volume). Animals (8–10 per experimental group per sex) were sacrificed after

3, 6, 12, and 18 months inhalatory exposure (7 hr/day, 5 days/wk) at 250-150, 50, 10, and 5 ppm EDC. Controls were exposed to ambient air under the same conditions. The animals used for the blood chemistry determinations at 3, 6, and 18 months were started on the experiment at 3 months age, whereas those employed for the 12th-month determinations were animals of the same strain, source, and stock as the others but were 14 months old when their exposure to EDC was initiated.

Clinical chemistry measurements were performed on blood samples obtained by heart puncture of ether-anesthesized animals using a SMA 12 Technicon autoanalyzer apparatus for blood urea nitrogen (BUN), serum glutamic-oxalacetic transaminase (SGOT), serum glutamic-pyruvic transaminase (SGPT), lactic acid dehydrogenase (LDH), bilirubin, alkaline phosphatase, γ-glutamil transpeptidase (γ-GT), cholesterol, creatine phosphokinase (CPK), uric acid, albumin, and total proteins. A Technicon SMA 7 apparatus was used for hemoglobin, hematocrit, erythrocyte volume, total red blood cell (RBC), and white blood cell (WBC) numbers. Platelet numbers were counted with a Coulter Counter LDT, whereas blood glucose was estimated enzymatically and leukocyte percentages microscopically after standard Giemsa staining. All samples were coded and read blindly. Statistical analysis was performed using a two-way analysis of variance and Tukey test (Kirk 1968).

Pharmacokinetics Studies

Animals employed were male Sprague-Dawley CD-COBS rats (190 ± 20 g obtained from Charles River Laboratories, Italy). Rats were housed in standardized conditions of temperature, humidity and noise isolation using 12-hour light-dark cycles; they were allowed free access to a pelletized diet and tap water until 12 hours before treatment. When the i.v. route was employed, EDC was freshly dissolved in sterile saline and injected in a volume of 3 ml/kg; when oral treatments were performed, corn oil was the vehicle, using a fixed volume of 1.5 ml/kg. A Rochester inhalation chamber was used for the inhalatory route; the desired EDC concentrations were obtained diluting a saturated EDC in air mixture (obtained through a Drechsel apparatus) in the chamber. The actual EDC concentrations in the ambient air were monitored at 15-30-minute intervals by gas chromatography using dimethylformamide as the trapping agent and chloroform as the internal standard.

Measurement of EDC in the Organism

Details of the methodology developed are presented elsewhere (Zuccato et al. 1980). Briefly, a head-space approach combined with a gas chromatography (GC) procedure was adopted, using methylene chloride as the internal standard. The GC apparatus used was a model 2150 Fractovap (C. Erba, Milan) equipped with a flame ionization detector, a 4-m silanized glass column (4-mm i.d.)

packed with Tenax GC (60-80 mesh, Applied Science Lab.); operating conditions were 150°C for the oven and 225°C for the detector and injector port with a 43 ml/minute flow for carrier N_2. Under the conditions used, the calibration curve was linear in the 0.025-100 µg EDC/ml blood range and reproducibility of the method was satisfactory, with variations between different determinations never exceeding 3%. The limit of sensitivity of the method was 25 ng EDC/ml blood or g tissue. For the analysis, 1 ml blood or 1-g aliquots of frozen tissues were homogenized at 4°C with distilled water (1:3, w/v) and 2 ml of the homogenate added to 2 ml of 10% citric acid in 8-ml glass prescription bottles equipped with a rubber septum and heated at 90°C for 30 minutes to allow homogenous diffusion of EDC to the upper gaseous compartment from which samples were withdrawn for GC analysis. Recovery of EDC from blood and tissues was satisfactory, ranging from 79.2 ± 3.1% for adipose tissue to 98.2 ± 1.9% for spleen tissue. The pharmacokinetic parameters were calculated using a peeling method with the aid of a Hewlett-Packard 9810 A computer.

RESULTS

Clinical Chemistry Parameters in Rats Exposed to Long-Term Inhalatory EDC

For the sake of simplicity, no detailed results will be presented for those parameters that did not show significant modifications in this study. In addition, we will describe the data obtained in rats initially exposed to EDC as young adults (i.e., 3 months old), considering separately those animals for which treatment was started at 14 months of age.

When the effects of inhalatory EDC on bone-marrow cellular production are considered, data obtained and schematically presented in Table 1 appear to indicate that the treatment employed was without any clearly recognizable detrimental effect. In fact, no statistically significant changes between treated and control animals were observed at any test period with regard to circulating RBC numbers. The mean erythrocyte volume and the hematocrit values were significantly increased at 18 months, but only in males exposed to 5, 10, and 150 ppm, whereas significant increases in hemoglobin concentration were seen only at 3 months and without significant dose-response relationship. With regard to total WBC numbers, in both sexes there was a significant increase at 3 months in rats exposed to 5, 10, and 50 ppm (but not at 250-150 ppm), whereas at 18 months there appeared to be a tendency towards dose-related decreases; however, these did not reach statistical significance. No significant changes in platelet numbers or in the relative percentages of lymphocytes, granulocytes, and monocytes could be detected. With regard to circulating protein levels and relative percentages, definite signs of deleterious effects induced by the EDC treatment employed could not be found. Total protein concentrations were not significantly different between treated and control

Table 1
Scheme of the Effect of Inhalatory EDC Exposure in Rats on Some Hematological Parameters

		Duration of exposure (months)[a]			
Parameter	Sex	3	6	18	12
RBC	M	=[b]	=	=	=
	F	=	=	=	=
Erythrocyte mean volume	M	=	=	=	S ↑ 5, 10, 150
	F	=	=	=	=
Hematocrit	M	=	=	=	S ↑ 5, 10, 150
	F	S[c] ↑ 150	=	=	=
Hemoglobin	M	S ↑ all doses	=		=
	F	no dr[d]	=	S ↓ 150	
WBC	M		=	N.S.[e] ↓ all doses	N.S. ↓ all doses
	F	S ↑ 5, 10, 50	=		N.S. ↓ 150
Platelets	M	=	=	=	=
	F	=	=	=	=

[a] For rats tested at 3, 6, and 18 months, EDC exposure (5, 10, 50, 150 ppm) was started at age 3 months; for rats tested at 12 months, EDC exposure was initiated at the age of 14 months.
[b] No difference from control.
[c] Statistically significant difference from controls at EDC concentration indicated.
[d] Dose-response relationship.
[e] Statistically not significantly different from controls at EDC concentrations indicated.

animals at all times investigated; percent albumin values were significantly increased in both sexes at 3 and 18 months but not at 6 months. In addition, there was no clear dose-response relationship and, indeed, at 18 months males exhibited an inverse dose-response relationship. The values in the highest-dose group were not significantly different from controls. Gamma globulin percentages were significantly lower at 3 months in the 150-ppm EDC group, however, there were no changes at 6 months, and at 18 months there was a significant decrease only at the 5- and 10-ppm doses. Beta globulins were significantly higher only at 3 months and then only at the 5-ppm exposure level; no changes were ever seen in alpha-1 globulin levels, whereas for alpha-2 globulins there was a significant, but not dose-related, decrease in all treated groups only at the third exposure month. These results are schematically presented in Table 2.

Table 3 (page 112) summarizes the results observed evaluating a series of other clinical chemistry parameters, which can be considered as indicators of the functional status of various major organs. It can be seen that when the serum levels of GOT and GPT enzymatic activities as well of γ-GT were measured, no clear indications of treatment-related abnormalities could be detected. In fact, no significant changes from control values were observed in either γ-GT or SGPT levels whereas as with regard to the levels of SGOT, there was an increase in treated males at the third month, but it was not significant and not dose-related.

Table 2
Scheme of the Effect of Inhalatory EDC in Rats on Serum Proteins

Parameter	Sex	Duration of exposure (months)			
		3	6	18	12
Total proteins	M	=	=	N.S. ↓ all doses	=
	F	=	=	=	=
% Albumin	M	S ↑ 10, 50, 150	=	S ↑ 5, 10, 50	S ↓ 5, 10, 50
	F	no dr	=	S ↑ all doses no dr	=
% Alpha-1 globulins	M	=	=	=	S ↑ 5, 10, 50
	F	=	=	=	S ↑ 10, S/150
% Beta-2 globulins	M	S ↓ all doses	=	=	S ↑ 5, 150
	F	no dr	=	=	S ↑ 5, 10
% Beta globulins	M	S ↑ 5	=	=	=
	F		=	=	S ↓ 5, 10
% Gamma globulins	M	S ↓ 150	=	S ↓ 5, 10	=
	F		=	S ↓ 50	=

For rats tested at 3, 6, and 18 months, EDC exposure (5, 10, 50, 150 ppm) was started at age 3 months; for rats tested at 12 months, EDC treatment was initiated at the age of 14 months. Symbols and abbreviations are the same as in Table 1.

A significant increase was present in females exposed to 5 and 150-250 ppm EDC at the same time period. No changes were observed at 6 or 18 months (Table 4, page 113). Alkaline phosphatase levels at 3 months were decreased significantly in all female groups, not significantly in males, and without dose-response relationship in either sex; at 6 and 18 months, the values of treated and control animals were not significantly different. With regard to bilirubin and cholesterol levels, again the results observed do not support clear treatment-associated effects, since only slightly and nonsignificantly increased (6 months) and decreased (18 months) levels could be measured.

In addition, these changes were present without dose-response relationship in only one sex. Slightly higher levels in CPK were seen at 18 months, but this increase was observable only in males exposed to 50 and 150 ppm and did not reach statistical significance. LDH levels were significantly higher at 3 months in both males and females of all treated groups, the increase being somewhat more evident in females, reaching a value of approximately 100% in the 150-250-ppm EDC group; a dose-response relationship was clear, however, only in males. The biological significance of this finding is rendered uncertain by the fact that at 6 months, even though there was still a tendency towards elevated levels, statistical significance was not reached. Similarly, at 18 months, significantly higher LDH levels were seen only in males exposed at 5, 50, and 150 ppm EDC, and no changes from controls were evident in females. Table 4 also shows

Table 3
Scheme of the Effect of Inhalatory EDC Exposure in Rats on Various Blood Chemistry Parameters

Parameter	Sex	Duration of exposure (months)			
		3	6	18	12
SGOT	M	N.S. ↑ all doses	=	=	S ↑ 5; S ↓ 50, 150
	F	S ↑ 5, 150	=	=	S ↑ 5, 10; S ↓ 50, 150
SGPT	M	=	=	=	S ↑ 50, 150
	F	=	=	=	N.S. ↑ all doses
γ-GT	M	=	=	=	S ↑ 50, 150
	F	=	=	=	N.S. ↓ 150
Bilirubin	M	=	N.S. ↑ 150	N.S. ↓ all doses	S ↑ 5
	F	N.S. ↑ 150	N.S. ↑ all doses	=	S ↓ 10, 50, 150, no dr
Cholesterol	M	N.S. ↑ 150	=	=	S ↓ 50, 150
	F	N.S. ↓ all doses	=	N.S. ↓ all doses	N.S. ↓ 150
Alkaline phosphatase	M	S ↓ all doses, no dr	S ↓ 5	=	
	F	S ↑ 10, 50, 150, no dr	N.S. ↑ all doses	S ↑ 5, 50, 150, no dr	S ↓ 5, 10, 50
LDH	M	S ↑ 10, 50, 150, no dr	S. ↑ 50, N.S. ↑ 150	=	S ↓ 5, 10, 150, no dr
	F		=		
CPK	M	N.S. ↑ 10, 50	S ↓ 5	N.S. ↑ 50, 150	=
	F	S ↓ 50	S ↓ 5, 10, 50	S ↑ 5	=
BUN	M	=	N.S./all doses	=	S ↑ 150
	F	S ↓ all doses, no dr	=	=	S ↑ 50, 150, dr
Uric acid	M	S ↑ 50	=	=	=
Glucose	F	=	=	N.S. ↑ 50, 150	=

For rats tested at 3, 6, and 18 months, EDC exposure (5, 10, 50, 150 ppm) was started at age 3 months; for rats tested at 12 months, EDC treatment was initiated at the age of 14 months. Symbols and abbreviations are the same as in Table 1.

Table 4
Values of Some Blood Chemistry Parameters in Rats Exposed to EDC for 18 Months

Parameter	Sex	\\ EDC concentrations (ppm)				
		0	5	10	50	150
SGOT (milliunits/ml)	M	124 ± 18	88 ± 19	124 ± 9.9	108 ± 13	110 ± 12
	F	136 ± 19	105 ± 14	119 ± 21	102 ± 14	121 ± 13
SGPT (milliunits/ml)	M	28.7 ± 2.0	39.8 ± 5.4	29.5 ± 3.3	25.2 ± 3.3	35.2 ± 6.1
	F	26.2 ± 3.3	35.5 ± 5.7	32.8 ± 3.1	26.6 ± 3.0	31.5 ± 2.8
γ-GT (milliunits/ml)	M	2.50 ± 0.6	2.88 ± 0.5	2.63 ± 0.7	3.00 ± 0.8	2.25 ± 0.4
	F	1.63 ± 0.4	2.25 ± 0.4	2.00 ± 0.3	2.13 ± 0.4	1.50 ± 0.5
Cholesterol (mg %)	M	126 ± 5.3	127 ± 12	119 ± 8.9	112 ± 7.7	122 ± 6.2
	F	161 ± 16	145 ± 15	134 ± 9	140 ± 7.5	147 ± 10
LDH (milliunits/ml)	M	905 ± 36	725 ± 24[a]	765 ± 25[a]	800 ± 11[a]	857 ± 26
	F	890 ± 48	737 ± 29[a]	897 ± 58	762 ± 33	767 ± 25
BUN (mg %)	M	17.4 ± 0.4	19.5 ± 0.5[a]	16.5 ± 0.7	15.6 ± 0.3	16.0 ± 1.6
	F	17.0 ± 0.9	19.0 ± 0.4	19.5 ± 0.6	17.5 ± 0.8	18.5 ± 1.3
Uric acid (mg %)	M	2.58 ± 0.3	3.29 ± 0.6	2.78 ± 0.3	2.52 ± 0.3	2.49 ± 0.2
	F	2.85 ± 0.2	2.58 ± 0.3	2.23 ± 0.1	3.30 ± 0.6	2.78 ± 0.3

EDC exposure was started in these rats at age 3 months.
[a] At least $p < 0.05$ vs controls.

that no clear changes could be seen regarding BUN levels in these animals. BUN values were significantly lower at 6 months in males exposed to 5 ppm and in females treated with 5, 10, and 50 ppm EDC but were not lower in either sex when the EDC concentration was 150 ppm. In addition, no significant changes were seen at 18 months, except for an increase in males exposed to 5 ppm. Serum glucose levels were also not clearly modified by the EDC treatment employed. The same conclusion would seem to apply to uric acid levels in the circulation, which decreased in females at 3 months (but with an inverse dose-response relationship) and slightly and nonsignificantly increased at 6 (but not 18) months in both sexes.

The levels of Na^+, K^+, Cl^-, Ca^{++}, and P_i were also measured in these animals. Although significantly different values from controls were occasionally seen (e.g., higher Na^+ levels in both sexes after 3 months at 150–250 ppm, lower P_i levels at 18 months in all treated males), none of these modifications were consistently present in both sexes nor was any dose-response relationship recognizable.

Similarly, no consistent and significant differences in any of a series of urinary parameters (pH, proteins, bilirubin, glucose, hemoglobin, erythrocytes, leukocytes, epithelial cells, casts, crystals, mucus, and microorganisms) were detected between treated and control animals when measured after 18 months exposure. In view of their greater age (14 months) at treatment initiation, the results obtained in rats after 1 year of inhalatory exposure to EDC should be analyzed separately, although for comparative purposes the data are also schematically presented in Tables 1–3. In these animals EDC appeared to exert a relatively greater toxic effect than it did in younger animals.

Briefly, no changes were seen in RBC and hemoglobin values, significant but not clearly dose-related increases being detectable only in males with regard to the hematocrit value and mean erythrocyte values. A nonsignificant, nondose-related tendency towards lower WBC numbers was also observed. With regard to serum proteins, no clear changes were seen, since the significant differences observed for the various parameters were either in opposite directions at the same test period, present only at low or intermediate doses, or not detectable in both sexes.

On the other hand, significant changes were seen in parameters indicative of liver and kidney function. In fact, SGPT levels were significantly and markedly increased in both males and females exposed to 50 and 150 ppm, and γ-GT levels were also significantly higher in females treated with the two higher EDC doses. In male rats, there was a tendency towards an increase in the circulating levels of this enzyme in all groups, however the changes did not reach statistical significance and a dose-effect relationship was not evident. When SGOT levels are considered, there were significant increases in both sexes at 5 and 10 ppm but significant decreases at 50- and 150-ppm concentrations (Table 5). Bilirubin, CPK, alkaline phosphatase, LDH, and glucose levels in the serum were not significantly modified, whereas cholesterol levels were significantly lower at 50 and 150 ppm in both sexes.

Table 5
Values of Some Blood Chemistry Parameters in Rats Exposed to EDC for 12 Months

Parameter	Sex	EDC concentrations (ppm)				
		0	5	10	50	150
SGOT (milliunits/ml)	M	58.2 ± 3.0	93.5 ± 13.5	69.2 ± 13.7[a]	22.6 ± 2.1[a]	26.0 ± 2.3[a]
	F	68.2 ± 6.7	102.4 ± 13.4[a]	112.7 ± 13.5[a]	33.6 ± 7.2[a]	22.6 ± 2.2[a]
SGPT (milliunits/ml)	M	22.9 ± 2.3	28.8 ± 3.6	23.2 ± 3.5	90.0 ± 9.3[a]	111.0 ± 13.4
	F	15.7 ± 1.0	23.4 ± 3.1	28.2 ± 4.5	143.0 ± 11.7[a]	110.1 ± 10.7[a]
γ-GT (milliunits/ml)	M	0.88 ± 0.2	1.63 ± 0.7	1.13 ± 0.3	1.50 ± 0.3	1.50 ± 0.4
	F	0.83 ± 0.3	0.63 ± 0.2	0.65 ± 0.4	1.63 ± 0.4	1.75 ± 0.2
Cholesterol (mg %)	M	136 ± 5	130 ± 4	125 ± 2[a]	102 ± 1[a]	102 ± 1[a]
	F	136 ± 4	146 ± 16	119 ± 5	103 ± 1[a]	101 ± 2[a]
LDH (milliunits/ml)	M	587 ± 14	670 ± 23[a]	600 ± 41	695 ± 21[a]	200 ± 23[a]
	F	617 ± 16	682 ± 27[a]	200 ± 28[a]	770 ± 18[a]	205 ± 32[a]
BUN (mg %)	M	10.3 ± 0.3	11.0 ± 0.5	10.2 ± 0.4	10.6 ± 0.5	15.5 ± 0.7[a]
	F	10.5 ± 0.3	10.8 ± 0.3	10.7 ± 0.5	10.7 ± 0.7	15.2 ± 0.8[a]
Uric acid (mg %)	M	0.80 ± 0.1	1.05 ± 0.1[a]	0.95 ± 0.1	1.50 ± 0.1[a]	1.90 ± 0.1[a]
	F	0.94 ± 0.1	1.08 ± 0.1	1.25 ± 0.1	1.63 ± 0.1[a]	3.41 ± 0.3

EDC exposure was started on these rats at age 14 months.
[a] At least $p < 0.05$ vs controls.

In addition, significantly higher levels of uric acid (at 50 and 150 ppm, dose-related, with increases up to approximately 300%) and blood glucose (at 150 ppm) were also observed. It should be mentioned, however, that the control values for both these parameters were significantly lower than seen in untreated rats tested at other times in this study. In this connection, no significant changes in any of the urinary parameters investigated could be detected, which is similar to what we observed in "younger" animals exposed for 18 months.

Finally in the "older" animals, significant, but not clearly dose-related changes in the circulating levels of K^+ and Ca^{++} were seen at EDC levels above 10 ppm. Specifically, K^+ concentrations increased (4.91 ± 0.20 and 6.26 ± 0.21 meq/ml in the control and 150-ppm male groups, respectively), whereas Ca^{++} levels decreased (9.98 ± 0.13 and 8.73 ± 0.13 meq/ml in the control and 150-ppm male groups, respectively). No significant changes in the serum concentrations of Na^+, Cl^-, and P^- were detected.

Pharmacokinetics of EDC in the Rat

This section describes a series of results obtained in investigations aimed at determining the distribution and persistence of EDC in the body when administered to rats over a range of dosages and concentrations. The intravenous, oral, and inhalatory routes were used to administer EDC. Although studies on the pharmacokinetics of a series of EDC metabolites have been conducted and are in progress, these results are not considered here. In addition, it should be noted that only the data on EDC levels in blood, liver, lung, and adipose tissue are presented. In fact, although determinations of the levels of this chemical were also performed using brain, kidney, and spleen tissue, data obtained for EDC concentrations in these tissues were superimposable with those observed in the blood (a district with which these organs appear to be in direct equilibrium) without influencing, therefore, the overall kinetic pattern of EDC in the body.

EDC distribution and persistence in the rat organism was first investigated employing i.v. injections at doses of 1, 5, and 25 mg/kg. Levels determined in the blood are shown in Figure 1; under these conditions the disappearance is rapid and biphasic, with levels above the sensitivity limit of our method detectable until 30 minutes, 60 minutes, and approximately 2 hours for the 1, 5, and 25 mg/kg doses, respectively. Note that the steepness of the second or beta phase decreases with the increase in the EDC dose injected, indicating a dose-dependence of the disappearance of the chemical under these conditions and suggesting that its elimination may be a saturable process. The kinetic parameters calculated for these conditions are in Table 6; in addition to the already described changes with dose in the T½ beta (and in the elimination rate constant K_{el} values), note that the time zero concentration (C_0) and area under the curve (AUC) values increase linearly and exponentially, respectively, with the rise in the dose.

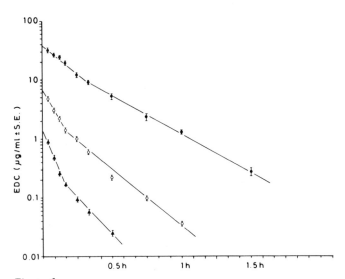

Figure 1
Blood levels of EDC in rats after i.v. administration. (●) 25 mg/kg i.v.; (○) 5 mg/kg i.v.; (▲) 1 mg/kg i.v.

EDC levels in blood, liver, lung, and epidydimal adipose tissue after single oral administration of 25, 50, and 150 mg/kg are shown in Figures 2, 3, and 4. In addition to the rapid absorption of the chemical, these figures also show that, as was expected, the liver is the tissue in which EDC accumulates most rapidly— peak concentrations were reached within 10 minutes of administration. Disappearance from this organ begins rapidly, following a biphasic, biexponential curve whose second component (the gamma phase of the curve) is practically equivalent to the disappearance curve of EDC from the general circulation, in which the peak is reached slightly later than in the liver. EDC in the lung appears to be in equilibrium with blood, although levels are at all times lower, presumably because of expiration of the chemical. At variance with these organs, accumulation in the adipose tissue is slower, with peak levels that are reached at 45-60 minutes but which are significantly higher (approximatley five times at the 50 and 150 mg/kg doses) than in blood. Disappearance from adipose tissue is monophasic in its appearance and essentially equivalent in its rate to what is seen for blood.

The kinetic parameters for the oral treatment are listed in Table 6, from which the dose-dependence of EDC elimination as well as EDC's high uptake in adipose tissue can be noted. AUC values that are approximately 12 and 4 times greater in blood or liver at 25 and 150 mg/kg, respectively, suggest a possible saturation of the adipose tissue at high EDC doses. When one relates the dose administered and peak blood level observed, the curve appears linear up to 50 mg/kg, with a perceptible decrease in steepness thereafter, possibly

Table 6
Pharmacokinetic Parameters of EDC in Rats

Route	Dose	Parameters[a]	Blood	Adipose Tissue	Lung	Liver
Oral	25 mg/kg	T½ β[b]	24.62	23.22	24.10	18.47
		AUC	446	5119	136	679
		peak level	13.29	110.67	2.92	30.02
		AUC blood : AUC tissue	—	0.09	3.28	0.66
	50 mg/kg	T½ β[b]	44.07	30.11	38.26	42.31
		AUC	1700	12,543	538	1897
		peak level	31.94	148.92	7.20	55.00
		AUC blood : AUC tissue	—	0.13	3.16	0.90
	150 mg/kg	T½ β[b]	56.70	57.63	44.57	66.47
		AUC	7297	29,468	648	5384
		peak level	66.78	259.88	8.31	92.10
		AUC blood : AUC tissue	—	0.25	11.26	1.35
Inhalatory	50 ppm	T½	12.69	22.63	11.26	10.72
		AUC	26	391	6	17
		C_0	1.42	10.24	0.39	1.02
		AUC blood : AUC tissue	—	0.07	4.33	1.53
	250 mg/kg	T½	22.13	28.12	15.53	17.51
		AUC	1023	13,558	279	694
		C_0	30.92	265.47	13.88	22.06
		AUC blood : AUC tissue	—	0.08	3.66	1.47
Intravenous	1 mg/kg	T½ β	7.30			
		AUC	9			
		C_0	1.50			

5 mg/kg	T½ β	9.49
	AUC	54
	C_0	8.00
25 mg/kg	T½ β	14.07
	AUC	595
	C_0	38.12

[a] T½ = min; AUC = μg ∘ min · ml^{-1}; C_0 and peak level = μg/ml or μg/g.
[b] In the case of liver, T½ values are for the gamma phase.

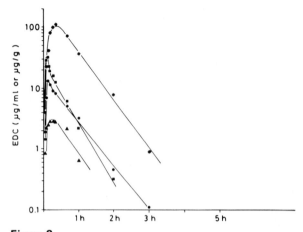

Figure 2
EDC levels after oral administration in rats; dose is 25 mg/kg. (*) Adipose tissue; (●) blood; (■) liver; (▲) lung.

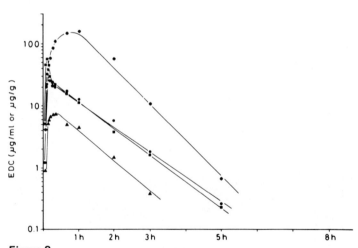

Figure 3
EDC levels after oral administration in rats; dose is 50 mg/kg. (*) Adipose tissue; (●) blood; (■) liver; (▲) lung.

indicating a relative saturation in gastrointestinal absorption of EDC at doses of 100-150 mg/kg. A limited series of experiments have been performed to examine whether repeated oral treatments significantly modified the distribution and kinetics of EDC in the rat. However, no significant differences in the kinetic parameters were seen between single and ten daily administrations of 50 mg/kg; nor were significant differences seen between male and female rats.

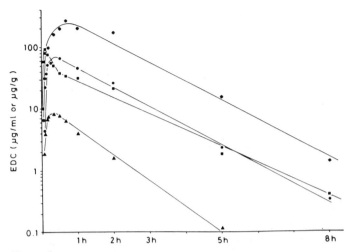

Figure 4
EDC levels after oral administration in rats; dose is 150 mg/kg. (*) Adipose tissue; (●) blood; (■) liver; (▲) lung.

The distribution and disappearance of EDC in the body were also investigated after inhalatory exposure, and Tables 7 and 8 present the blood and tissue levels determined in rats after exposure to this chemical for different times. It can be observed that steady-state concentrations in the body are reached relatively slowly—in 2-3 hours, depending on the level of EDC in the atmosphere. A very clear dose-dependence of the EDC in body levels can also be noticed under these conditions, with differences in EDC concentrations on the order of 20-30 times between exposures to 50 and 250 ppm when the blood, liver, lung, and adipose tissues are considered. The highest absolute levels of EDC were again measured in the adipose tissue, with concentrations that were between 8 and 9 times greater than those measured in the blood. As previously seen for the i.v. and oral routes, after inhalation the data obtained with regard to the brain,

Table 7
EDC Tissue Levels After 50-ppm Inhalatory Exposure

Time of exposure	EDC µg/g or µg/ml ± S.E.			
	blood	liver	lung	adipose tissue
30 min	0.48 ± 0.05	0.32 ± 0.02	0.14 ± 0.02	2.91 ± 0.22
1 hr	0.92 ± 0.09	0.67 ± 0.04	0.27 ± 0.03	7.49 ± 0.60
2 hr	1.34 ± 0.09	0.84 ± 0.09	0.34 ± 0.03	10.31 ± 0.94
4 hr	1.34 ± 0.11	1.14 ± 0.17	0.42 ± 0.05	11.08 ± 0.77
6 hr	1.37 ± 0.11	1.02 ± 0.10	0.38 ± 0.02	10.19 ± 1.00

Table 8
EDC Tissue Levels After 250-ppm Inhalatory Exposure

Time of exposure	EDC µg/g or µg/ml ± S.E.			
	blood	liver	lung	adipose tissue
30 min	6.33 ± 1.04	3.82 ± 0.78	2.19 ± 0.21	26.75 ± 3.12
1 hr	11.65 ± 1.12	7.34 ± 0.74	6.40 ± 0.20	82.64 ± 2.19
2 hr	23.64 ± 0.91	16.39 ± 1.17	14.07 ± 0.47	151.53 ± 12.17
3 hr	29.36 ± 1.01	20.83 ± 2.21	14.47 ± 1.12	252.18 ± 14.62
6 hr	31.29 ± 1.19	22.49 ± 1.12	14.14 ± 0.90	273.32 ± 12.46

kidney, and spleen were essentially superimposable on the EDC blood levels and are therefore not presented in detail in this paper.

The disappearance of EDC from the organism after a steady-state condition has been reached by maintaining the rats for 6 hours in an atmosphere containing 50 or 250 ppm, is shown in Figures 5 and 6. At both exposure levels, the highest rate of elimination is for the lung and the slowest is for the adipose tissue, from which EDC disappearance, although again monoexponential (as for the other tissues), is significantly more sluggish. From the pharmacokinetic parameters calculated for this experimental condition (Table 6), the dose-dependence of EDC disappearance rates can easily be noted, although the dose influence is less marked than observed in the case of the oral route. The marked dose-dependence of the AUC and C_0 values alluded to above is evident from the data of Table 6 and from the fact that the EDC dose does not markedly

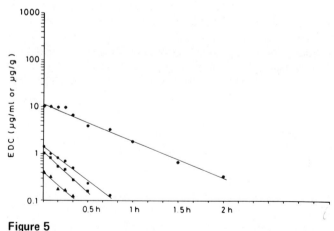

Figure 5
Levels of EDC in rats after inhalatory exposure to 50 ppm. Levels were measured at the termination of a 5-hr exposure period. (*) Adipose tissue; (●) blood; (■) liver; (▲) lung.

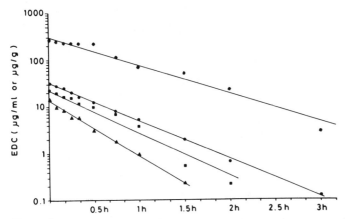

Figure 6
Levels of EDC in rats after inhalatory exposure to 250 ppm. Levels were measured at the termination of a 5-hr exposure period. (*) Adipose tissue; (●) blood; (■) liver; (▲) lung.

influence the relative distribution of the chemical in the various organs, as indicated, for instance, by an AUC blood-to-AUC adipose tissue ratio that is practically constant despite a fivefold increase in exposure levels.

DISCUSSION

Considering first the clinical chemistry data, it would appear that long-term exposure (1 year or longer) to inhalatory EDC in concentrations up to 150 ppm is not associated with marked and widespread toxicity in the rat. Indeed, in animals initially exposed to this chemical when 3 months old, the treatment employed was well tolerated for at least 18 months, as judged by the analysis of a series of standard parameters indicative of the function of various major organs. Evidence suggesting (but not proving) a somewhat greater toxicity of EDC, involving the liver and kidney, was obtained when this treatment was employed on animals that were older (i.e., 14 months) at treatment initiation. Liver toxicity in the latter group of animals after 12 months of treatment is suggested in the first place by the significantly higher SGPT levels in the 50- and 150-ppm groups and by the observation of significant decreases in SGOT values at the same EDC concentrations, accompanied by significant increases at lower exposure levels.

In further support of a hepatotoxicity of EDC under these conditions could also be the tendency towards a dose-related increase in γ-GT levels in all treated groups, although the change reached statistical significance only in females exposed to 50 and 150 ppm EDC. The possibility of toxicity for the kidney rests on the observed increases in BUN and uric acid levels, which were significant in animals treated with 150 ppm and 50-150 ppm, respectively.

However, no changes in any urinary parameter were seen in these rats. Of uncertain interpretation are the dose-related decreases in cholesterol levels seen at the two higher EDC doses and the dose-related increase in K^+ and decrease in Ca^{++} values seen at doses above 10 ppm. A detailed histological investigation currently in progress at the Oncology Institute of the University of Bologna should shed light on these problems.

It should be noted, however, that findings of others support the possibility suggested by these results, namely, that the liver and kidney can be prime targets for EDC chronic toxicity. The presence of slight to moderate fatty degeneration and occasional necrosis in these organs was described in animals of different species, dead after exposure to 500 ppm EDC for few weeks (Heppel et al. 1946; Hofman et al. 1971); liver impairment was observed by Alumot et al. (1976) in rats fed a diet containing 200-250 ppm for 2 years. Lioia and Fondacaro (1959) report abnormal liver function and histology in rabbits exposed to 3000 ppm for short periods. Analogous to what was observed with other chlorinated hydrocarbons, human liver and kidney abnormalities associated with the ingestion of or occupational exposure to EDC have been described by various investigators (NIOSH 1976).

Although various reports are available on the acute or short-term effect of exposure to inhalatory EDC in animals, as summarized in a recent publication by the U.S. Department of Health, Education, and Welfare (DHEW) (NIOSH 1976), a paucity of information exists on the effect of chronic treatments with this chemical. In addition, available studies have generally employed relatively limited number of animals and shorter periods of exposure. Heppel et al. (1946) did not observe effects on clinical examination or at necropsy in limited numbers of rats and guinea pigs exposed to 100 ppm EDC for 4 months. Similarly, Spencer et al. (1951) report that the exposure of rats to 100 and 200 ppm for 6 months was not associated with any adverse effect, as judged by general appearance, behavior, mortality, growth, organ function, and blood chemistry. Similarly treated guinea pigs had, on the other hand, increased liver-to-body weight ratios and reduced growths; the limited numbers of animals involved prevents conclusions regarding possible species differences in sensitivity. In these studies and in others (NIOSH 1976) EDC appeared to exhibit a relatively steep dose-response curve, since exposure to 300-400 ppm EDC could not be protracted for more than 3-6 weeks because of high levels of mortality. Also in this study, exposure to 250 ppm was accompanied by significant mortality (Maltoni et al., this volume), forcing the use of 150 ppm as the highest exposure dose for these investigations.

Results presented here thus appear to be in general agreement with previously available information, allowing us to extend the period during which exposure to inhalatory EDC (in concentrations up to 150 ppm) is not associated with marked and generalized toxicity in the rat to at least 12-18 months. At the same time, the detection of functional signs suggesting effects on the liver and kidney indicates that the levels used in the carcinogenicity bioassay of inhalatory EDC were biologically active, a fact that could give greater confidence to the negative results obtained in this testing (Maltoni et al., this volume).

In considering the second aspect examined in these investigations, it should be noted that although the in vivo biotransformation of EDC has been the object of a number of studies by different groups (Reitz et al., Rannug, Anders and Livesey, all this volume), to the best of our knowledge no previous data on the distribution and kinetics of this compound in the body based on direct determinations of the chemical in tissues exist in the literature. There is the elegant study of Filser and Bolt (1979) in which the kinetics of a series of halogenated ethylenes in rats was investigated through the use of animals exposed to different gaseous levels in a closed inhalation system. However, in that study the kinetic behavior of the xenobiotics in the tissues was not directly measured; a total body kinetics was only extrapolated from the measurements of the decline in the atmospheric concentration of the chemical. The "tissue" curves presented in the Filser and Bolt study are thus based on theoretical considerations, assuming the various body tissues to be kinetically identical and, more importantly, without taking into account metabolic processes.

Data described earlier indicate that a series of general similarities can be recognized when the kinetics and distribution are determined after EDC treatment by the i.v., oral, and inhalatory routes. In addition to kinetic curves that are similar, another general similarity for the three routes has to do with the relative distribution of EDC in the different tissues. Namely, in all cases the highest quantities of EDC are seen in the adipose tissue and for each route the lowest AUC values are found in the lung, most probably because of its expiratory role. An influence of the dose on the kinetic parameters, recognized for other halogenated ethylenes in recent publications by other groups (Gehring et al. 1977, 1978; Bolt 1978; McKenna et al. 1978; Reichert and Henschler 1978; Filser and Bolt 1979), is also present in at least two of the treatment routes employed and, by inference, also with inhalatory treatments.

With the i.v. method of administration, our results in fact support this influence for EDC disappearance from the circulation, as directly reflected by the diminishing values of the rate constant K_{e1} (0.277, 0.142, and 0.078 min^{-1} for the blood at 1, 5, and 25 mg/kg, respectively) with the increase in the EDC quantity administered, a finding which suggests that the process can be saturable. The same also appears to be true for the oral treatment (K_{e1} values of 0.046, 0.017, and 0.010 for blood were computed for doses of 25, 50, and 150 mg/kg), a condition in which the data further suggest that a saturation may also exist with regard to the gastrointestinal absorption of EDC. In fact, the curve relating the observed peak blood levels and dose administered is linear up to a dose of 50 mg/kg, but at higher doses the curve exhibits a clear downward inflection.

Thus for both the i.v. and oral routes, the AUC values (which provide an estimate of the total time of exposure of a tissue to this chemical) increase almost exponentially up to a point (viz., 50 mg/kg in the case of oral treatments) but only linearly above the critical dose. Since only two inhalatory dose levels have been explored so far, it is still impossible to state whether under these conditions an EDC concentration exists above which the increase in the exposure levels results in proportionally lower rises in AUC values. Our data only allow us to point out that within the range of inhalatory levels explored, the rate of disappearance of EDC (once exposure is terminated) is dose-dependent (for blood, K_{el} of 0.055 and 0.031 at 50 and 250 ppm).

After mentioning that the route employed influences at least some kinetic parameters for selected organs (e.g., higher EDC peak levels and AUC values are seen after oral treatment as compared with inhalatory treatment in the liver, whereas the C_0 values in the adipose tissue are greater after inhalation), of possible relevance is the finding that when the oral dose is increased sixfold from 25 to 150 mg/kg, one observes an approximately 16-fold increase in the blood AUC; passing from 50 to 250 ppm atmospheric concentration, the increase in the AUC for the same region is approximately 40-fold. At equivalent zero time or peak levels, such as can be obtained with 50 mg/kg os and 250 ppm inhalation (see Table 6), the AUC values were higher in six out of the seven organs investigated when the treatment was administered orally, with differences in the range of 70–270%, depending on the organ considered. The only exception was the adipose tissue, for which no significant differences in AUC values were seen at these two doses. It should be emphasized, however, that the inhalatory AUC values considered earlier are those measured after termination of inhalation exposure to EDC, the comparison being between an acute oral treatment and an acute inhalatory treatment giving equivalent initial levels of this chemical in the circulation.

Although the mathematical analysis is still incomplete (especially for the inhalatory treatment) our data for the i.v. treatment seems to fit a two-compartment model of the type

$$\rightarrow \boxed{1} \rightleftarrows \boxed{2}, \qquad\qquad (1)$$
$$\downarrow$$

in which the first compartment is represented by the blood and the tissues that rapidly equilibrate with the circulation and the second compartment by the tissues in which equilibrium is reached more slowly or which have a greater affinity for EDC, such as the adipose tissue.

Under the conditions of the investigation, the general structure of the model for the inhalatory exposure is similar, at least in the specific experimental conditions investigated, which involve a first period of exposure to EDC so as to reach steady-state levels and determination of the kinetic parameters at the termination of such exposure. This condition can explain the absence of the rapid alpha disappearance phase in our inhalatory studies, which is at variance with what is seen in the i.v. studies. The alpha phase is, in fact, heavily influenced by the equilibration between compartments one and two, an equilibrium already reached in our inhalatory conditions. With regard to the oral route, in which the results obtained indicate that EDC in the liver has a kinetic behavior somewhat different from what is seen with the other routes, a tentative model might be represented by a three-compartment model of the following type:

$$\begin{array}{c} \boxed{1} \rightarrow \\ \uparrow\downarrow \\ \leftarrow \boxed{2} \rightleftarrows \boxed{3} \, . \end{array} \tag{2}$$

In this hypothetical model the first compartment would consist of the liver and intestines and compartments two and three of the tissues reaching equilibrium with the blood rapidly and slowly, respectively. A more formal and complete analysis of these aspects will be presented elsewhere (F. Spreafico et al., in prep.), and these models should still be considered as tentative.

Interest in the long-term toxicity of EDC has been renewed in recent years by the results obtained by the National Cancer Institute (NCI 1978) supporting the carcinogenic potential of this compound for rats and mice treated by the oral route. Maltoni et al. (this volume) did not find statistically significant increases of neoplasms in rats and mice submitted to long-term inhalatory exposure to this chemical at levels up to 150 ppm. There are some differences in the experimental conditions of the two studies with regard to variables that may be important determinants of the apparently contrasting results, variables ranging from the possible purity of the EDC employed (see Hooper et al., Reitz, Appendix, all this volume) to the animal strain used. Also, it is well known that marked differences in metabolic activity towards various xenobiotics can exist not only among different species but also among different inbred or outbred strains within a given species (Hucker 1970; Thorgeirsson and Nebert 1977), as was shown by Bolt (1978) for the significantly different liver microsomal metabolizing capacity for vinyl bromide between Sprague-Dawley and Wistar rats.

It is not the purpose of this paper to discuss this contrast in findings; however, since different routes of treatment were employed for the two carcinogenicity bioassays, we may ask if the resulting different kinetics and distribution could have played a role. In answering this question, we note that at this time kinetic data are available for only some of the doses employed in the NCI study and in the study of C. Maltoni et al., in which the treatments were at 45 and 95 mg/(kg · day). For other dose levels, extrapolated rather than experimentally determined data have to be employed for making comparisons. Although preliminary results would appear to indicate that no marked differences in kinetic profiles exist between single and subacute (10 days) treatments for both routes, the possibility cannot be discounted that EDC kinetics and distribution in the various tissues may be significantly modified after chronic treatments.

With these important provisos, when the AUC for blood, liver, lung, and adipose tissue were estimated for a 100 mg/kg oral dose and a 7-hour, 150-ppm inhalatory exposure with EDC, the values computed were greater by at least 15-25% for the inhalatory treatment when the liver and blood are considered, and by over 50% in lung and adipose tissue. In other words, it would seem that the total EDC quantity to which the tissues were exposed in the inhalatory bioassay were not significantly inferior to those reached with the oral carcinogenicity investigations. However, considering that evidence does exist supporting the concept that the carcinogenic activity of EDC is mediated by biotransformation products of this chemical (see Rannug and Beije 1979), the possibility that different quantities of biologically active and inactive metabolites are formed after oral and inhalatory treatments is still open.

The data presented here do not directly clear up this problem, although it was found that the curve of EDC concentration in the liver after ingestion shows a higher peak, a significantly steeper ascending phase, and a slower disappearance rate than after inhalation, even if the total AUC are equivalent or indeed higher for a 7-hour inhalation period. On this basis, a mechanism similar to that advanced for vinyl chloride by Hefner et al. (1975) and Watanabe et al. (1976) could in principle be postulated. According to this hypothesis, the liver metabolic capacity for EDC is saturable (Filser and Bolt 1979), so that when confronted with lower quantities of the chemical only nononcogenic metabolites are produced, but when confronted with higher EDC quantities, oncogenic metabolites are produced (such as chloroethylglutathione, a biotransformation product of EDC known to be a potent mutagen [Rannug and Beije 1979]).

In the absence of direct data on the levels and relative biological activity of the metabolites formed after oral and inhalatory exposure to EDC, such a mechanism remains purely hypothetical and its relevance in carcinogenicity testing remains unproven, especially considering the possibility that a significant biotransformation of EDC may take place at extrahepatic sites in organs possessing different saturation rates and exhibiting different kinetics of EDC accumulation. Perhaps other factors are behind the contrasting results that have been obtained, such as the presence of chemical impurities in the EDC employed for bioassay, which was of technical grade in the NCI study. The kinetic data presented in this paper should help, however, in the planning of future and we hope, definitive investigations.

ACKNOWLEDGMENT

Part of this work was supported by the Chemical Manufacturing Association, Washington, D.C., and by AMPE, Brussels.

REFERENCES

Alumot, E., E. Nachtomi, E. Mandel, P. Holstein, A. Bondi, and M. Herzberg. 1976. Tolerance and acceptable daily intake of chlorinated fumigants in the rat diet. *Food Cosmet. Toxicol.* 14:105.

Bolt, H. M. 1978. Pharmacokinetics of vinyl chloride. *Gen. Pharmacol.* **9**:91.
Filser, J. G. and H. M. Bolt. 1979. Pharmacokinetics of halogenated ethylenes in rats. *Arch. Toxicol.* **42**:123.
Gehring, P. J., P. G. Watanabe, and J. D. Young. 1977. The relevance of dose-dependent pharmacokinetics in the assessment of carcinogenic hazard of chemicals. *Cold Spring Harbor Conf. Cell Proliferation* **4**:187.
Gehring, P. J., P. G. Watanabe, and C. N. Park. 1978. Resolution of dose-response toxicity data for chemicals requiring metabolic activation: Example—vinyl chloride. *Toxicol. Appl. Pharmacol.* **44**:581.
Hefner, R. E., P. G. Watanabe, and P. J. Gehring. 1975. Preliminary studies on the fate of inhaled vinyl chloride monomer in rats. *Ann. N.Y. Acad. Sci.* **246**:135.
Heppel, L. A., P. A. Neal, T. L. Perrin, K. M. Endicott, and V. T. Posterfield. 1946. The toxicology of 1,2-dichloroethane (ethylene dichloride). V. The effect of daily inhalations. *J. Ind. Hyg. Toxicol.* **28**:113.
Hofman, H. T., H. Birnstiel, and P. Jobst. 1971. On the inhalation toxicity of 1,1- and 1,2-dichloroethane. *Arch. Toxicol.* **27**:248.
Hucker, H. B. 1970. Species differences in drug metabolism. *Annu. Rev. Pharmacol.* **10**:99.
Kirk, R. E. 1968. *Experimental design: Procedures for the behavioural sciences.* Wadsworth, Belmont, California.
Lioia, N. and S. Fondacaro. 1959. Toxicity of 1,2-dichloroethane. III. Liver function in experimental poisoning. *Folia Med.* (Naples) **42**:1524.
McKenna, M. J., J. A. Zempel, E. O. Madrid, W. H. Braun, and P. J. Gehring. 1978. Metabolism and pharmacokinetic profile of vinylidene chloride in rats following oral administration. *Toxicol. Appl. Pharmacol.* **45**:821.
NCI (National Cancer Institute). 1978. *Bioassay of 1,2-dichloroethane for possible carcinogenicity.* DHEW publication number 78-1361. Government Printing Office, Washington, D.C.
NIOSH (National Institute for Occupational Safety and Health). 1976. *Occupational exposure to ethylene dichloride (1,2-dichloroethane).* DHEW publication number 76-139. Government Printing Office, Washington, D.C.
Rannug, U. and B. Beije. 1979. The mutagenic effect of 1,2-dichloroethane on *Salmonella typhimurium.* II. Activation by the isolated perfused rat liver. *Chem.-Biol. Interact.* **24**:265.
Reichert, D. and D. Henschler. 1978. Uptake and hepatotoxicity of 1,1-dichloroethylene by the isolated blood-perfused rat liver. *Int. Arch. Occup. Environ. Health* **41**:169.
Spencer, H. C., V. K. Rowe, E. M. Adams, D. D. McCollister, and D. D. Irish. 1951. Vapor toxicity of ethylene dichloride determined by experiments on laboratory animals. *Arch. Ind. Hyg. Occup. Med.* **4**:482.
Thorgeirsson, S. S. and D. W. Nebert. 1977. The Ah Locus and the metabolism of chemical carcinogens and other foreign compounds. *Adv. Cancer Res.* **24**:149.
Watanabe, P. G., G. R. McGowan, and P. J. Gehring. 1976. Fate of [^{14}C] vinyl chloride after single oral administration in rats. *Toxicol. Appl. Pharmacol.* **36**:339.
Zuccato, E., F. Marcucci, and E. Mussini. 1980. GLC determination of ethylene dichloride (EDC) in biological samples. *Anal. Letters* (in press).

COMMENTS

WARD: Have you done any experiments with multiple doses on a daily basis, to mimic the chronic studies?

SPREAFICO: We have investigated EDC kinetics after one single treatment or exposure, as well as after 2 weeks (5 days/wk). The kinetic parameters after the tenth "dose" were not different in respect to the first one. It is possible, however, that differences are only seen after a more prolonged period of exposure than we have so far investigated. We simply don't know.

WARD: In multiple exposures, the lower doses may behave more like the single high doses.

SPREAFICO: It's possible. So far, within the limits of what we have done, we haven't seen significant differences.

PLOTNICK: You would presume that if the compound was a reasonably good or effective inducer, you would have noted that in reduced blood levels or tissue levels.

SPREAFICO: As mentioned before, your hypothesis, which incidentally is examined in my discussion (Spreafico et al., this volume), could be right but could also be wrong. It is difficult to speculate in the absence of data. We intend to follow up on this point, but again we cannot say anything beyond what is already mentioned.

PLOTNICK: Are there any indications in these tests of sleeping times?

ANDERS: Most of low-molecular-weight halogenated compounds are poor inducers, if inducers at all. If a compound doesn't produce enzyme induction by 5 days, it's unlikely to prove to be an inducer with prolonged administration.

SPREAFICO: Very frankly, I don't know, and it is also difficult to speculate safely on such a point. We normally think about microsomal metabolism, but here one might have combinations of microsomal and nonmicrosomal metabolism. If the latter is true, our scale of times in these speculations may be entirely wrong. We have thought of these problems but don't have enough evidence on which to base a "safe" speculation. We have a lack of real data, therefore it's difficult to make assumptions. You can extend your reasoning only up to a certain point, but you have to qualify that many of these reasonings are based on unproven assumptions.

MALTONI: I have some reservations about the matter of low doses being transformed, detoxicated, etc. because the time of treatment is more important than the level of dose. In our studies on vinyl chloride (VC), we repeated animals for 100 hours on different schedules—4 hours in 1 day for 25 weeks or 1 hour daily for 4 days weekly for 25 weeks. The 100 hours times 10,000 ppm would give you a value of one million for the total quantity of VC given. If you expose animals 4 hours daily, 5 days weekly for 52 weeks at 50 ppm for a year, you get a value of 52,000. And if you expose animals in the same way at 25 ppm, for 1 year, you get 26,000. The result in the first exposure (1 million) is (+), whereas for 52,000 it is +++, and for 26,000 it is ++. And so it means that during this 100 hours the liver should have been very much saturated.

HOOPER: But the 10,000-ppm dose is not fully absorbed because there is a limit to what the lung can absorb.

MALTONI: Yes, however, you get saturation in the liver.

REITZ: Furthermore, I think the metabolism creating a reactive intermediate is saturated. We know that already because the tumor incidence at the same concentrations leveled out with your VC. Also, we don't know that metabolism plays the same role in VC exposure as it does with EDC.

MALTONI: I did not like to come up with the fact that a low exposure is innocent and a very high exposure sometimes would be more dangerous. I want just to point out this. It is a matter of environmenal hygiene that I want to emphasize.

SPREAFICO: And I would like to emphasize again that what I explained is an hypothetical mechanism presented in the attempt to reconcile the conflicting data emerging from the NCI bioassay (NCI 1978) and Dr. Maltoni's (Maltoni et al., this volume) experiments. "Hypothesis" is the critical word. Secondly, it seems to me that many problems raised with NCI bioassay are still unresolved and, at least in my mind, they cast heavy doubts on the validity of the NCI results.

MALTONI: What did you see in spleen, Federico [Spreafico]? It is my impression that there may be some cellular depletion of spleen correlated to treatment. This is just my impression, and I do not yet have a number. This does not deal with cancer in our own view but I would like to know, do you have some cellular depletion in spleen?

SPREAFICO: Cesare [Maltoni], I do not understand what you are asking me. The levels of EDC in the spleen are equal to those in blood after both oral and inhalatory exposure. Regarding our clinical chemistry data, we have no easy signal change in the blood, or whatever, that gives you an indication of functional or anatomical derangement in the spleen. If your impression is true, one could get some data studying the immunological capacity of these rats.

HOOPER: I still think we want to look at the life-table data to resolve the apparently discrepant results from the gavage and inhalation studies. Another possibly minor point, a fly in the ointment (or oil in the rat), is that the proper inhalation experiment to parallel the gavage study would be to administer corn oil by gavage while EDC is given by inhalation. There may be more lipid in 1 ml/day of corn oil sufficient enough to increase the fat deposition in the mammary tissues thus affecting the appearance of tumors in the gavage study. A more likely explanation is that there are strain differences in the sensitivity of mammary tissue between Osborne-Mendel and Sprague-Dawley female rats.

SPREAFICO: I would say no. Actually, the quantity of oil one employs is very low—and it doesn't really change the histology, nor the total amount of adipose tissue.

REITZ: Didn't you report in some of your data, though, that the pharmacokinetic parameters were unchanged if you administered EDC in water at 25 mg/kg?

SPREAFICO: The data that we have using water as the EDC vehicle are limited to only one exposure route, i.e., i.v., and we have relatively less information on this treatment modality.

PLOTNICK: When you were discussing the levels in the spleen, you said that they paralleled the levels in the blood. That's a 1:1 relationship. But you're only talking about EDC again.

SPREAFICO: Right.

PLOTNICK: You're not talking about EDC plus the metabolites?

SPREAFICO: Correct.

PLOTNICK: The spleen may be filled with metabolites.

SPREAFICO: It is possible.

PLOTNICK: These metabolites may be more toxic than the test compound.

SPREAFICO: Yes, your point may be true, although I emphasized in the very beginning that we measured only EDC, which is possibly just the tip of the iceberg.

PLOTNICK: Yes, but at a point of 3-6 hours afterwards, you're going to have significant amounts of metabolites from the liver redistributed to other organs.

SPREAFICO: Yes, and possibly even faster than that. But, again, until you really measure the levels of metabolites and you know their biological reactivity, you are in for a surprise. So I wouldn't really care to speculate beyond what I've already said.

WARD: There are thousands of experiments in the literature in which two different rat strains or even different routes of exposure are used and the results don't agree. There's no reason that we should seek the ultimate reason why there are differences because all these discrepancies may be red herrings that just don't mean anything. The reason may just be strain differences, route differences, etc. So, the best comparison would be the same strain, the same lab, and the same conditions.

SPREAFICO: I could not agree more. But this is a very general problem, and I don't know how appropriate it is for us to go into it now.

WARD: Dr. Van Duuren will present an EDC skin painting study in mice in which EDC induced lung cancers (see Van Duuren, this volume). That's similar to the results of our mouse study.

With the Shimkin strain-A mouse lung tumor bioassay, there are really no false positives, but there are false negatives. So the lung tumor seems to be a valid end point when you survey all the types of chemicals that induce lung tumors in mice.

References

NCI (National Cancer Institute). 1978. *Bioassay of 1,2-dichloroethane for possible carcinogenicity.* Technical report series number 55. DHEW publication number (NIH) 78-1361. Government Printing Office, Washington, D.C.

Pharmacokinetics and Macromolecular Interactions of Ethylene Dichloride: Comparison of Oral and Inhalation Exposures

RICHARD H. REITZ, TONY R. FOX,
JEANNE Y. DOMORADZKI, JOHN F. QUAST,
PAT LANGVARDT, AND PHILIP G. WATANABE
Toxicology Research Laboratory
Health and Environmental Sciences
Dow Chemical Company
Midland, Michigan 48640

Two long-term bioassays of ethylene dichloride (EDC; 1,2-dichloroethane) for possible carcinogenicity have been carried out. In the study conducted by C. Maltoni (this volume), rats and mice were exposed to EDC vapor at concentrations as great as 150 ppm. No specific types of tumors were produced by EDC. A small increase in the incidence of mammary tumors in some groups of female rats was attributed to the fact that these groups of animals survived longer than controls. A second study commissioned by the National Cancer Institute (NCI) investigated the effects of EDC administered chronically by gavage in corn oil. In this study a variety of tumors, which appeared to be treatment-related, were observed in rats and mice (Ward, this volume). The apparent contradiction in the results of these two bioassays prompted us to carry out studies to see whether these differences could be related to the actions of EDC at the molecular level.

There are many processes that occur in a living organism after administration of a foreign chemical such as EDC. Many of these processes dramatically affect the toxicity of particular chemicals. For example, as outlined in Figure 1, a chemical may be excreted unchanged or after conversion to nontoxic metabolites such as glucuronides. Alternatively, the chemical may be transformed by metabolic activation to a reactive metabolite. The reactive metabolite can undergo detoxification to produce an inactive metabolite or can bind to various other sites in the cell. Most of these are noncritical (covalent binding, nongenetic), but a few of the sites may have a critical role. Thus if EDC bound to DNA to produce altered bases (covalent binding, genetic), and these altered bases escaped the DNA repair systems and caused mispairing in replication, there might be a finite chance of transforming the target cells to malignant clones. Since the dynamic interplay of these processes may vary according to the route of administration of a given chemical, a comparative study of the pharmacokinetics and macromolecular interactions of EDC after oral or inhalation exposure was undertaken.

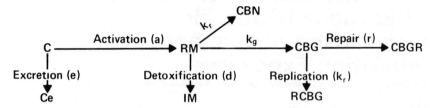

Figure 1
Processes that may affect the toxicity of a chemical in vivo. (C) Chemical; (RM) reactive metabolite; (Ce) excreted chemical; (IM) inactive metabolite; (CBN) covalent binding, nongenetic; (GBG) covalent binding, genetic; (CBGR) repaired covalently bound genetic material; (RCBG) retained genetic program, critical and noncritical.

COMPARATIVE STUDY OF ORAL AND INHALATION EXPOSURES

In many respects our studies are similar to those already reported by F. Spreafico (this volume) and we obtained very similar results. Thus, we are in the position of being able independently to confirm some of his fine work. In addition, there are several further conclusions that can be drawn from the studies reported here. One of the most important of these is in the area of possible species differences. Osborne-Mendel rats were employed in our studies, as well as the NCI bioassay, and preliminary experiments indicated that there were no major differences between the oral pharmacokinetics of EDC in this strain and those reported by Dr. Spreafico for Sprague-Dawley rats (F. Spreafico, pers. comm.).

In the NCI study, Osborne-Mendel rats developed squamous cell carcinomas of the forestomach and hemangiosarcomas of the spleen, liver, and other organs (NCI 1978). We studied the effects of either 150 ppm EDC inhalation (6 hr) or administration of 150 mg/kg EDC (1.5% solution in corn oil) by gavage. The inhalation exposure corresponds to the highest level in Dr. Maltoni's study, whereas the oral dose was chosen to correspond to the highest dose administered to male rats in the NCI study. As reported by Dr. Ward (this volume), this dose was later reduced because it produced early mortality, so that the time-weighted average dose for male rats in the NCI study was somewhat lower (95 mg/kg).

Pharmacokinetics of EDC in Blood

Our first study dealt with the pharmacokinetics of EDC in blood after inhalation. Computer modeling of the data gave the curve shown in Figure 2. Steady-state levels of 8-9 μg EDC/ml blood were reached in 2-3 hours and remained constant thereafter until the termination of the exposure at 6 hours. Immediately following exposure, blood levels fell rapidly. A semilogarithmic plot of blood levels versus time after exposure termination is shown in Figure 3.

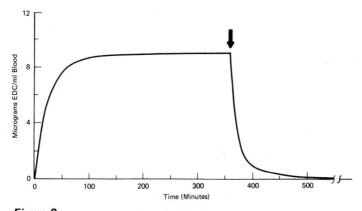

Figure 2
Blood levels of EDC observed during and following a 6-hr inhalation exposure to 150 ppm EDC. Data from four male Osborne-Mendel rats were fitted to a two-compartment open model as described by Gehring et al. (1976). The computer plot of this model is shown. (→) Exposure termination.

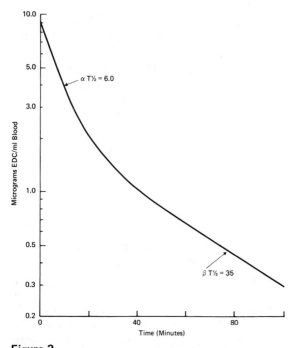

Figure 3
Semilogarithmic plot of EDC blood levels vs time after exposure termination (min). Experimental data from four male Osborne-Mendel rats were fitted by computer modeling and the computer fit is shown.

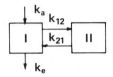

Figure 4
Two-compartment open model used to describe the pharmacokinetics of EDC in Osborne-Mendel male rats (Gehring et al. 1976).

Elimination of EDC appeared to be biphasic, with an initial alpha phase and a slower beta phase. The half-lives for these two phases were 6 minutes and 35 minutes, respectively. These data may be modeled as a two-compartment open system, as described by Gehring et al. (1976). This model is shown in Figure 4. The kinetic parameters estimated for this model are: $k_e = 0.092$, $k_{12} = 0.019$, $k_{21} = 0.025$. The volume of distribution for this model system was 513 ml/kg. Using these parameters, we calculated the total area under the curve (AUC) as 3018 μg min/ml. The predicted blood concentration of EDC 18 hours after termination of inhalation exposure is well below detection limits (less than 10^{-8} μg/ml).

For comparison, Spreafico's pharmacokinetic modeling of EDC after oral administration of EDC gave a volume of distribution of 390 ml/kg, with a half-life of 57 minutes for the terminal phase of elimination. His calculated AUC value was 7297 μg min/ml, and his model also predicts essentially complete elimination of EDC from the body in 24 hours (Spreafico, pers. comm.).

The most striking difference between the oral pharmacokinetics reported by Spreafico (pers. comm.) and our inhalation data is that the peak blood levels of EDC are much higher after oral exposure than after inhalation exposure. Otherwise, the similarities are marked. This difference was also reported.

Use of Radioactive Labeling to Trace EDC Metabolism

The next aspect of our studies dealt with the fate of EDC after administration by the two routes. For these studies we employed [1,2-^{14}C]EDC prepared by New England Nuclear. After exposure to the radioactive chemical, animals were housed in all-glass metabolism cages designed to allow separate collection of urine and feces. Air flowing through these cages was sequentially passed through traps designed to recover quantitatively any unmetabolized EDC (charcoal trap) as well as any $^{14}CO_2$ (2-methoxyethanol/ethanolamine, 3:7). Urine was collected on dry ice and kept frozen until analysis. Forty-eight hours after the start of the experiment, the animals were sacrificed and the tissues were analyzed for radioactivity. The results of these experiments are summarized in Table 1.

Table 1

Fate of [^{14}C]EDC in Rats 48 Hours After Oral (150 mg/kg) or Inhalation (150 ppm, 6-hr) Exposure

	Oral		Inhalation	
	μmole/kg	% metabolites	μmole/kg	% metabolites
Body burden (A) (total radioactivity)	1539 ± 391	—	512 ± 135	—
Charcoal trap (B)	447 ± 60	—	9.4 ± 0.4	—
Total metabolites (A − B)	(1092)	(100)	(503)	(100)
Urine	926 ± 348	85.7	432 ± 121	84.4
CO_2 trap	83.1 ± 11.9	7.7	36.1 ± 6.89	7.0
Total carcass (48-hr after exposure)	46.9 ± 10.1	4.3	22.7 ± 3.38	4.4
Feces	23.6 ± 14.7	2.1	8.90 ± 2.84	1.7
Cage wash	12.5 ± 6.37	1.1	3.34 ± 1.34	0.7

Values are mean ± S.D., with $n = 4$ for each route of exposure, and are μmole equivalents of EDC, based on the specific activity of [^{14}C]EDC (3.2 mCi/mM).

In the oral balance study, 1520 μmoles [^{14}C]EDC (150 mg/kg body weight) were administered. Total radioactivity recovered (body burden) was equivalent to 1539 μmole equivalents (101%). It is not possible to determine percent recovery after inhalation exposure, but the body burden at the end of the 6-hour exposure was 512 μmole equivalents, about one-third that seen after oral dosing.

During the first 48 hours after administration of the oral dose, 447 μmole equivalents EDC were retained in the charcoal trap designed to recover unmetabolized EDC (29% of the body burden). In contrast, only about 1.8% of the body burden was recovered as apparently unchanged EDC after inhalation exposure (9.4 μmole equivalents EDC). One possible explanation for this difference may be that the peak blood levels of EDC were much higher after oral exposure than after inhalation exposure (about 70 μg/ml vs 8-9 μg/ml). However, this possibility cannot be evaluated without a more complete study of the dose-dependency of metabolism in each case.

Radioactivity recovered in the CO_2 traps and in the urine was composed of metabolites derived from [^{14}C]EDC (all of the urinary radioactivity is nonvolatile). In addition, because of the rapid clearance of EDC from blood, it was assumed that radioactivity remaining in the body after 48 hours, or recovered in the feces or cage wash, was primarily metabolites. Thus, total metabolites after exposure by the two routes may be estimated by subtracting the amount of radioactivity recovered in the charcoal traps from the total radioactivity (Table 1). This suggests that about twice as much EDC is metabolized after oral exposure as compared with inhalation exposure (1092 vs 503 μmole equivalents).

The primary route of elimination appears to be urinary excretion of nonvolatile metabolites (Table 1). Significant radioactivity also appeared in the CO_2 trap, with much smaller amounts detected in feces, cage wash, or carcass. After normalization for the different amounts of metabolites formed, the distribution of radioactivity after the two routes of administration was virtually identical (Table 1, columns 2, 4). Furthermore, preliminary experiments designed to characterize the urinary metabolites suggest that two major metabolites were formed in each case. The metabolites appear to be identical after either route of administration and are formed in approximately the same ratio in each case. Studies to identify the chemical structures of the urinary metabolites by gas chromatography-mass spectroscopy are underway.

It is important to note that about 96% of the radioactivity was eliminated from the body in 48 hours, which is in agreement with the pharmacokinetic models. This is significant because it indicates that it is unlikely that unforeseen toxicity owing to bioaccumulation of the parent compound will occur after repeated exposures to EDC. The small amount of radioactivity that remained in the body (4%) may be related to two factors. First, a considerable portion of the EDC was converted to $^{14}CO_2$. This suggests that radioactive one-carbon fragments may enter normal biosynthetic pools and be incorporated into undamaged cell components. Second, the radioactivity may represent covalent binding of reactive EDC metabolites; this possibility will be discussed later.

Distribution of radioactivity into particular tissues was examined in these experiments. Tissues that developed tumors in the NCI experiments (forestomach, spleen, liver) as well as nontarget tissues (e.g., kidney, lung, stomach, and remaining carcass homogenate) were surveyed. The results are presented in Table 2. There was no striking difference between target and nontarget tissues by either route of administration. A remarkably consistent pattern was observed between tissues from animals exposed orally and animals exposed by inhalation. In each case, the levels are about twofold higher after oral exposure to EDC. Thus, studies of the fate and distribution of EDC suggest that the two routes of administration were nearly equivalent and cannot provide an explanation for the contradictory results in the two bioassays.

Macromolecular Binding of EDC

Banerjee and Van Duuren (1979) reported that microsomal preparations from rat liver metabolically activate EDC to a reactive species, which covalently binds to macromolecules. Rannug et al. (1978) have reported a second pathway for metabolic activation involving cytosolic enzymes and dependent upon glutathione (GSH). In collaboration with Dr. Fred Guengerich (Vanderbilt University), we have carried out studies that confirm the in vitro binding results of both groups (unpubl. results). Since macromolecular binding may be related to toxic effects of chemicals, including carcinogenicity, it was important to carry out

Table 2
Distribution of Radioactivity in Selected Tissues of Rats 48 Hours After Exposure to [^{14}C]EDC by the Oral or Inhalation Routes

	Nanomole equivalents/g tissue	
	oral (150 mg/kg)	inhalation (150 ppm, 6 hr)
Liver[a]	154 ± 26	75 ± 13
Kidney	120 ± 20	77 ± 8.6
Lung	51 ± 16	35 ± 5.4
Spleen[a]	59 ± 19	38 ± 6.0
Forestomach[a]	108 ± 31	37 ± 7.2
Stomach	62 ± 21	30 ± 6.4
Carcass	23 ± 4.4	13 ± 4.0

Results are reported as nanomole equivalents of EDC/g tissue ± S.D. (n = 4). Radioactivity was determined by combustion of tissue samples, trapping $^{14}CO_2$.

[a]Site where malignant tumors were in the observed NCI bioassay after EDC was given by gavage. No excess tumors were reported in the inhalation bioassay (Maltoni, this volume).

in vivo studies of macromolecular binding of EDC after exposure by the two routes.

In these studies, animals were administered [^{14}C]EDC and then sacrificed either 4 hours (oral) or 6 hours (inhalation) after the beginning of the exposure. Tissue samples were homogenized, precipitated with trichloroacetic acid (TCA), and then extensively washed with TCA and methanol. Finally, the insoluble residue from this procedure is digested in NaOH and counted in a scintillation counter. Radioactivity measured by this procedure is assumed to be covalently bound to DNA, RNA, or protein (Jollow et al. 1973). The results of this study are listed in Table 3. Macromolecular binding was diffuse, rather than localized. That is, there was no apparent difference between the group of target and nontarget tissues.

Also, in this experiment, there was no striking difference between the levels of macromolecular binding after oral exposure as compared with inhalation exposure. Levels of binding were slightly higher after inhalation exposure, whereas most of the other parameters studied revealed slightly higher effects of EDC in the oral experiments. The reason for this change is not clear, but it is apparent that if gross macromolecular binding were an indicator of carcinogenicity, oral exposure should be slightly less effective than inhalation in inducing tumors, not more. Hence, these studies also failed to provide an explanation for the contradictory results of the two bioassays.

Table 3
Macromolecular Binding in Selected Tissues of Rats After Exposure to [^{14}C]EDC by Oral or Inhalation Routes

	Nanomole equivalents EDC/g tissue	
	oral (150 mg/kg)	inhalation (150 ppm, 6 hr)
Liver[a]	175 ± 24	268 ± 45
Kidney	183 ± 25	263 ± 48
Spleen[a]	65 ± 21	130 ± 22
Lung	106 ± 34	147 ± 16
Forestomach[a]	160 ± 19	71 ± 19
Stomach	90 ± 2	156 ± 29

Results are reported as nanomole equivalents of EDC/g tissue ± S.D. ($n = 4$). Radioactivity was determined by scintillation counting of digests. Animals were sacrificed 4 hr after oral dosing or immediately following a 6-hr inhalation exposure.

[a]Site where malignant tumors were observed in the NCI study after EDC was given by gavage. No excess tumors were reported in the inhalation bioassay (Maltoni, this volume).

PROSPECTIVE STUDIES OF EDC ACTIVITY

This perplexing problem has two remaining areas that we plan to study. The first of these involves the pathway for metabolic activation discussed by Dr. Rannug (1978). This seems to be particularly pertinent, since only the pathway involving GSH conjugation has given rise to mutants in the Ames test (although both mixed function oxidase and GSH S-transferase catalyze covalent binding of EDC to DNA or protein in vitro). We suspect that there may be more involvement of the GSH pathway after oral exposure to EDC, since EDC absorbed from the gut would be taken directly to the liver, a rich source of GSH and GSH transferase enzymes. If this is so, this should be reflected in a higher level of EDC adducts in DNA purified from target organs after oral administration.

The second area of study involves determination of the underlying mechanisms of tumorigenesis. Although the direct reaction of some chemicals with DNA has been correlated with the induction of tumors, it is becoming increasingly clear that other factors can also be involved in tumor formation. For instance, repeated subcutaneous injection of salt, glucose, or water can induce tumor formation (Grasso and Golberg 1966). Chronic physical irritation can greatly potentiate the effect of previously applied carcinogens (Berenblum 1944), as does the repeated application of certain tumor promoters, such as the phorbol esters (Boutwell 1978). All these treatments result in a stimulation of normal DNA synthesis in the affected tissues. One consequence of increasing DNA replications is that the relative rates of DNA synthesis and DNA repair are altered. This may be important, because there is evidence that DNA repair systems are less effective in correcting small amounts of genetic damage when extensive DNA replication is taking place (Berman et al. 1978; McCormick 1979). Consequently, chronic cellular regeneration, such as may take place during repeated administration of toxic doses of a chemical, could lead to increased spontaneous rates of tumor formation.

Either of these mechanisms could give rise to a positive response in an animal bioassay such as those of the NCI (1978). However, although the end result—tumor formation—is the same, there are fundamental differences between the two processes, with important implications for safety evaluation:

1. Low levels of DNA alkylation (if they escape repair and are fixed by replication) may constitute some finite amount of risk. In contrast, low levels of cytoplasmic damage, if not sufficient to cause cell death and subsequent regeneration, are generally reversible and should not contribute added cancer risk.
2. Low levels of DNA alkylation are very difficult to detect in affected tissues. However, cytoplasmic damage sufficient to stimulate regeneration is readily visualized by conventional histopathology and may also be studied by measuring [^{3}H]thymidine incorporation into DNA.

We plan to carry out a series of in vivo experiments with EDC administered by the two routes to assess the relative importance of both mechanisms in each case. Nongenetic toxicity may be particularly important, because as reported by Spreafico and Maltoni (both, this volume), EDC was well tolerated at all inhalation doses up to 150 ppm, whereas early mortality was quite evident in the NCI gavage study.

In conclusion, it appears that the greatest understanding of species effects, route effects, and dose dependency in the carcinogenic process will depend upon careful integration of three areas of study: pharmacokinetics, molecular mechanisms, and animal bioassays.

NOTE ADDED IN PROOF

During the meeting at which the papers in this volume were given, it became apparent that there was considerable uncertainty about the purity of the EDC employed in the NCI bioassay. Consequently, a sample of this material was obtained from the NCI and analyzed at the Dow Chemical Company and elsewhere. The results of this analysis may be found in the appendix to these proceedings.

REFERENCES

Banerjee, S. and B. L. Van Duuren. 1979. Binding of carcinogenic halogenated hydrocarbons to cellular macromolecules. *J. Natl. Cancer Inst.* 63:707.

Berenblum, I. 1944. Irritation and carcinogenesis. *Arch. Pathol.* 38:233.

Berman, J. J., C. Tong, and G. M. Williams. 1978. Enhancement of mutagenesis during cell replication of cultured liver epithelial cells. *Cancer Lett.* 4:277.

Boutwell, R. K. 1978. Biochemical mechanisms of tumor promotion. In *Mechanisms of tumor promotion and cocarcinogenesis* (ed. T. J. Slaga, A. Sivak, and R. K. Boutwell), vol. 2, p. 49. Raven Press, New York.

Gehring, P. J., P. J. Watanabe, and G. E. Blau. 1976. Pharmacokinetic studies in the evaluation of the toxicological and environmental hazards of chemicals. In *Advances in modern toxicology* (ed. M. A. Mehlman, R. E. Shapiro, and H. Blumenthal), vol. 1, part 1. Hemisphere Publishing Co., Washington, D.C.

Grasso, P. and L. Golberg. 1966. Subcutaneous sarcoma as an index of carcinogenic potency. *Food Cosmet. Toxicol.* **4**:297.

Jollow, D. J., J. R. Mitchell, W. Z. Potter, D. C. Davis, J. R. Gillette, and B. B. Brodie. 1973. Acetaminophen-induced hepatic necrosis II. Role of covalent binding in vivo. *J. Pharmacol. Exp. Ther.* **187**:195.

McCormick, J. J. 1979. "Evidence that DNA excision repair processes in human fibroblasts can eliminate potentially cytotoxic and mutagenic lesions." Paper read at the 10th Annual Meeting of the Environmental Mutagen Society, New Orleans, Louisiana, March 8-12.

National Cancer Institute. 1978. *Bioassay of 1,2-dichloroethane for possible carcinogenicity*. NCI carcinogenesis technical report series no. 55. DHEW publication number (NIH) 78-1361. Government Printing Office, Washington, D.C.

Rannug, U., A. Sundvall, and C. Ramel. 1978. The mutagenic effect of 1,2-dichloroethane on *Salmonella typhimurium* I. Activation through conjugation with glutathione in vitro. *Chem. Biol. Interact.* **20**:1.

COMMENTS

WARD: How about the clearance of radioactivity from the different tissues? Is that different in your study?

REITZ: We have not measured clearance from tissues.

WARD: For example, there are some studies on aromatic amines in progress now showing that some metabolites tend to build up in those target sites of the aromatic amines where you get cancer.

PLOTNICK: You have concentrations in certain organs with amines, particularly the lung, that, for some reason, nobody has ever been able to explain. But would you like to speculate on the possible alkylated intermediate or alkylated DNA product?

REITZ: Not until we have some more data.

PLOTNICK: The other point I have with respect to chronic exposure is the following. Looking at the data on the GSH conjugate, which obviously is mutagenic, and the other pathway, which I think you and I presume will be through a nonspecific dehalogenation followed by an aldehyde, and alcohol aldehyde, presumably haloacetic acid, which do you think would be probably the most important?

It's my personal view that, at least in the chronic exposure situation, you're going to deplete liver GSH early, if this is a major method of biotransformation on elimination. With continuous exposure those levels are not going to be there in your secondary pathway. What we call the secondary pathway may be the predominant pathway.

REITZ: That may or may not be so. We found in other studies in our laboratory that in some cases where you get GSH depletion after exposure to chlorinated solvents the animals apparently have a compensatory mechanism. We can see GSH depletion after 1 week, 2 weeks, or 3 weeks. But then, if we go on and look at those animals after 6 months of continuous exposure, the GSH levels are identical to those in controls.

So I'm not sure that we can say that this pathway would not become increasingly important, instead of less important on chronic exposure. Obviously, that's someting that needs to be carefully investigated.

AMES: Just one other point about that other pathway. Chloroacetaldehyde (2-chloro-1-acetaldehyde) is an extremely potent mutagen. If the pathway goes through chloroethanol and then chloroacetaldehyde, that is known to interact with adenine and make flourescent products.

We tried chloroethanol a while back in the *Salmonella* system (with microsomes). It was not very effective, as I remember, and I don't know whether we were seeing the metabolism. Human alcohol dehydrogenase is known to be fairly nonspecific, and could convert alcohols such as chloroethanol to aldehydes. I'm not sure we were getting that to work very well in the system, so that the mutagenicity test may not be giving just the right result.

I think it's very important to do the kind of studies that you and Dr. Spreafico are doing. There is one thing to keep in mind in DNA damage studies. Carbon tetrachloride (CCl_4) may work through generation of a free radical and then lipid peroxidation and production of damaging hydroxyl radicals and singlet oxygenetics. This is an indirect mechanism, and therefore you wouldn't see the CCl_4 attached to the DNA because the damage would be indirect. One ought to keep in mind that there may be pathways through free radicals, giving indirect, oxidative damage, and we're just not going to see that by a direct DNA adduct analysis. Also, those kinds of radical-initiated events aren't showing up in *Salmonella*; we miss CCl_4 completely.

REITZ: That may not be a contradiction at all. In fact, that would be entirely consistent with the operation of several types of processes that could show up as positives in bioassays, which, after all, are just looking at the end result—did the animal get an increased incidence of tumors or not? Now, if there are epigenetic mechanisms that play significant roles, then what you've described for CCl_4 is exactly that type of mechanism.

AMES: I think it's genetic, but it's indirect. CCl_4 may be active because it starts off lipid peroxidation in the membrane and that initiates a whole cascade of free radicals and peroxides and hydroxyl radicals, and then those hit the DNA.

REITZ: I see what you mean.

AMES: What you will see is that you have DNA damage, but you won't see the CCl_4 bound to the DNA. So there could be DNA damage, but one would miss it if you were at the right things.

REITZ: Exactly. This is one of the reasons we try to assess genetic damage by measuring both the incorporation of adducts, as well as the DNA repair that occurs after DNA damage.

AMES: Perhaps measuring repair would be a good way.

REITZ: That's an excellent way of doing that.

AMES: All of the heavily chlorinated chemicals that I think may be working through free radicals are not showing up in our test, maybe because of short-lived active forms or other reasons.

ANDERS: The evidence for free radicals as the primary damaging species in CCl_4 is perhaps a little shaky. Although I won't argue with you that the lipid peroxidation is taking place, the evidence that lipid peroxidation is responsible for cell death is not as solid.

A question back to you: Where would ethylene turn up in your traps, if anywhere?

REITZ: Ethylene would probably have showed up on the charcoal.

ANDERS: Do you have any data on how much of the radioactivity in the charcoal trap is ethylene versus EDC?

REITZ: Well, being familiar with your own work, we'd assume that the amount of ethylene would be small. We're hoping to desorb that charcoal and do gas chromatographic analysis of the radioactivity, but we haven't completed those studies yet, so I can't answer the question.

ANDERS: One more point: It is important to measure the rate of disappearance of alkylated or bound material. There are several studies showing that it disappears at a much slower rate in target tissue than it does in nontarget tissue.

REITZ: I agree that those would be very interesting experiments to do. The data we have at present are too limited to say much about clearance from target and nontarget tissues, so I really can't answer your question.

HOOPER: If you can find a host of compounds that are noncarcinogenic but induce tissue or cell damage, cell divisions, etc., who's to argue against the epigenetic mechanisms, if they produce cancer?

REITZ: I think there may be a variety of reasons why carcinogenicity has not been demonstrated for every substance capable of causing tissue damage. First, not every substance has been evaluated under truly chronic administration (i.e., for most of the lifetime of the animal). Acute or subchronic treatment would probably not result in tumor production. In addition, the severity of tissue damage varies greatly from chemical to chemical, and chemicals that cause weak tissue damage may escape detection. Furthermore, there are great differences in the spontaneous tumor rates among the various organs, strains, and species of animals. A given degree of tissue damage may be more significant at a site where spontaneous tumor formation is present. Finally, we should recognize that no chemical will be purely genetic or purely epigenetic in its action. Most chemicals will produce a mixture of genetic and epigenetic effects, with one or the other aspect predominating. We would not expect chemicals that are primarily epigenetic in character to induce any tumors as long as tissue damage is absent. However, when tissue damage is present, even very small amounts of genetic damage could be significant. This could lead to different carcinogenic potencies in chemicals producing equal tissue damage. However, we do not believe that chemicals that induce tumors primarily through epigenetic mechanisms do not present the same risk to human populations as agents that act through primarily genetic mechanisms. So I think some sort of classification is very important for risk estimation.

AMES: One very quick point: Half the workers are smokers, and they're getting all of these promoters and whatnot into them. If all kinds of chemicals turn out to be synergistic with smoking, you may find there is one risk level for smokers and one for the nonsmokers.

Teratogenicity and Reproduction Studies in Animals Inhaling Ethylene Dichloride

K. SURYANARAYANA RAO, JANIS S. MURRAY,
MARY M. DEACON, JACQUELINE A. JOHN,
LINDA L. CALHOUN, AND JOHN T. YOUNG
Toxicology Research Laboratory
Health and Environmental Sciences U.S.A.
Dow Chemical Company
Midland, Michigan 48640

Ethylene dichloride (EDC; 1,2-dichloroethane) has been used as a precursor in the production of vinyl chloride monomer and as an industrial solvent. EDC is acutely toxic at low levels when ingested, inhaled, and absorbed through the skin or mucous membrane. Oral ingestion studies with EDC in rats and rabbits has shown the LD_{50} to be 680 mg/kg and 860 mg/kg, respectively (NIOSH 1977). Spencer et al. (1951) observed no adverse effects among rats and rabbits inhaling 200 ppm (157, 7-hr exposures) or 400 ppm (167, 7-hr exposures) EDC, respectively. Overt signs of toxicity were observed among rats exposed to 400 ppm EDC; no female rat survived more than ten exposures in 14 days.

The potential effects of EDC on human fertility and reproduction are unclear. In a two-generation reproduction study in rats, Vozovaya (1974) reported that exposure to 57 mg/m^3 (14 ppm) EDC for 6 months caused decreased fertility and increased perinatal mortality in the first but not the second generation. In another study by the same investigator, changes in the estrus cycle were observed among female rats exposed to 5 mg/m^3 (1.25 ppm) EDC for 4 hours/day for 1-9 months (Vozovaya 1971). In a subsequent study, the same investigator reported that exposure to an unspecified level of EDC resulted in an increased embryonic death rate and a fivefold increase in the incidence of preimplantation loss (Vozovaya 1976). EDC was reported to pass through the placental barrier of rats and accumulate in the tissues of the fetus, particularly the liver (Vozovaya and Malyarova 1975).

The purpose of the studies presented herein was to evaluate the effects of inhaled EDC on embryonal and fetal development in rats and rabbits and on the reproductive capacity of rats.

MATERIALS AND METHODS

EDC used for these studies was obtained from the Dow Chemical Company, Freeport, Texas. The test material was analyzed prior to use and found to be

99.9% pure. Female Sprague-Dawley rats (Spartan Research Animals, Inc., Haslett, Michigan) weighing approximately 250 g and New Zealand white rabbits (Langshaws Rabbitry, Augusta, Michigan) weighing 3.5-4.5 kg were used for the teratology study.

The rats were bred by the supplier; the day on which sperm was seen in a vaginal smear was considered to be day 0 of pregnancy. Rabbits were allowed at least 3 weeks for acclimation to the laboratory before being placed on study. Female rabbits were artificially inseminated (Gibson et al. 1966) and immediately thereafter were given a single 50 I.U. (international units) intravenous injection of human chorionic gonadotropin to induce ovulation. The day of insemination was considered day 0 of gestation. For the reproduction study male and female Sprague-Dawley rats, 6-7 weeks of age, were used. These rats were allowed to acclimate for at least 10 days before being placed on study. All animals were housed singly in wire-bottomed cages and were maintained on Purina Laboratory Chow and tap water free choice. During the exposure period, food and water were withheld and animals were group housed (rats) or housed two per cage (rabbits). During the periods between exposures, all animals were maintained in a room that was designed to maintain a temperature of 72°F and a relative humidity of 45% with a 12-hour light-dark cycle.

Chambers, Vapor Generation, and Analysis

Inhalation exposures were conducted in 4.3-m^3 Rochester-type stainless steel and glass chambers under dynamic airflow conditions. The temperature and humidity in the chambers were controlled by a system designed to maintain a temperature of approximately 70°F and a relative humidity of approximately 50%. The concentration of EDC in a chamber was generated by metering the material with a precision pump at a calculated rate into a vaporization vessel heated to 90°C. The vapor was then swept into the main chamber airflow, via a compressed air stream, where it was diluted to the desired concentration and mixed with incoming air by turbulence. Nominal concentrations (the ratio of the rate of EDC vaporized to the rate of airflow through the chamber) were calculated on a daily basis. EDC concentration in the chamber was determined two to three times per hour by infrared spectrophotometry using a Miran I Variable Filter Infrared Analyzer at a wavelength of 8.1 μm. Distribution of test material within the chamber was verified to be within 10% of the target concentration prior to initiation of animal exposures.

Experimental Design

For the teratology study, groups of 16-30 bred rats were exposed to filtered room air, 100 or 300 ppm EDC for 7 hours/day on days 6 through 15 of gestation. Concurrently, groups of 19-21 bred rabbits were exposed in the same chambers to filtered room air, 100 or 300 ppm EDC for 7 hours/day on days 6 through 18 of gestation. Control animals were placed in chambers identical to those used for the EDC exposures.

The mated rats and rabbits were observed daily, weighed periodically during the experimental period, and sacrificed by exposure to carbon dioxide on day 21 or 29 of gestation for rats and rabbits, respectively. The number of corpora lutea and the number and position of live, dead, and resorbed fetuses were recorded. After being weighed, measured for length, and sexed (rats only), all the fetuses were examined for external alterations and cleft palate. One-third of the fetuses of each litter were examined immediately for evidence of soft-tissue alterations by dissection under a low-power stereomicroscope (Staples 1974). Rabbit fetuses were sexed on the basis of internal genitalia. The head of each rat fetus examined for soft tissue alterations was placed in Bouin's solution and subsequently examined by the razor section technique of Wilson (1965). All the fetuses in each litter were cleared and stained with alizarin red-S (Dawson 1926) to permit examination for skeletal alterations.

For the reproduction study, the F_0 rats were randomized into test groups consisting of 30/sex in the control group and 20/sex in each of the treated groups at 25, 75, or 150 ppm EDC. These rats were exposed to the respective concentrations of EDC for 60 days (prebreeding period), 6 hours/day, 5 days/week. During the rest of the period, the exposure was 6 hours/day, 7 days/week. Control animals were not placed in a chamber because of lack of chamber space. Like exposed animals, controls were group-housed in exposure cages and deprived of food and water for the exposure period. These animals were placed in a holding room supplied with filtered room air at ambient temperature and humidity. Maternal animals were not exposed from gestation day 21 through the fourth day postpartum to allow for delivery and rearing of the young. After 60 days of exposure, the F_0 males and females of each respective treatment group were bred, one to one, to produce the F_{1A} generation. Seven days following sacrifice of the last F_{1A} litter, the F_0 animals were bred again to produce the F_{1B} litters. During the mating period, daily vaginal smears were prepared from each female rat to identify the day 0 of gestation. Nonpregnant females were removed from exposure for 6 days, so as to receive the same number of total exposure days as pregnant animals. Male rats continued to be exposed daily after the first 60 days.

During the study, all animals were observed daily for changes in appearance and demeanor. Body weight and food consumption were monitored weekly. After mating, the females were observed for parturition. The date of parturition and the number of live and dead newborn pups were recorded on the day of parturition (day 0). The number of live pups and their sex was noted on days 1, 7, 14, and 21 after delivery. Each litter was weighed on days 1, 7, and 14. On day 21, individual body weights and litter weights of the weanlings were recorded.

Gross Necropsy and Histopathology

At approximately 21 days of age, all F_{1A} and F_{1B} pups were sacrificed and a gross pathologic examination was made. Organ weights of the kidneys and liver from five randomly selected pups of different litters, sex, and dose levels were recorded. Sections of these organs, as well as any other target organs observed during gross examination, were preserved for microscopic examination. The tissues from pups on the lower dose levels were examined only if histopathological changes were observed at the highest dose level.

A gross pathological examination was conducted on the parental animals after weaning of F_{1B} pups. Organ weights of liver and kidneys were recorded. The following organs or tissues, as well as any other target organs observed during gross examination, were collected and preserved in buffered 10% Formalin: liver, kidneys, ovaries, uterus, and testes. Microscopic examination of these tissues from ten randomly selected rats per sex in the control and top dose groups were made by a veterinary pathologist. The tissues from rats on the lower dose levels were examined only if histologic changes were observed at the high dose level.

Statistical Evaluation

Analysis of food consumption and body and organ weights were made by one-way analysis of variance and Dunnett's test (Steel and Torrie 1960). Fertility index was analyzed by the Fisher's exact probability test (Siegel 1956). The Wilcoxon test as modified by Haseman and Hoel (1974) was used to evaluate the incidences of fetal alterations, survival indices, and resorptions. The level of significance chosen for all cases was $p < 0.05$.

RESULTS

Teratology Study

Extensive evidence of maternal toxicity among bred rats exposed to 300 ppm EDC was observed. Death occurred in 10 out of 16 rats exposed to 300 ppm EDC (Table 1). Lethargy, ataxia, decreased body weight and food consumption, and some evidence of vaginal bleeding was observed prior to death. There were no deaths at the 0- or 100-ppm dose levels. One rat at 300 ppm EDC exhibited implantation sites at cesarean section; however, all of the implantations were resorbed. Exposure of pregnant rats to 100 ppm EDC had no apparent effect on the mean litter size, the incidence of resorptions, or the fetal body measurements (Table 1). The rats exposed to 100 ppm EDC gained statistically significantly more weight than the controls during gestation.

Among the offspring of rats exposed to 100 ppm EDC, no major malformations occurred at an incidence that was significantly different statistically from that of the controls (Table 2). Skeletal examination revealed a significant

Table 1
EDC Rat Teratology: Observations at Cesarean Section

	Exposure level of EDC (ppm)		
	0	100	300
Number deaths/number females	0/30	0/30	10/16
Number litters	22	15	1[a]
Implantation sites/dam[b]	14 ± 2	15 ± 2	14
Fetuses/litter[b]	13 ± 3	14 ± 2	—[a]
Implantations resorbed (%)	21/299 (7)	7/221 (3)	14/14 (100)[c]
Litters with resorptions (%)	7/22 (32)	4/15 (27)	1/1 (100)
Litters totally resorbed	0/22	0/15	1/1
Resorptions/litters with resorptions	21/7	7/4	14/1
Sex ratio, M:F	49:51	44:56	—[a]
Fetal body weight (g)[d]	5.51 ± 0.42	5.65 ± 0.26	—[a]
Fetal crown-rump length (mm)[d]	43.7 ± 1.1	43.3 ± 1.0	—[a]

Rats were exposed to 0, 100, or 300 ppm EDC for 7 hr/day on days 6–15 of gestation.
[a] There is only one litter that was totally resorbed at the 300-ppm dose level.
[b] Mean ± S.D.
[c] Statistically significant from the control value by a modified Wilcoxon test, $p < 0.05$.
[d] Mean of litter ± S.D.

decrease in the incidence of bilobed thoracic centra among litters of rats exposed to 100 ppm EDC. This decreased incidence of a minor skeletal variant is indicative of normal variation in this species and has no toxicological significance.

Among bred rabbits, there were three maternal deaths at 300 ppm EDC and four deaths at 100 ppm EDC (Table 3). Gross necropsy did not reveal any treatment-related pathological changes. The incidence of pregnancy was not affected by exposure to EDC. Exposure of pregnant rabbits to 100 or 300 ppm EDC had no apparent effect on the mean litter size, incidence of resorptions, or the fetal body measurements (Table 3). Body weights of EDC-exposed rabbits were generally comparable with controls.

Exposure of rabbits to EDC did not significantly alter the incidence of major malformations (Table 4). Among litters exposed to 300 ppm EDC, only one fetus exhibited multiple external malformations, which included acephaly, omphalocele, kyphosis, and bilateral ectrodactyly and anonychia. Soft-tissue examination revealed a missing thymus, diaphragmatic hernia, and multiple heart anomalies. Upon skeletal examination, this same fetus showed delayed ossification of ribs and vertebrae, bilobed and unfused thoracic centra, and misshapen sternebrae. One fetus exposed to 100 ppm EDC showed misshapen vertebrae, hemivertebrae, delayed ossification of thoracic vertebrae, and unfused thoracic centra. Among the control litters, a single fetus exhibited a missing left

Table 2
EDC Rat Teratology: Incidence of Anomalies

	Exposure level of EDC (ppm)		
	0	100	300
External examination[a]	278/22	214/15	0/0
Soft-tissue examination[a]	90/22	67/15	0/0
Skeletal examination[a]	278/22	214/15	0/0
Bones of the skull[a]	188/22	147/15	0/0
Total major malformations[b]	2 (2)	1 (1)	
External examination[b]			
Omphalocele[c]	0 (0)	1 (1)	
Hypoplastic tail[c]	1 (1)	0 (0)	
Soft tissue examination[b]			
Missing innominate artery[c]	1 (1)	0 (0)	
Dilated ureter	1 (1)	2 (1)	
Hemorrhage-kidney	0 (0)	1 (1)	
Skeletal examination[b]			
Skull			
Delayed ossification	16 (8)	9 (6)	
Sternebrae			
Delayed ossification	4 (3)	7 (4)	
Ribs			
Extra	5 (3)	3 (2)	
Lumbar spur(s)	63 (17)	33 (12)	
Vertebrae			
Delayed ossification	49 (14)	36 (9)	
Bilobed thoracic centra	11 (7)	1 (1)[d]	

Rats were exposed to 0, 100, or 300 ppm EDC for 7 hr/day on days 6–15 of gestation.
[a] Number of fetuses/number of litters.
[b] Number of fetuses (number of litters) affected.
[c] Considered to be a major malformation.
[d] Statistically significantly different from the control value by a modified Wilcoxon test $p < 0.05$.

kidney. Examination of the rabbit fetuses for evidence of skeletal alterations revealed a statistically significant decrease in the incidence of 13 ribs and lumbar spurs among litters at the 100-ppm dose level. A decrease in the incidence of lumbar spurs was also observed among litters of rabbits exposed to 300 ppm EDC. These alterations are considered to be minor skeletal variants. The decreased incidence of these alterations are indicative of the normal variation in this species and has no toxicological significance.

Table 3
Rabbit Teratology: Observations at Cesarean Section

	Exposure level of EDC (ppm)		
	0	100	300
Number deaths/number females	0/20	4/21	3/19
Number litters	14	12	13
Implantation sites/dam[a]	9 ± 2	7 ± 3	8 ± 2
Fetuses/litter[a]	7 ± 3	6 ± 3	6 ± 2
Implantations resorbed (%)	20/124 (16)	11/86 (13)	17/101 (15)
Litters with resorptions (%)	9/14 (64)	6/12 (50)	8/13 (62)
Litters totally resorbed	0/14	0/12	0/13
Resorptions/litters with resorptions	20/9	11/6	17/8
% Dead fetuses	0 (0/101)	1 (1/75)	1 (1/85)
Fetal sex ratio, M:F	47:53	52:48	49:51
Fetal body weight (g)[b]	34.8 ± 6.0	35.8 ± 7.5	37.3 ± 6.9
Fetal crown-rump length (mm)[b]	89.7 ± 7.2	92.1 ± 6.0	93.1 ± 6.4

Rabbits were exposed to 0, 100, or 300 ppm EDC for 7 hr/day from days 6–18 of gestation. No values are statistically significantly different from the control values by the appropriate statistical test, $p < 0.05$.
[a] Mean ± S.D.
[b] Mean of litter ± S.D.

Reproduction Study

Inhalation of 25, 75, or 150 ppm of EDC did not alter the physical appearance in any of the rats during the study. However, during the seventh week of the study, a syndrome similar to sialodacryoadenitis spread among both control and treated animals, subsiding after the eighth week of the study. During this period the affected animals showed symptoms of red crusty material around the eyes and nose, as well as conjunctivitis. During the latter part of the study, three animals (one control female, one 25-ppm female, and one 25-ppm male) died spontaneously. Gross and microscopic examination of these animals indicated that the deaths resulted from spontaneous lesions and probably were not related to EDC exposure.

Body weights of the F_0 males and females were unaffected during the first 60 days of premating exposure. Additionally, female body weights during gestation and lactation of F_{1A} and F_{1B} litters were comparable between control and treated groups. A significant increase in food consumption was apparent in the latter part of the study in males at the 150-ppm exposure level. Significant decreases in food consumption were observed sporadically at the 75- and 150-ppm exposure levels. These changes were of an inconsistent nature, hence are not considered treatment-related. In females, a significant decrease in food

Table 4
Rabbit Teratology: Incidence of Anomalies

	Exposure level of EDC (ppm)		
	0	100	300
External examination[a]	101/14	75/12	85/13
Soft tissue examination[a]	44/14	32/12	36/13
Skeletal examination[a]	101/14	75/12	85/13
Total major malformations[b]	1 (1)	1 (1)	1 (1)
External examination[b]			
Multiple malformations[c]	0 (0)	0 (0)	1 (1)[d]
Left front paw rotated laterally and forward	1 (1)	0 (0)	1 (1)[d]
Soft tissue examination[b]			
Left kidney missing[c]	1 (1)	0 (0)	0 (0)
Enlarged adrenal gland	0 (0)	0 (0)	1 (1)
Ectopic umbilical artery	0 (0)	0 (0)	1 (1)
Skeletal examination[b]			
Skull			
Delayed ossification	8 (4)	2 (2)	3 (2)
Misshapen bones	1 (1)	0 (0)	1 (1)
Sternebrae			
Delayed ossification	61 (13)	46 (11)	54 (12)
Misshapen	0 (0)	1 (1)	2 (2)
Ribs			
Delayed ossification	0 (0)	0 (0)	1 (1)[d]
Extra	55 (13)	24 (10)[f]	44 (12)
Lumbar spur(s)	18 (10)	2 (2)[f]	6 (4)[f]
Asymmetric rib	1 (1)	0 (0)	2 (2)
Vertebrae			
Hemivertebrae[c]	0 (0)	1 (1)[e]	0 (0)
Misshapen[c]	0 (0)	1 (1)[e]	0 (0)

Rabbits were exposed to 0, 100, 300 ppm EDC for 7 hr/day on days 6–18 of gestation.
[a] Number of fetuses/number of litters examined.
[b] Number of fetuses (number of litters) affected.
[c] Considered to be a major malformation.
[d] These anomalies are from a single fetus, which exhibited the multiple anomalies.
[e] These anomalies are from a single fetus from a dam exposed to 100 ppm EDC.
[f] Statistically significantly different from the control value by a modified Wilcoxon test, $p < 0.05$.

consumption was observed only during the first week of the study at the 75- and 150-ppm exposure levels. This is attributed to the initial stress of exposure. No other significant deviations in food consumption were observed among female rats.

Reproduction data from F_{1A} and F_{1B} litters are presented in Tables 5 and 6, respectively. The fertility index of any group exposed to EDC by inhalation was not significantly different from the controls. At birth, the average number of pups per litter was not statistically different from controls in both F_{1A} and F_{1B} litters except in the F_{1A} 75-ppm exposure level. The gestation survival index (the percentage of pups alive at birth) and the 1-day, 7-day, 14-day, and 21-day survival indices of the treated groups were all comparable to or higher than the control group survival. The sex ratio of the 21-day-old pups was comparable between the treated and the control group. The exposure to EDC did not affect the neonatal body weight or growth of pups to weaning in F_{1A} or F_{1B} litters. External and internal examination (at weaning) of these pups did not reveal any unique treatment-related anomaly.

No significant deviations in organ weights of F_0 adults and F_{1A} weanlings of either sex were observed in any of the treated groups. However, the mean

Table 5
Effect of EDC on Rat Reproductive Indices—F_{1A} Litters

	EDC (ppm)			
	0	25	75	150
Number females	30	20	20	20
Fertility index[a]	70% (21/30)	75% (15/20)	80% (16/20)	75% (15/20)
Gestation days[b]	22 ± 0.4	22 ± 0.5	22 ± 0.6	22 ± 0.5
Number pups at birth[b] (live and dead combined)	13 ± 3	12 ± 3	10 ± 3[c]	12 ± 0.5
Gestation survival index[d]	94%	96%	94%	96%
1-Day survival index[e]	95%	97%	99%	98%
7-Day survival index[e]	90%	95%	91%	95%
14-Day survival index[e]	89%	95%	90%	92%
21-Day survival index[e]	87%	95%	90%	81%
Sex ratio on day 21, M:F	49:51	43:57	48:52	53:47

Male and female rats were exposed to 0, 25, 75, and 150 ppm EDC by inhalation for 6 hr/day for 176 days.
[a] Number of females delivering a litter/number of females placed with males.
[b] Mean ± S.D.
[c] Significantly different from control vaues by the appropriate statistical test, $p < 0.05$.
[d] Percentage of newborn pups that were alive at birth.
[e] Percentage of live-born pups that survived for 1, 7, 14, or 21 days.

Table 6
Effect of EDC on Rat Reproductive Indices—F_{1B} Litters

	EDC (ppm)			
	0	25	75	150
Number females	30	20	20	20
Fertility index[a]	87% (26/30)	85% (17/20)	90% (18/20)	85% (17/20)
Gestation days[b]	22 ± 0.4	22 ± 0.4	21 ± 0.5	22 ± 0.4
Number pups at birth[b] (live and dead)	13 ± 3	12 ± 2	12 ± 2	14 ± 2
Gestation survival index[c]	94%	93%	95%	98%
1-Day survival index[d]	95%	100%	100%	98%
7-Day survival index[d]	88%	99%[e]	98%[e]	95%
14-Day survival index[d]	85%	97%[e]	98%[e]	85%
21-Day survival index[d]	84%	97%[e]	98%[e]	85%
Sex ratio on day 21, M:F	45:55	45:55	51:49	46:54

Male and female rats were exposed to 0, 25, 75, and 150 ppm EDC by inhalation for 6 hr/day for 176 days.
[a] Number of females delivering a litter/number of females placed with males.
[b] Mean ± S.D.
[c] Percentage of newborn pups that were alive at birth.
[d] Percentage of live born pups that survived for 1, 7, 14, or 21 days.
[e] Significantly different from control values by the appropriate statistical test, $p < 0.05$.

kidney weight (absolute and relative) of the F_{1B} male weanling pups at the 25-ppm dose level of EDC was significantly higher than the control group means (Table 7). No significant differences in the kidney weights were observed in the 75- or 150-ppm groups. In the absence of a dose-related effect, the isolated significant increase in kidney weight at the low-dose level was considered a random variation unrelated to EDC exposure. No differences in liver weight, relative or absolute, were noted in the male weanling F_{1B} pups. The treated F_{1B} female weanlings were all comparable to the control mean values in liver and kidney weights.

No treatment-related histopathologic changes were apparent in any of the groups. Practically all (control and treated) adult F_0 male and female rats had some degree of chronic renal disease with no alteration in severity resulting from treatment. There were no histopathologic changes attributed to EDC exposure in the kidneys and livers of F_{1A} or F_{1B} weanlings.

Table 7
Effect of EDC on Organ Weights (Mean ± s.d.) of F_{1B} Weanling Rats

	ppm	N	Body weight (g)	Liver g	Liver g/100g	Kidneys g	Kidneys g/100g
Males	0	5	42 ± 9	2.11 ± 0.50	5.01 ± 0.58	0.49 ± 0.10	1.17 ± 0.12
	25	5	42 ± 6	2.28 ± 0.33	5.45 ± 0.13	0.63 ± 0.05[a]	1.53 ± 0.18[a]
	75	5	40 ± 5	2.06 ± 0.37	5.18 ± 0.48	0.56 ± 0.10	1.42 ± 0.14
	150	5	36 ± 6	19.2 ± 0.25	5.36 ± 0.62	0.45 ± 0.06	1.25 ± 0.08
Females	0	5	36 ± 10	1.84 ± 0.54	5.15 ± 0.49	0.47 ± 0.11	1.34 ± 0.15
	25	5	38 ± 13	2.11 ± 0.72	5.55 ± 0.59	0.60 ± 0.13	1.62 ± 0.36
	75	5	39 ± 6	2.08 ± 0.35	5.34 ± 0.38	0.57 ± 0.11	1.45 ± 0.09
	150	5	37 ± 9	1.93 ± 0.54	5.15 ± 0.46	0.49 ± 0.10	1.34 ± 0.10

Male and female rats were exposed to 0, 25, 75, and 150 ppm EDC by inhalation for 6 hr/day for 176 days.
[a] Significantly different from the control mean by the Dunnett's test, $p < 0.05$.

DISCUSSION

EDC was not teratogenic in rats inhaling 100 ppm or in rabbits inhaling 100 or 300 ppm of the compound for 7 hours each day during the period of major organogenesis. No major soft-tissue or skeletal malformations were seen at a significantly increased incidence over controls among litters of rats exposed to 100 ppm EDC. In rabbits, only one malformed fetus at the 300-ppm dose level and some skeletal alterations in a single fetus exposed to 100 ppm of the test compound were observed. No alterations were significantly increased in incidence over control values.

In the teratology study, severe maternal toxicity was observed among the rats exposed to 300 ppm EDC; two-thirds of the animals died during the exposure period. Only one of the surviving females was pregnant at the time of cesarean section, and all of her implantations were resorbed. The embryotoxicity seen in this animal is considered secondary to the maternal toxicity observed. No fetuses were available for examination at this exposure level, therefore, no conclusions may be drawn concerning the teratogenic potential of inhaled EDC in the rat at 300 ppm. No signs of toxicity were observed among rats exposed to 100 ppm of the test compound. Maternal toxicity was noted in rabbits, as evidenced by maternal deaths at both dose levels.

No evidence of embryotoxicity or fetotoxicity was observed among rats exposed to 100 ppm EDC or among rabbits exposed to 100 or 300 ppm EDC for 7 hours/day. No adverse effects on the incidence of pregnancy, the mean litter size, the incidence of resorptions, or the fetal body measurements were observed among the litters of exposed animals.

In the reproduction study, incidence of external or internal malformations among treated F_{1A} and F_{1B} pups was not statistically significantly different from controls. No unique malformation was observed in treated pumps that was not seen historically in this laboratory in Sprague-Dawley rats.

Organ weights monitored at terminal sacrifice of F_0 animals did not reveal any treatment-related effect. Likewise, organ weights of F_{1A} or F_{1B} pups at weaning were comparable between control and treated pups. However, a significant increase in the kidney weight (relative and absolute) was observed in F_{1B} male weanlings at the 25-ppm dose level and not at the higher levels. Because of the absence of dose-response and the lack of histopathologic changes in these animals, the isolated effect on kidney weight was considered unrelated to EDC exposure. Gross and microscopic examination of tissues from F_0 adults and F_{1A} and F_{1B} weanlings did not reveal any changes that were considered to be related to exposure to EDC by inhalation.

The results presented here are at variance with the work of Vozovaya (1974), who showed decreased fertility in rats exposed to 14 ppm EDC. However, in this study exposure of rats to over tenfold-higher concentration (150 ppm) of EDC was without effect on reproduction. No explanation for the discrepancy in the results of the two studies is apparent at the present time.

In conclusion, our data suggest that a teratogenic effect was not discerned in rats exposed to 100 ppm or rabbits exposed to 100 or 300 ppm EDC during organogenesis. There was no evidence of an embryotoxic or a fetotoxic effect at these dose levels. Maternal toxicity was seen in rabbits exposed to 100 or 300 ppm EDC and in rats at 300 ppm but not at 100 ppm of the test material. In addition, chronic exposure of male and female rats to EDC (up to 150 ppm, 6 hr/day) did not result in any adverse effect on reproduction over one generation within two litters.

REFERENCES

Dawson, A. V. 1926. A note on the staining of the skeleton of cleared specimens with alizarin red-S. *Stain Technol.* **1**:123.

Gibson, J. P., R. E. Staples, and J. W. Newberne. 1966. Use of the rabbit in teratogenicity studies. *Toxicol. Appl. Pharmacol.* **9**:398.

Haseman, J. K. and D. G. Hoel. 1974. Tables of Gehan's generalized Wilcoxon test with fixed point sensoring. *J. Stat. Comp. and Simulation* **3**:117.

NIOSH (National Institute of Occupational Safety and Health). 1977. *Registry of toxic effects of chemical substances,* vol. 2. DHEW, Cincinnati, Ohio, entry NKI05250.

Seigel, S., ed. 1956. The case of two independent samples. In *Nonparametric statistics for the behavioral sciences,* p. 96. McGraw-Hill, New York.

Spencer, H. C., V. K. Rowe, E. M. Adams, D. D. McCollister, and D. D. Irish. 1951. Vapor toxicity of ethylene dichloride determined by experiments on animals. *AMA Arch. of Industr. Hyg.* **4**:482.

Staples, R. E. 1974. Detection of visceral alterations in mammalian fetuses. *Teratology* **9**:A-37.

Steel, R. G. D. and H. H. Torrie, ed. 1960. Analysis of variance. In *The one-way classification in principles and procedures of statistics,* p. 99. McGraw-Hill, New York.

Vozovaya, M. A. 1971. Variations in the estral cycle of rats during the chronic combined action of gasoline and dichloroethane vapor. *Akush. Ginekol.* (Mosc.) **47**(12):65.

_____. 1974. Development of progeny of two generations obtained from females subjected to action of dichloroethane. *Gig. Sanit.* **7**:25.

_____. 1976. The effect of small concentrations of benzene and dichloroethane separately and combined on the reproductive function of animals. *Gig. Sanit.* **6**:100.

Vozovaya, M. A. and L. K. Malyarova. 1975. Mechanism of action of ethylene dichloride on the fetus of experimental animals. *Gig. Sanit.* **6**:94.

Wilson, J. G. 1965. Methods for administering agents and detecting malformations in experimental animals. In *Teratology: Principles and techniques* (ed. J. G. Wilson and J. Warkany), p. 262. University of Chicago Press, Chicago.

COMMENTS

WARD: Is EDC found in the fetus after the mother is exposed?

RAO: No. Fetuses were not exposed to the EDC, and we did not look for EDC levels in the fetus.

WARD: Wouldn't that be a reason why it's not teratogenic?

RAO: No. It is not conventional in teratology studies to monitor the concentration of the chemical in the fetuses. Based on the simple chemical structure, I have no reason to expect there is any placental barrier for a chemical like EDC.

SIMMON: You discounted one grossly abnormal fetus that you observed in the high-treatment dose.

RAO: Yes.

SIMMON: How many headless rabbits would you have to have in the treated group before you would consider it significant?

RAO: I would say if I noticed at least one more I would have associated this anomaly with EDC exposure.

SIMMON: How many headless rabbits have you seen in control populations before?

RAO: We have not seen that in our laboratory. In fact, I have searched the literature. As I told you, Tony Palmer has published an extensive list of spontaneous incidences in his rabbit population in England (Palmer 1968), and he has seen this.

REITZ: I would simply interject here that we could take some guidance from standard statistical procedures. If something were statistically significant it should not be ignored. But there are statistical procedures that are designed to eliminate the chance observations from those that are frequent.

RAO: There is a 10% incidence of spontaneous anomalies in the rabbit pups, compared to 3% in the case of rats. So for some reason there is a relatively higher incidence of anomalies in rabbits, mainly because the rabbit population is not well defined, that is, rabbits are not bred under controlled conditions.

INFANTE: Are the rib changes you reported considered abnormalities or are they variations of normalities? Is there some kind of consensus or agreement on this among teratologists?

RAO: We consider such changes as variations of the normal ossification of the bones. They are not anomalies. The bone is there, it's just a question of the intensity and degree of ossification being delayed for some reason. But even on that, we saw a decrease not an increase.

INFANTE: Then if these are normal variations, and you saw a significant excess of variation, then what would you conclude? Would you still conclude that there's no teratogenic effect?

RAO: Variation of a normal ossification is not a teratogenic response. Assuming we have seen an increased incidence of the decreased ossification, then I would consider it some kind of an embryotoxic effect or a fetotoxic effect. As a result of the fetal toxicity, overall development of the fetus is delayed. But it is not a teratogenic response, per se.

PARKER: In some of the studies that we have done, we have been able to associate a decreased birth weight for the delayed ossification, but I didn't notice in your results that there was a decreased birth weight for the delayed ossification.

RAO: No. Again, it all depends upon the intensity and the degree of the decrease in ossification. The overall size, either birth weight or the length of the pup, was not affected; the entire length apparently is normal. It is just the ossification was slightly delayed.

Again, these variations can vary from mild to moderate to severe. It's very hard to judge, and it's a very subjective thing. Any time we find them we record them and then try to evaluate them based on the total package of data, rather than one particular variation or anomaly.

INFANTE: If you have, say, an excess of normal variation of the ribs, spurs or whatever, what would you conclude about that? I'm not a teratologist, and when I see these studies I don't know what it means.

RAO: I consider such variations some kind of a fetotoxic effect, that's all. It tells me nothing more than fetal toxicity, as a result of the gross effect on the overall development of the fetus. It has nothing to do with organ differentiation. Organ differentiation—the bone formation, the skeletal formation, etc.— has already occurred in these pups. It is just that the ossification process, which is occurring normally, is slightly displaced in time.

Let me give an example: We did a study employing (2,4-dichlorophenoxy)acetic acid (2,4-D) wherein we actually allowed one-half of the litters to be sacrificed by cesarean section on day 21; the other half were allowed to litter normally.

There was no difference in growth or apparance of the pups in both groups that grew up to weaning.

AMES: Have you tried to write the Russians? Also, why do you think that in your study it seems pretty clear there's not much of an effect compared to the Russian results (Vozovaya 1971, 1974, 1976; Vozovaya and Malyarova 1975)?

RAO: I have no plans to write the Russians.

FABRICANT: I'm interested in the dominant lethal data that you presented. What was the percentage of postimplantation mortality at the different concentrations that you looked at? And at what stage during gestation was it determined?

RAO: Okay. Let me tell you, in the conventional dominant lethality you expose only the males.

FABRICANT: I've done it.

RAO: Okay. You are familiar with that. In this particular case, we did not do that. We exposed both males and females. In the reproduction studies we don't look for implantations because these animals were allowed to deliver normally. Unless you sacrifice the animals, you don't know how many implantations they have.

However, you can see that the total numbers of pups born alive on day 1 were comparable between control and EDC-treated groups. Assuming that the chemical had significant dominant lethality, then I would have expected a decrease in the number of pups that were born, but I didn't see it.

FABRICANT: What's the litter size that you saw?

RAO: I believe it was in the range of 7-9.

INFANTE: If in fact you did have small litters in your exposure group, how would you know if that effect was a dominant lethal effect or if it was a transplacental effect since you exposed both sexes?

RAO: Litter size in EDC-treated groups was comparable with controls in our study.

KARY: We might perhaps go back to a question on foreign literature, in why there might be some differences. One of the problems that we have, of course, is communication. It's difficult to get answers to questions from countries behind the Iron Curtain.

Another problem that we have is verifying exposure levels. One of the first questions that arises when we assess foreign literature is: What were the exposure levels, and how are they monitored? We usually never have an answer to that question. And it's critical.

INFANTE: I had another question in terms of methodology, not in terms of your study. I think in the first part you mentioned that you sacrificed the animals at day 21 because of concern for cannibalizing.

RAO: That's right.

INFANTE: But in the other study, where you ran the F_{1A} and F_{1B} generations, these pups were suckled, were they not?

RAO: That's correct.

INFANTE: Then why weren't you concerned about cannibalization in the second case? I don't understand.

RAO: In the first study, we are interested in picking up the teratogenic effects. In the second study, teratogenesis is of secondary importance. In the second study the most important and critical information we are looking for is fertility, that is, whether exposure of both males and females is going to affect the fertility of that generation and future generations. That's why these two studies are always done—to evaluate different types of parameters, to see the different locus of action.

Also the second study by itself will not answer whether or not the chemical has any teratogenic effects. If we had done just the reproduction study, it would only prove that EDC did not affect the fertility of the exposed pairs. But still it doesn't answer the question about any teratogenic effects. That's why we do the other study.

We did not see any cannibalization in the second study. The total number of pups from day 1 to day 21 is practically the same.

INFANTE: One more question: Is there any concern about looking for postnatal developmental defects, or is this outside of pure teratology? I know, for example, at a meeting last year on pesticides, Renata Kimbrough reported the results from one study in which the pups were normal at birth (Kimbrough et al. 1974), however, a certain percentage of them would not expand their lungs. If you were doing classical teratology, you would miss that effect, because the pup appeared normal. I'm wondering if there is any concern about revamping methodologies for a certain number of these studies to look for postnatal behavioral changes or metabolic effects.

RAO: Let me first answer the lung problem. If the animals do not have the capability to expand their lungs, chances are they will die within days or hours as a result of hypoxia.

Coming back to your question about the behavioral aspect—this science is quite new and the methodology is still being developed. We don't know the implications of those techniques that are developed. At this stage these techniques are not sufficiently sophisticated to have any relevance in terms of extrapolation to human beings.

KARY: A point of clarification—perhaps contrast would be a better word. A teratology study does not seek to answer the same thing that a postnatal study would. They are really, technically, two different things, so that if you were looking for postnatal effects, you would not revamp the teratology study, you would design a postnatal study.

BUSEY: Which you did, in fact, do.

RAO: Yes. In the reproduction study, we didn't look at the behavioral aspect because of the technology available for the test procedures.

BUSEY: But you would have picked up the phenomena that Dr. Infante was talking about.

RAO: Yes.

References

Kimbrough, R., T. B. Gaines, and R. E. Linder. 1974. 2,4-Dichlorophenyl-*p*-nitrophenylester (TOK) effects on the lung maturation of lung fetus. *Arch. Environ. Health* **28**:316.
Palmer, A. K. 1968. Spontaneous malformations of the New Zealand white rabbit: The background to safety evaluation tests. *Lab. Anim.* **2**:195.
Vozovaya, M. A. 1971. Variations in the estral cycle of rats during the chronic combined action of gasoline and dichloroethane vapor. *Akush. Ginekol.* (Mosc.) **47**(12):65.
⎯⎯⎯⎯⎯. 1974. Development of progeny of two generations obtained from females subjected to action of dichloroethane. *Gig. Sanit.* **7**:25.
⎯⎯⎯⎯⎯ 1976. The effect of small concentrations of benzene and dichloroethane separately and combined on the reproductive function of animals. *Gig. Sanit.* **6**:100.
Vozovaya, M. A. and L. K. Malyarova. 1975. Mechanism of action of ethylene dichloride on the fetus of experimental animals. *Gig. Sanit.* **6**:94.

An Investigation of Possible Sterility and Health Effects from Exposure to Ethylene Dibromide

GARY TER HAAR
Toxicology and Industrial Hygiene
Ethyl Corporation
Baton Rouge, Louisiana 70801

Ethylene dibromide (EDB; 1,2-dibromoethane) has been manufactured for over 50 years. Levels of exposure in the workplace have decreased from perhaps an average of 5 ppm to less than 1 ppm. Ambient levels are well under 0.0001 ppm. Studies of workers exposed to EDB have shown no evidence of reduced sperm counts. In addition, the size of EDB workers' families are similar to those expected.

Epidemiological studies of four worker populations have not shown an increase in cancer that could be attributed to EDB. These detailed studies of worker exposure to EDB have not uncovered chronic effects on humans. As levels of EDB in the ambient atmosphere are four to five orders of magnitude lower than those in the workplace, there should be no measurable risk to the population at large from EDB.

EDB is a dense, colorless, nonflammable liquid. Its vapor pressure of 11.0 mm of Hg at 25°C is rather high even though it is a relatively large molecule (m.w., 187.87). It is an organic chemical with a fairly simple structure:

$$\begin{array}{c} \text{Br} \quad \text{Br} \\ | \quad | \\ \text{H}-\text{C}-\text{C}-\text{H} \\ | \quad | \\ \text{H} \quad \text{H} \end{array}$$

Manufacture of EDB involves simple addition of elemental bromine to ethylene, followed by a purification step. The production of EDB makes up the largest single use of bromine, and the product is usually manufactured at plants located near the natural bromide brines as obtained from wells in Michigan and Arkansas.

USES OF EDB

EDB was first produced on a commercial scale in the mid-1920s for sale to the producers of lead antiknock compounds. The bromine atoms in EDB react with the lead atoms during gasoline engine combustion to produce lead bromide salts, which temporarily vaporize and clear from the engine before solidifying. This use of EDB as a lead scavenger has always been its largest use, reaching over 200 million pounds each year since 1954. Usually EDB is blended as a supplement with other scavengers, mainly ethylene dichloride (EDC; 1,2-dichloroethane). The scavengers are used to form more volatile lead compounds during combustion, and these are more completely removed from the combustion chamber. The introduction of any alternative formulation of antiknock additive would require additional time-consuming testing and research in the current model lead-tolerant emission control systems before EPA could accept its use under the terms of the recent amendments of the Clean Air Act. Thus, regulating EDB out of gasoline might result in eliminating the use of lead antiknock additives.

Lead antiknocks are widely recognized as energy savers. The elimination of EDB as a scavenger in leaded fuels would severely reduce the utility of this energy-saving additive. The result would be a significant increase in the nation's crude oil usage. Furthermore, the oil industry would be unable to produce the required volumes of gasoline during the next 3 or 4 years without lead antiknocks being available. This is because additional, very expensive, refinery process equipment would have to be installed to produce the necessary volume of high-octane blending stocks. Restrictions on EDB, would have an extremely adverse impact on petroleum utilization, gasoline engine performance, and, in a number of respects, on the nation's economy.

EDB has other important commercial uses and has been used in federal, state, and international quarantine treatments since the early 1950s. In the Animal and Plant Health Inspection Service (APHIS), it is recognized as a basis for the importation of many fruits and vegetables from foreign countries, for movement interstate from areas regulated for certain insect pests (e.g., Mexican fruit fly in Texas and tropical fruit flies in Hawaii), and for the treatment of certain agricultural exports from the U.S. to foreign countries in accordance with their import requirements. In addition to federal quarantine regulations, a number of states require the use of EDB as a basis for the movement of some agricultural products to prevent pest spread.

Other uses of EDB include fumigation of grain, including some for export, and spot fumigation of milling machinery. EDB is a widely used and extremely significant soil fumigant. About 5 million pounds of EDB are used annually in the U.S. for soil fumigation—an essential component of many significant agronomic, fruit, and vegetable crop production systems. It is one of the most effective and economical nematocides available. Removal of this pesticide would result in decreased agricultural commodity productivity.

LEVELS OF EDB IN AIR

We should consider five general areas when we look at EDB concentrations in air: EDB manufacturing sites, refineries, gasoline stations, fumigation, and ambient.

While information is sketchy for worker exposure to EDB prior to 1970, data from the Dow Chemical Company (Rausch 1978) indicates a range of 1-24 ppm for 38 samples taken in 1949 and 1952. Data in 1971 and 1972 indicate levels were of the order of a few ppm, generally less than 5 ppm. These data probably represent general practice before 1975.

There has been an increasing effort since 1975 to lower worker exposure to EDB. Ethyl Corporation (Ter Haar 1978) reported a range of nondetected to 4.50 ppm for 95 samples. Sixty-six of these samples were below 0.5 ppm, 8 between 0.5 and 1 ppm, and 21 between 1 and 4.5 ppm. Dow (Rausch 1978) reports 0.8-5.0 ppm and Great Lakes Chemical Corporation (Hunt 1977) reports similar data, with nearly all samples below 5 ppm and more than half below 1 ppm. All manufacturing areas today would probably be below 1 ppm on an 8-hour time weighted average.

Workers in refineries have very little opportunity for exposure to EDB. The antiknock fluid is the only source of EDB in a refinery and its use is closely controlled. As a result, EDB exposures at gasoline bulk loading stations and refineries are 0.0001 ppm or less.

Similarly, employees in gasoline stations have minimal exposure. Thirty-four samples showed a range of 0.00002-0.00006 ppm. This was data generated by EPA and published in 1975 and 1976 (Going and Long 1975; Going and Spigarelli 1976).

Exposure of field applicators to EDB is largely dependent on the applicators' technique. In a report to the Occupational Safety and Health Administration (OSHA) and the Environmental Protection Agency (EPA) (Hunt 1977), Great Lakes Chemical Corporation stated that most exposure to EDB occurred during calibration, chemical transfer, and equipment repairing procedures. Findings also indicated that exposure can occur during active application, but significant exposure only occurred as a result of poor technique or nonadherence to label direction. Values for personnel exposure ranged from nondetected to 0.08 ppm for typical exposure. Under misuse conditions, exposures ranged from 0.06-0.52 ppm.

Ambient levels range from less than 0.00001 ppm to 0.00005 ppm according to the EPA work. Leinster et al. (1978) report an average of 0.000005 ppm for samples taken near a busy road in London. The data summarized above indicate clearly that the only area where meaningful exposure to EDB might occur is in the manufacturing or possibly fumigation and soil application areas.

ACUTE HUMAN EFFECTS FROM EDB

EDB has been used as an ingredient in antiknock compounds for more than 50 years both in the U.S. and abroad with no report of any lasting adverse effect on the health of the worker or user. Workers have suffered accidental exposure to EDB with minor chemical burns or other temporary effects.

In 1976, a record was made of Ethyl employees with known work exposure to EDB together with their mortality experience. Fifty-three workers were identified with exposures varying from 3 months to 10 years. There was one death in this group with the cause being listed as cancer of the kidney. This study does not reveal any abnormal mortality experience in those Ethyl workers with known exposure to EDB.

CHRONIC EFFECTS OF EDB

In hens, EDB causes decreased egg size, impaired uptake of labeled proteins by ovarian follicles, infertility of eggs, and cessation of egg laying (Bondi et al. 1955; Alumot et al. 1968; Alumot and Harduf 1971). In male chicks, sperm count was not affected, and body and testes weight remained normal (Alumot et al. 1968).

In large doses a reversible effect upon spermatogenesis has been demonstrated in rats and bulls without concurrent adverse effect on reproduction (Amir and Volcani 1965; Amir 1973, 1975). There is an adequate factor of safety between these reversible effects seen in bulls and worker exposure.

Sperm production remained normal (Bondi and Alumot 1969) in two rams dosed chronically at 2–5 mg EDB/(kg body weight · day) for 119 days, then at a lethal level of 10 mg/kg for 12 days. Spermatogenesis remained unaffected even at the lethal dose level.

Cows dosed (Bondi and Alumot 1966) with 2 mg EDB/kg body weight beginning during the second or third month of pregnancy and continuing through three lactation periods showed no effects and offspring were normal. Fertility was unimpaired. Female calves did not differ from controls when dosed with EDB (unspecified dose) from birth through first parturition. Fertility was unaltered as measured by insemination and parturition records.

On the basis of the literature data cited earlier, there is some reason to suspect reproductive effects from EDB. Ethyl's medical department arranged in 1977 to have sperm counts taken on 59 employees at Ethyl Corporation's plant at Magnolia, Arkansas. These workers have had some work exposure to EDB, other brominated hydrocarbons, and bromine. This was followed up in 1978 with 24 additional samples. Table 1 shows these results.

Eighteen of those submitting specimens in 1978 also submitted specimens in 1977. Of these, 10 (55.6%) had higher counts in 1978 than in 1977 and 8

Table 1
Sperm Count of EDB Workers

Sperm counts	1977	1978
0–20 million	11.9% (one zero count)	20.8% (no zero count)
20–40 million	15.3%	12.5%
40–100 million	49.1%	37.5%
Over 100 million	23.7%	29.2%

(44.4%) had lower counts in 1978 than in 1977. The sperm levels found in workers exposed to EDB are not different than those reported in the literature (Nelson and Bunge 1974; Smith and Steinberger 1977).

We divided the 1977 population into two groups, with exposures to EDB estimated to be below 0.5 ppm and between 0.5 and 5 ppm. There were 40 men whose estimated exposure was less than 0.5 ppm. Of these, 10% had counts less than 20 million, 10% between 20 million and 40 million, 55% between 40 million and 100 million, and 25% over 100 million. Of those whose exposure was estimated to be 0.5-5 ppm, the percentages were 16, 26, 37, and 21 respectively.

This treatment of the data and the similarity of the distribution of sperm counts to literature values indicate that EDB at the levels found in our workplace have not had an adverse effect on the sperm count of the workers.

To further study the question regarding the reproductive system, the major producers of EDB commissioned Equitable Environmental Health, Inc., to conduct a retrospective study of inferred fertility of married men occupationally exposed to EDB. Data on the number of children born to EDB workers at Ethyl, Dow, Great Lakes, Associated Octel, and Houston Chemical were compared with national standards. Levels of exposure were estimated at less than 0.5 ppm and between 0.5 and 5.0 ppm. Wong et al. (1978) reported on the U.S. workers in this study. Table 2 shows a summary of the data for three of the plants. Clearly there is no effect of EDB on this population with regard to children born. The Houston Chemical plant presented a different picture. Table 3 shows these results.

The authors concluded (Wong et al. 1978) with respect to this group that "the observed-to-expected births ratio is consistently and significantly low, suggesting a possible antifertility effect." Exposure levels on an individual basis

Table 2
Observed and Expected Births (Dow, Ethyl, Great Lakes)

Exposure groups	Observed births	Expected births
White < 0.5 ppm	20	22.27
White 0.5-5 ppm	28	27.36
All nonwhites	1	2.04

Table 3
Observed and Expected Births (Houston Chemical)

Race	Observed births	Expected births
White	11	22.20
Nonwhite	2	3.32

were not available from the plant, so it is impossible to address a possible dose-response relationship. However, for the entire group at this plant, the average exposure level lies below 5 ppm. Thus the antifertility observation cannot be attributed to a difference in exposure level. However, it should be pointed out that the prevalence of vasectomies at this plant was the highest among the four, and was also slightly higher than that in the country." The high incidence of vasectomies in the Houston Chemical's work force indicates a strong desire for birth control in this population and is a likely explanation of the somewhat reduced birthrate.

The data for the British group was handled differently (pers. comm.). The 82 married male employees were divided into four exposure groups: non-exposure to EDB ($N = 41$); occasional exposure ($N = 17$); regular exposure ($N = 13$); and irregular exposure ($N = 11$).

For each exposure group, the number of person-years of observation since joining the company (onset of exposure) was computed by age of father and by parity at the time of joining the company (Table 4). The indirect method of standardization was employed to adjust for the confounding effects of paternal age and parity on birth rates. The age- or parity-specific rates per person-year for the nonexposed group were used as the standard. The indirect method of standardization was chosen, because, among the four, the non-exposed group is the largest, and, accordingly, its rates would be more stable than those in the other groups. The standard birth rates from the nonexposed group were applied to the age- or parity-specific person-years in each of the three exposed groups to obtain the expected births. The expected births are the number of births one would expect to observe in the exposed group, had the latter been experiencing the same fertility rates as the nonexposed group. As an index, the ratio of the actually observed births in each exposed group to the expected births was formed. These indirectly standardized fertility ratios are

Table 4
Observed and Expected Births (Associated Octel)

Exposure	Observed births	Expected births	Observed to expected
Adjusting for age			
Occasional	14	17.19	0.81
Regular	8	7.16	1.12
Irregular	13	15.43	0.84
Combined	35	39.78	0.88
Adjusting for parity			
Occasional	14	9.71	1.44
Regular	8	7.74	1.03
Irregular	13	9.56	1.36
Combined	35	27.01	1.30

adjusted for age or parity. A ratio larger (smaller) than one indicates higher (lower) fertility of the exposed group than the nonexposed group. It was not feasible to carry out the analysis adjusting for both paternal age and parity simultaneously because of the extremely small numbers in the exposed groups. Treating the number of observed births as a Poisson random variable, none of the ratios is significantly different from one at the 0.05 level. This statistical test is only an approximate one, since the standard rates are based on a fairly small sample ($N = 41$).

Worker exposure to EDB began over 50 years ago. The Dow Chemical Company studied workers exposed to EDB at two plants (Rausch 1978) and Associated Octel in England also studied workers exposed to EDB at two plants (Ter Haar 1978). These reports cover about 450 workers. The Dow data for Unit 1 goes back to the 1920s; the data for the other plant, to 1942. Tables 5 and 6 show these data. The workers from Unit 1 showed a slight excess of cancer but there was exposure to several other chemicals during their working history. The plant started in 1942 involved almost pure EDB exposure. Cancer in this group was less than the rate for the control population.

Octel also studied workers at two plants. At the Amlwich plant, records begin in 1952. Data for this plant compare favorably with annual death rates for males from this area of Wales called Gwynedd. Table 7 shows these results.

Octel also studied workers at a fairly remove site near the sea in southwest England. Exposure was solely to EDB manufacture. No other chemical processes

Table 5
Observed and Expected Deaths for Selected Causes by Unit, 1940–1975

	Unit 1		Unit 2 (less arsenical employees)	
	observed	expected	observed	expected
All causes	20	20.4	15	13.0
Malignant neoplasms				
Total	1	3.8	5	2.2
Respiratory system	0	1.3	0	.6
Digestive system	1	1.0	2	.7
All other sites	0	1.5	3	.9
Diseases of cardiovascular system	11	9.6	6	7.0
Emphysema, chronic bronchitis, and asthma	0	.4	1	.2
Influenza, pneumonia	0	.4	2	.3
All external causes	3	2.9	0	1.3
All other causes	2	3.3	1	2.0
Death certificates not obtained	3	–	0	–

Expected deaths based on U.S. white male mortality rates.

Table 6
Observed and Expected Deaths for Selected Causes by Duration of Exposure and Interval Since First Exposure, Unit 1 and Unit 2 Less Arsenicals Employees, 1940–1975

Duration of exposure and interval since first exposure	Total deaths observed	Total deaths expected	Total malignancies observed	Total malignancies expected
< 1 Year exposure				
Total	2	5.1[a]	0	.9
< 15 Years	0	1.6	0	.2
15–24 Years	2	1.5	0	.3
25+ Years	0	1.9	0	.4
1–5 Years exposure				
Total	23	19.8	3	3.5
< 15 Years	8	7.1	1	1.1
15–24 Years	10	7.2	1	1.3
25+ Years	5	5.4	1	1.2
6+ Years exposure				
Total	10	8.5	3	1.6
< 15 Years	2	1.9	0	.3
15–24 Years	6	3.5	3	.7
25+ Years	2	3.2	0	.7
Total years exposure				
Total	35	33.4	6	6.0
< 15 Years	10	10.6	1	1.6
15–24 Years	18	12.2	4	2.2
25+ Years	7	10.5	1	2.2

[a] Expected number of deaths not additive due to rounding errors.

Table 7
Deaths per 1000 in Males 45–64 Years Old

Cause	Studied group	Gwynedd
All deaths	8.3	14.2
Ischemic heart disease	3.5	5.5
Cerebrovascular disease	1.2	1.5
Malignant neoplasm		
All	2.7	3.7
Bronchus	1.5	1.4
Stomach	0.3	0.5
Intestine	0.3	0.7
Suicide	0.3	0.2

Table 8
A Comparison Between the Mortality Rates of the Studied Group and the General Population of the Locality

Age (years)	Studied population					Southwest England (death rate/1000 men)			
	number man years	number deaths		death rate/1000 man-years		1961		1970	
		total	cancer	total	cancer	total	cancer	total	cancer
25–44	1797	3	0	1.67	0	1.62	0.30	1.38	0.32
45–64	1440	15	1	10.42	0.69	12.49	3.43	12.95	3.44
65–74	177	9	2	50.85	11.30	52.65	11.04	49.53	11.65
>74	47	7	1	148.94	21.28	136.52	16.15	135.78	16.99

were involved. The plant began operation in 1940 and closed in 1973. Table 8 shows these results. Again the workers exposed to EDB have a similar mortality experience to the control group. In particular, cancer incidence is similar.

EDB has been manufactured for many years. This has given the industry the opportunity to evaluate long-term low-level effects. The studies available to date do not support the hypothesis that EDB causes sterility or other reproductive effects in workers, nor is there evidence that exposure to EDB increases the likelihood of cancer.

REFERENCES

Alumot, E. and Z. Harduf. 1971. Impaired uptake of labeled proteins by the ovarian follicles of hens treated with ethylene dibromide. *Comp. Biochem. Physiol.* 6:39861.

Alumot, E., E. Nachtomi, O. Kempenich-Pinto, E. Mandel, and H. Schindler. 1968. The effect of ethylene dibromide in feed on the growth, sexual development and fertility of chickens. *Poult. Sci.* 47:1979.

Amir, D. 1973. The sites of spermicidal action of ethylene dibromide in bulls. *J. Reprod. Fertil.* 35:519.

———. 1975. Individual and age differences in the spermicidal effect of ethylene dibromide in bulls. *J. Reprod. Fertil.* 44:561.

Amir, D. and R. Volcani. 1965. Effects of dietary ethylene dibromide on bull semen. *Nature* 206:99.

Bondi, A. and E. Alumot. 1966. *Effect of ethylene dibromide fumigated feed on animals.* Final technical report for special foreign currency research under PL-480 (to USDA), August.

———. 1969. *Mechanism of action of halogenated hydrocarbons used as fumigants on animals.* First annual report of research under PL-480, A10-MQ-8 (to USDA), August.

Bondi, A., E. Olomuck, and M. Calderon. 1955. Problems connected with ethylene dibromide fumigation of cereals—II. Feeding experiments with laying hens. *J. Sci. Food Agric.* 6:600.

Going, J. E. and S. Long. 1975. *Sampling and analysis of selected toxic substances. Task II-Ethylene dibromide.* EPA 560/6-75-001.

Going, J. E. and J. L. Spigarelli. 1976. *Sampling and analysis of selected toxic substances. Task IV-Ethylene dibromide.* EPA 506/6-76-021.

Hunt, W. C. 1977. *Submission to Occupational Safety and Health Administration and Environmental Protection Agency, EDB Criteria Document and EDF Petition.* Great Lakes Chemical Corporation, December 12.

Leinster, P., R. Perry, J. Young. 1978. Ethylene dibromide in urban air. *Atmos. Environ.* 12:2243.

Nelson, C. M. K. and R. G. Bunge. 1974. Semen analyses: Evidence for changing parameter of male fertility potential. *Fertil. Steril.* 25:503.

Rausch. D. 1978. *Submission to Docket Officer, Docket H-111. Occupational Safety and Health Administration. Comments concerning a standard for occupational exposure to ethylene dibromide.* Dow Chemical USA, May 15.

Smith, K. D. and E. Steinberger. 1977. What is oligospermia? In *The testes in normal and infertile men* (ed. P. Trever and H. R. Nankin). Raven Press, New York.

Ter Haar, G. 1978. *Submission to Docket Officer, Docket H-111. Occupational Safety and Health Administration. Response to the request for comments and information on occupational exposure to ethylene dibromide.* Ethyl Corporation, May 1.

Wong, W., H. M. Utidjian, and V. S. Karten. 1978. Retrospective evaluation of reproductive performance of workers exposed to ethylene dibromide (EDB). *J. Occup. Med.* **21**:*98.*

COMMENTS

SIMMON: How long have workers been exposed to EDB?

TER HAAR: About 20 or 30 years. EDB has been manufactured for 50 years.

PLOTNICK: I'm not sure that I can agree with your conclusion that humans are less susceptible than animals. If EDB is a human carcinogen, the levels at which workers have been exposed have never gotten anywhere near, in most cases, the 20-ppm standard for very long periods of time.

TER HAAR: That is true, but we aren't talking about a 20-ppm standard. The NCI study (1978) is at 10 ppm, and it shows a very high incidence of nasal tumors. The data indicate that 1-5 ppm is not an unusual level of exposure.

WARD: The humans aren't constantly exposed like the animals are.

TER HAAR: The human exposures are averages for 8 hours per day.

WARD: So it averages 1 ppm or a little over that?

TER HAAR: Or more.

WARD: In the NCI inhalation EDB study, the mice may have responded like humans. There are almost no nasal tumors in the low-dose mice (10 ppm). In the rats, there are some nasal tumors. The mice have lung tumors induced at the lower dose, but they responded with very few nasal tumors.

PLOTNICK: Rats at 20 ppm had an excess of splenic angiosarcomas, a very rare tumor.

TER HAAR: Yes, and we would pick that up in the humans, I would expect, if it occurred.

CAPALDI: Do you have any idea what the exposure levels for humans might have been in 20 or 30 years?

TER HAAR: Well, our best guess is that the average level is not markedly different from 5 ppm, but with more excursions. We're very conscious today to keep the workers from excursions; for instance, a worker would cross a board over a tank filled with EDB. We used to send a worker across the tanks routinely to see if the tops were rusted out. I believe that the exposures under these conditions would have been high; they could easily have been 30 or 40 ppm.

HOOPER: Gary [Ter Haar], how many workers are at risk in these kinds of studies?

TER HAAR: The Dow study (Rausch 1978) had 450 workers total. The total number of workers worldwide, is probably less than 1000.

AMES: In this study, how many workers were studied who had been exposed for 15 or 20 years? And did you split them into smokers and nonsmokers? Smoking is such an important confounding variable.

TER HAAR: I think they did consider smoking.

HOOPER: Has sufficient time elapsed (15-20 years) since the end of the workers' 15-year exposure to overcome the long latency of some of these cancers? Can we expect to begin seeing cancers in this time period?

TER HAAR: The exposures began in the Octel work (Ter Haar 1978) around 1940. One Dow study goes back to 1920 and the other one to 1942.

AMES: When one looks at these epidemiological studies, the problem is that by the time you sort out the smokers and the nonsmokers and have people in the 15-20 year range, you'd end up with such a small n that even with more than doubling of risk you would probably never pick it up. That's the trouble, it is so hard to get some degree of confidence in negative epidemiological studies.

TER HAAR: I think it also depends on the kind of tumor. We'd pick up a nasal tumor immediately. I think we can absolutely say we've had no nasal tumors in the workers.

AMES: It may not be the same kind of tumor.

TER HAAR: Possibly so. And I think we'd have picked up the angiosarcomas if we had them in these workers. If it's a lung tumor, it might be more difficult.

INFANTE: I just want to say that I view these studies with sympathy—sympathy from the standpoint that it's very difficult to identify an occupational cohort with a large enough sample to follow over a long enough period to be able to evaluate carcinogenicity. These studies simply do not answer the question: Is EDB carcinogenic or isn't it? If you look at the two studies (see Table 5), total malignant neoplasms in one study is only 3.8. In fact we know that in males the highest site you should expect is the respiratory system, and, even if there were no effect at all, you haven't even identified one respiratory system cancer.

Now, if you want to talk about nasal cancer, you're talking about less than 2% of total respiratory system cancers, so if you ever identified one you'd probably have something like 100-fold risk in this population.

TER HAAR: You'd have a 100-fold risk, but you'd only have one nasal tumor.

INFANTE: That's right. And what would you say about that?

TER HAAR: I would say that that was evidence that it was carcinogenic. But we don't put 1000 workers at great risk.

INFANTE: But the point is that you cannot make a health evaluation on the basis of these studies; they're too insensitive because of the small numbers. If there are 400 or more workers, I wonder what the latency period is. The first 15 years tell you little about carcinogenic effect. Even if you did have an effect, you would have to look at it. And to look at total cancers of only 3.8 expected in one group and 1.2 expected in the other, simply does not tell you anything about cancer in this population. If, in fact, there were a positive effect. I could accept it, but, in fact, there is no effect at all and you can't say anything about it.

TER HAAR: I can say someting about it.

INFANTE: But you can't conclude that there is inconsistency between the animal and the human data because you have no human data upon which to make the evaluation.

TER HAAR: EDB is far less potent in humans than it is in animals.

INFANTE: But you don't know that.

TER HAAR: I think the data teach us that.

AMES: I'm not sure you can say it's far less potent until one finally analyzes the thoroughness of the test.

TER HAAR: Well, Dow has been through the math and has reached the conclusion that it's far less potent.

GOLD: Are there plans to follow up these workers and examine mortality figures?

TER HAAR: Many of these workers are retired so I expect a large portion of this population can be followed because the exposures of these people were for long periods of time. We have one starting in 1920 and two in the '40s. These are not our workers. I don't know what Dow plans to do with them, but I would assume they're going to follow them. I would suppose the same for Octel.

GOLD: In the fertility study it would be more accurate to compare workers' births to regional rates rather than to national rates. People in the southwest-central region, where the study plants are located, tend to have more children than the national average. Also, national rates are based on unmarried and married women. Both of these factors would tend to underestimate the number of expected births.

SPREAFICO: Did you have a greater abortion rate as a percentage of the population in Houston?

TER HAAR: We did not ask the workers that question. It is a sensitive question, and unless you have a really driving reason to know this information, a worker resents that type of question, just like he resents having sperm counts taken.

SPREAFICO: You didn't go through hospital records, did you?

TER HAAR: No. If you wanted to really look at that question, each person would have to be interviewed by a doctor and have a life history taken on him and his family. If you really went into detail on a questionnaire I think you could get the information you wanted without offending the workers, but that was not done in this particular case.

WARD: Don't you perform detailed physical exams and recordkeeping for these people?

TER HAAR: Yes. But we do not ask them if their wife had an abortion.

FABRICANT: Do you have any plans to do any cytogenetics on these workers?

TER HAAR: No. We haven't considered it.

INFANTE: With dibromochloropropane (DBCP; 1,2-dibromo-3-chloropropane) the XYY test was positive. I'm wondering, if you're able to obtain the sperm samples from these individuals, whether you might not consider doing that kind of analysis for these workers.

TER HAAR: NIOSH looked at these samples in some detail from the Houston population.

INFANTE: But they didn't do chromosomal Y nondisjunction in sperm.

TER HAAR: That might be true.

INFANTE: In your sample (Table 1), how long were these 59 individuals exposed, and were they still exposed at the time these sperm samples were taken?

TER HAAR: They were still exposed at the time the sperm samples were taken. All of them would have been exposed for at least 1 year, and probably not many of them more than 5 years.

FABRICANT: Has dominant lethal testing been done?

TER HAAR: Yes.

HOOPER: You argue that the potency of EDB in rats is much greater than that in people, that, in effect, people are less senistive to EDB than rats. I'm not sure that is true, and the following example may help to illustrate this. If workers are exposed to time-weighted average of 1 ppm, they have a daily exposure of 1.3 mg/(kg · day) (7 mg/m^3 × 15 m^3/day ÷ 70 kg). This is close to the calculated dose of EDB to give half the rats and mice cancer if given daily over their lifetimes. If we assume for the sake of argument that workers are like rats and mice, we would expect half of them to get cancer from a lifetime exposure at this dose level. If workers were exposed for only one-fifth or one-half of their lifetime, the risks would be lower, but still considerable. The important point is not the exact increase in risk from this exposure, but that epidemiology requires a large increase in risk before it can be considered significant, usually in sample sizes much larger than 450 people. Thus, even if workers were exposed for some fraction of their lifetime to dose levels that would give half the animals cancer, I'm not sure you would detect the effect in this study. Thus, EDB could be as effective in humans as in rodents in producing cancer, and yet we would not see its cancer effect in this study using this simple risk calculation.

WARD: Are the workers actually exposed, or is EDB just in air today? The 1 ppm EDB is in the air, but are they breathing it in? Do they wear masks or other types of gear to prevent their inhalation of EDB?

TER HAAR: Most of the levels are less than 1 ppm, regardless of whether they've got the respirators on or not. We let the pump run all day if we're making a measurement of the person to determine what his exposure is. We don't turn off the pump when he's got the respirator on. If he does a certain job where his instructions call for respiratory protection, for instance cleaning out a tower, the pump would still run to sample his atmosphere while he was doing that particular work. Therefore the levels that our people are exposed to on an average for 8 hours have not exceeded 4.5 ppm in the air. The actual exposure these people received was considerably less than that, because there are times when they would have had their respirators on. The levels in the plant would on the average be of the order of .1 ppm or less.

INFANTE: This Dow study (Rausch 1978) is comparable to doing a carcinogenesis bioassay study with one animal. You simply don't have the population followed for a long enough period. You can see that from the expected cases. I don't know how you can conclude that there is a difference in carcinogenic risk in animals and humans because I don't know how you can extrapolate quantitatively from animals to men. You simply don't have the methodology to do that yet. I realize that people do it, but we're talking about science, and I think there is simply not a scientific means to do quantitative risk assessment with the methodology available today, either from epidemiologic studies extrapolated to the universe or from animal studies.

GOLD: Do you have information about the exposure levels of fumigation workers?

TER HAAR: The exposure levels for the fumigation workers ranged from nondetectable to about 0.08 ppm for typical exposures. Under misuse conditions, levels as high as .5 ppm were detected. If the worker takes some equipment apart and doesn't wear respiratory protection during that time, he can get 20–30 ppm for short periods of time.

AMES: Of all the workers exposed in the past, the ones who might have gotten the highest dose could be the chemical workers.

TER HAAR: I think that's right.

AMES: In any case, now they're getting much lower doses and people are much more aware of it.

TER HAAR: I believe the average exposures are probably about one-fifth of what they were 6 or 7 years ago. The big difference is that peak exposures are avoided. For example, we are very careful about sending people into areas where their eyes burn unless they have respiratory protection.

GOLD: Do you or anyone else have any data on ethylene dichloride (EDC; 1,2-dichloroethane) prevailing exposure levels?

JACOBS: I don't have anything on exposure during production of EDC.

TER HAAR: Levels are somewhat higher than they are for EDB, about 5 ppm, similar to what prevailed before we really tightened exposure to EDB.

JACOBS: Just for comparison, the vapor pressure of EDC at 30°C is about 100 mm whereas the vapor pressure of EDB is 17.4 mm at 30°C.

TER HAAR: Our levels of exposure to EDC are somewhat higher than EDB. The worker doesn't like EDB, so it's not as hard to convince him to do certain things to protect himself from EDB as it is from EDC. There's frequently difficulty in convincing people to do the right things, but it's less difficult when the chemical will acutely hurt you.

PLOTNICK: I'd feel more comfortable scientifically with a positive result in the epidemiologic study of workers, even though the negative number is small. The least that we can say here is that there hasn't been anything overt that has shown up. However the study population is very small, and when you've got a risk of 1%, a 1% difference in a couple of workers isn't going to show up.

INFANTE: The point is the EDB study is not a negative study, it's a no-decision study. When you find two populations that have no lung cancer in men, that should be some indication, if you know nothing about methodology at all, that perhaps this population hasn't been followed very long or that it's a small sample size.

In doing these studies, you have to set them up to try to estimate, say, for the excessive risk that you might expect—what is power in the text? For example, what is the 80% chance of finding an effect if, in fact, there is one there? And I think you would find that there simply is not the sample size requirement, without even looking at any of the other methodology.

BUSEY: The same kind of argument can be used against the NCI study (1978), because the incidence of spontaneous neoplasms in the control was considerably lower in some instances than what would be expected. The same argument can be used there. Where do the pooled controls come from? Was the same pathologist used?

HOOPER: The pooled control rats were of the same strain, were housed in the same room, concurrently for at least 1 year, and were diagnosed by the same pathologist. But the important result is that tumor incidences were significant (Fisher exact, $p < 0.05$) whether pooled or matched controls were used. Thus, for example, even though the incidence of mammary fibroadenomas or adenocarcinomas in female Osborne-Mendel rats is quite low (0/20), using pooled control incidences (5/59) still produced a significant response when compared to an incidence of 24/50 in the dosed group.

BUSEY: Yes, and that makes the study a little bit suspect, using the same argument that Dr. Infante made.

HOOPER: Yes.

AMES: Epidemiologists have to make a more quantitative analysis of the limits of any study. Could it have detected the difference between cigarette smokers and nonsmokers. I think that's a key point. You can't always do an enormous epidemiological study because maybe there are only a couple of hundred men who have been exposed to the chemical. So, by making it more quantitative, we can start putting limits and determining the power of the test.

INFANTE: That's right and I know the difficulties, but you have to have something there to work with. When I was at NIOSH I tried for a year-and-a-half to find a cohort that was exposed to trichloroethylene and also had adequate records, long enough latency, and a large enough sample size. We simply could not find an adequately sized population.

If you don't have the numbers, you might find a positive, but if nothing turns out positive what can you say? Maybe, given another sample size and with another 15 years' followup, we might be able to make some kind of an interpretation of the study.

KELLAM: What's the level of sensitivity of the EDB study, with what number of people?

INFANTE: Without knowing what numbers or expecteds are involved here, I would say that if you're talking about nasal cancer you probably couldn't have detected a 20-fold risk, if in fact one had been present, simply because the background rate of nasal cancer is so low. The study didn't even pick it up for the respiratory system, and didn't pick up the lung cancer in this population.

AMES: If it is a very rare cancer, then the risk to the worker, whether it's 20-fold or 200-fold, can be low. Lung cancer, which one would like to look at it, is hard to get at.

TER HAAR: When the cancer assessment was made, the conclusions were a phenomenal rate of cancer for EDB workers. There must be something wrong with that assessment if it doesn't begin to predict these kinds of numbers and it's off by orders of magnitude.

PLOTNICK: That was based upon that NCI oral gavage?

TER HAAR: True.

PLOTNICK: Yet their maximum tolerated dose (MTD) was 90%?

TER HAAR: Yes.

PLOTNICK: That's what they used as the basis for the assessment, but you can't start with a number that you can't rely on.

TER HAAR: Somebody must have thought they could rely on it because they did it.

MALTONI: If that figure is correct, it means that this worker is being exposed to a dose that is not very high. If there were a positive risk, it would be a very, very long time before the tumor would appear.

So you are comparing a group of animals that have been exposed to a rather heavy dose and you are trying to find a set of tumors within a reasonably short period. If we assume that 1 year of age in rats corresponds to about 30 years, then by extrapolation you have to wait up to an age of 70-75 years in man to compare a low dose of vinyl chloride. And you could only evaluate results if the number of people older than that age range is large enough to be studied.

HOOPER: Are there any studies underway for sperm abnormalities among EDB workers?

PLOTNICK: There was an evaluation study done at NIOSH on sperm count and morphology, and it is my understanding that the results were negative.

TER HAAR: That was the Houston chemicals group.

AMES: In a way, it's all somewhat academic now, because people are much more careful about it. But isn't the key issue that we are careful enough now, and what kind of threshold limit values should one allow in the future? The epidemiology doesn't have the power to tell us that very well. Do we go along with the rats and mice and try to do a calculation that allows for a safety factor, or whatever, and say, "Okay, that looks about right" or "That doesn't look about right." Are we willing to accept one case of cancer in 1000 or one case in 10,000? Where does one draw the line? You're not going to pick up those kinds of tumor rates by epidemiology, anyway. These kinds of questions are difficult.

INFANTE: We don't know what the tumor incidence for these substances in humans is yet. I'd like to know. It would be nice because then we could put these things out there and say, "Ah, this is more carcinogenic than that one, on the basis of whatever." It would make all of our jobs much simpler. But, not having the methodology to do it, I don't think we should be flying by the seat of our pants and making up numbers.

AMES: It depends partly on how much it would cost. If you can keep the workers below 1 ppm, how much would it cost to keep them below .5 ppm or .2 ppm? That's part of the question. I don't know if it's difficult to lower it much more.

PLOTNICK: Industry says that 1 ppm can be reached without any trouble and without much additional cost. Anything below that would be exceedingly expensive. That is industry's position. There was the same kind of thinking with vinyl chloride, and it didn't cost them as much as they had anticipated.

TER HAAR: I think it cost them a whole pile of money.

PLOTNICK: Yes, but the NIOSH recommendation for EDC has been dropped from 145 mg/m^3 to 1 mg/m^3. That's a 145-fold decrease, and that's what is disturbing industry.

TER HAAR: The worst part of that recommendation is that it is for 15 minutes. If it were for 8 hours, it could be handled.

PLOTNICK: Is that 0.13 ppm?

TER HAAR: The 15-minute limit is the greater problem. Controlling transients at 1 mg/m^3 for 15 minutes, is very expensive. But to have to sample all day long every 15 minutes and run those samples to make sure OSHA is not going to cite you some time during the day is absolutely impossible.

GOLD: Is there a great deal of EDB used for fumigation?

TER HAAR: It is used mostly in gasoline, but there are some uses for it in fumigation where it is particularly needed. For instance, by law some of the citrus fruits cannot go to Japan unless they've been fumigated with EDB.

JACOBS: I think 90% of production is used for gasoline.

PLOTNICK: EDB has also been picked up to replace dibromochloropropane (DBCP; 1,2-dibromo-3-chloropropane) in a number of fumigation products.

TER HAAR: Yes.

GOLD: I do think that is one of the key questions: Is there some less toxic substance that could replace EDB?

JACOBS: If tetraethyl lead disappears, the only use of EDB left is as a fumigant. The use of tetraethyl lead is decreasing and will essentially disappear from the U.S. market.

References

NCI (National Cancer Institute). 1978. *Bioassay of 1,2-dibromoethane for possible carcinogenicity.* Technical report series number 86. DHEW publication number (NIH) 78-1336. Government Printing Office, Washington, D.C.

Rausch, D. 1978. *Submission of Docket Officer, Docket H-111. Occupational Safety and Health Administration. Comments concerning a standard for occupational exposure to ethylene dibromide.* Dow Chemical USA, May 15.

Ter Haar, G. 1978. *Submission to Docket Officer, Docket H-111. Occupational Safety and Health Administration. Response to the request for comments and information on occupational exposure to ethylene dibromide.* Ethyl Corporation, May 1.

Carcinogenicity and Metabolism of Some Halogenated Olefinic and Aliphatic Hydrocarbons

BENJAMIN L. VAN DUUREN
Laboratory of Organic Chemistry and Carcinogenesis
Institute of Environmental Medicine
New York University Medical Center
New York, New York 10016

Since the discovery of the carcinogenicity of vinyl chloride (VC) (Viola et al. 1971; Maltoni and Lefemine 1975), there has been an extraordinary upsurge in research, not only regarding ethylene dichloride (EDC; 1,2-dichloroethane) but also a series of related saturated and unsaturated halogenated short-chain hydrocarbons. This was to be expected because of the vast amounts of these compounds manufactured and used in the United States and abroad. Their uses range from small drycleaning industries to the large automobile, aircraft, and food-processing industries. In addition, they are used for a variety of purposes in agriculture. This state of affairs has raised grave concerns about their environmental dispersion, environmental persistence, and, hence, potential hazards to the ecosystem, particularly their effects on human health. The greatest concerns have been human health hazards in the chemical manufacturing industries and those industries where they are used directly. It is in these settings that the highest exposure levels to humans are obviously expected.

The organic chemistry and physical properties of these compounds have been well documented in many organic chemistry textbooks. Thus, research over the past 7 years has focused on the pharmacokinetics and toxicology of these compounds with special emphasis on potential animal and human carcinogenicity.

This discussion deals with a few selected compounds, their carcinogenicity, mutagenicity, and mode of action. Extensive reviews dealing with saturated and unsaturated halogenated hydrocarbons have appeared recently (Fishbein 1979a, 1979b).

TRICHLOROETHYLENE: CARCINOGENICITY, MUTAGENICITY, AND MODE OF ACTION

Carcinogenicity

Because of the unique structural similarity of trichloroethylene (TCE) oxide to the α-chloroethers, some of which are potent direct acting alkylating carcinogens

(Van Duuren et al. 1968), it was suggested at a conference of the New York Academy of Sciences in 1974 that TCE is a potential carcinogen (Van Duuren 1975).

This suggestion was based on the likelihood that TCE would be metabolized to its epoxide and that this epoxide would be the activated carcinogenic intermediate. A similar epoxide intermediate was suggested for VC (Van Duuren 1975). These epoxide intermediates were, at the time, unknown compounds. VC was by then a well-known carcinogen (Maltoni and Lefemine 1975), but there was no information concerning the carcinogenicity of TCE. The suggestion that TCE is carcinogenic was based on its structure-chemical reactivity relationship to the potent carcinogen, *bis*(chloromethyl) ether (Van Duuren et al. 1969). TCE oxide, VC oxide, as well as *bis*(chloromethyl) ether, have a common structural feature, namely, one or two carbon atoms in each of the three compounds bear both an ether oxygen and a chlorine atom as shown in Figure 1. This implies that these epoxides are reactive alkylating agents, similar to *bis*(chloromethyl) ether. It should be noted, however, that they differ in hydrolysis mechanisms and products, and VC oxide rearranges spontaneously to chloroacetaldehyde (Van Duuren 1975).

Subsequently, a report appeared on the carcinogenicity testing of this compound under sponsorship of the National Cancer Institute (NCI) (Public Health Service 1976a). In their bioassay, TCE was fed at high doses to B6C3F1 hybrid mice and Osborne-Mendel rats of both sexes. Hepatocellular carcinoma was observed in both sexes of mice in significant incidences ($P<0.01$). In the Osborne-Mendel rats, the tumor incidences observed, when compared to no-treatment controls, could not be ascribed to TCE, that is, it was not carcinogenic in this strain of rat. This report (Public Health Service 1976a) was widely publicized and criticized (Henschler et al. 1977) for a variety of reasons. The reasons

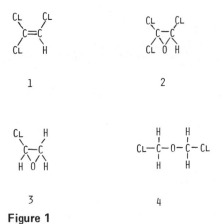

Figure 1
Structural similarity between the epoxides of TCE and VC and the carcinogen *bis*(chloromethyl) ether. (*1*) TCE; (*2*) TCE oxide; (*3*) VC oxide; and (*4*) *bis*(chloromethyl) ether.

most frequently cited were the high dosages used, the use of a liver tumor-susceptible strain of mice, and the lack of significant liver tumor incidences in Osborne-Mendel rats. Nevertheless, the report led some industries that were using TCE to discontinue its use and to use substitutes such as methylene dichloride, methylchloroform, and tetrachloroethylene (also known as perchloroethylene) (Price et al. 1978) on which carcinogenicity tests had not yet been reported by 1975, when the carcinogenicity results of the NCI were first made public (Public Health Service 1975). This development was unfortunate and costly. Tetrachloroethylene, for example, was subsequently shown to be hepatocarcinogenic by feeding in B6C3F1 mice, but not in Osborne-Mendel rats (Public Health Service 1977a).

In the mid-1970s, we undertook studies on the metabolism, carcinogenicity, and mode of action of a series of halogenated hydrocarbons of which TCE was one. Four bioassays were used to test TCE for carcinogenicity in ICR/Ha Swiss female mice. This is a random-bred strain of mice with a relatively low incidence of spontaneous tumors. The tests used were two-stage carcinogenesis on mouse skin, that is, a single application of TCE followed by repeated applications of the tumor promotor, phorbol myristate acetate (PMA; Chem. Abstr. Registry No. 16561-29-8; Van Duuren et al. 1978); repeated mouse skin application; once weekly subcutaneous injection; and once weekly intragastric intubation. The duration of the four tests ranged from 342 to 622 days and the median survival times in all tests were good, except in those tests where tumors developed. The complete details of these tests together with all or some of the same tests using other halogenated hydrocarbons were recently published (Van Duuren et al 1979). TCE had negative results in all four tests. The two-stage carcinogenesis and subcutaneous injection are, in our experience, particularly sensitive. The latter test is not regarded as a reliable indicator of carcinogenicity by some researchers (Clayson 1962), probably because of its sensitivity and confusion concerning the induction of sarcomas by inert solids and emulsions by subcutaneous implantation or injection, respectively. The two-stage carcinogenesis test proved valuable in indicating carcinogenicity for several other compounds that are potent carcinogens according to tests performed by other workers and by researchers in our own laboratory (see Table 1). In another laboratory, long-term feeding experiments with mice also were negative (C. Maltoni, pers. comm. and this volume).

TCE oxide, a possible carcinogenic intermediate, was tested by us as an initiating agent in two-stage carcinogenesis. This test was negative also (Van Duuren et al. 1979).

Mutagenicity

TCE has been reported to be mutagenic in *Escherichia coli* K-12 (Greim et al. 1975) in the presence of a microsomal activating system; however, the purity of the TCE used was not specified. The matter of the purity of TCE used in

Table 1
Initiation-Promotion for Screening Potential Carcinogens

Compound and dose (mg)	Two-stage carcinogenesis			Other bioassays	
	days on test	days to first tumor	mice with papillomas/ total papillomas	species and route	tumors
Allyl chloride (94)	505	197	7/10, $p < 0.025$ (Van Duuren et al. 1979)	mice, feeding	squamous carcinoma, forestomach (Public Health Service 1978b)[b]
DBCP (69)	499	257	6/6, $p < 0.05$ (Van Duuren et al. 1979)	mice, skin	squamous carcinoma, forestomach (Van Duuren et al. 1979)
				mice and rats, feeding	squamous carcinoma, forestomach (Public Health Service 1978a)[b]
VDC (121)	510	271	8/9 (1),[a] $p < 0.005$ (Van Duuren et al. 1979)	mice, inhalation	adenocarcinoma, kidney (Maltoni 1977)
ECH (2)	385	92	9/12 (1),[a] $p < 0.005$ (Van Duuren et al. 1974)	mice, s.c. injection	sarcomas, injection site (Van Duuren et al. 1974)
				rats, inhalation	carcinoma, nasal passages[c] (Public Health Service 1977b)

			Controls
PMA, 0.0025, 120 mice	342–468	141	9/10 (1)[a] (Van Duuren et al. 1979)
PMA, 0.0050, 90 mice	428–576	449	6/7 (1)[a] (Van Duuren et al. 1979)

Thirty female ICR/Ha Swiss mice per group were used, except where noted. Compounds applied once only in acetone followed by 2.5 μg PMA (for ECH initiation) or 5.0 μg PMA (for allyl chloride, DBCP, VDC initiation) in acetone, three times weekly for the duration of the test. Median survival times were greater than the duration of the test in all cases. All four compounds were inactive by repeated skin application in mice.

[a] Number of animals with squamous carcinoma of the skin.
[b] See text.
[c] Epidemiologic evidence indicates that ECH is a human carcinogen (Enterline 1980).

mutagenicity and carcinogenicity studies is described below, since this matter has complicated the interpretation of mutagenicity as well as earlier carcinogenicity studies.

Transforming Ability of TCE and Analogs

A recent report (Price et al. 1978) described the transformation of Fischer rat embryo cells (F1706) in culture by TCE, 1,1,1-trichloroethane, tetrachloroethylene, and methylene dichloride. The latter three compounds have been proposed and used as industrial replacements for TCE. In this assay, all four compounds caused transformation; however, the purity of the chemicals used was not specified.

Mode of Action of TCE

Several reports have appeared describing the metabolism of TCE in animals and man. Trichloroacetic acid (TCA) and trichloroethanol or glucuronides have been found in the urine of humans and dogs exposed to TCE by inhalation (Powell 1945). An epoxide was proposed as an intermediate by the same author. In a subsequent study (Daniel 1963), rats were fed [^{36}Cl] TCE and the metabolites were isolated from the urine. Both TCA and trichloroethanol were found. The specific activities of the two metabolites isolated were approximately equal to that of the administered TCE. This result suggested that there is a pathway via trichloroacetaldehyde, as suggested by Powell (1945), and that no exchange of chloride with the body chloride pool occurred. Daniel (1963) also found that trichloroethanol is not the precursor of TCA, that is, these two metabolites are formed via independent pathways. Leibman (1965) showed that TCE is converted in vitro to trichloroacetaldehyde (chloral) in liver microsomes of rats, rabbits, and dogs. The same author also examined the effects of activators and inhibitors of the enzyme system. Byington and Leibman (1965) presented similar evidence concerning the metabolism of TCE in rat liver. Other possible intermediates have been proposed, for example, a chloronium ion intermediate (Van Dyke and Chenoweth 1965).

Several metabolic pathways have been suggested for TCE (Bonse and Henschler 1976; Banerjee and Van Duuren 1978); the exact pathways are yet to be determined. The known metabolic products and possible intermediates are given in Figure 2.

The known formation of chloral from TCE in vitro is of interest, since chloral, under basic conditions, is known to undergo cleavage of the C-C linkage (Johnson et al. 1951) to yield chloroform, which is a liver carcinogen in B6C3F1 mice (Public Health Service 1976b; Reuber 1979). However, it is not known whether this reaction occurs in the microsomal system in vitro or in vivo.

Because of the electron-withdrawing effect on the chlorine atoms, chloral behaves as an electrophile and is known to react with a number of nucleophiles;

KNOWN METABOLITES

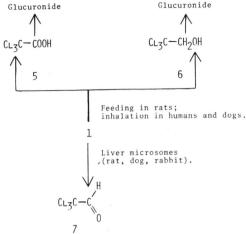

Figure 2
Known metabolites and proposed intermediates in the metabolism of TCE, structure *1* (see text). The three known metabolites are TCA, structure *5*; trichloroethanol, structure *6*; and trichloroacetaldehyde, structure *7*. The proposed intermediates are TCE oxide, structure *2*; 1,1,2-trichloro-1,2-dihydroxyethane, structure *8*; and the cyclic chloronium ion of TCE, structure *9*. Two of the proposed intermediates, *8* and *9*, are shown in brackets because they are expected to be transitory intermediates only.

with water it yields chloral hydrate [$Cl_3C-CH(OH)_2$], and it reacts also with bisulfites, ammonia, hydroxylamine, and formamide (Johnson et al. 1951).

TCE oxide was synthesized and characterized in our laboratory (Kline and Van Duuren 1977). Its rate of hydrolysis, together with that of a series of related epoxides, was also determined. TCE oxide hydrolyzes rapidly in phosphate buffer at pH 7.4 with a half-life of 1.3 minutes (Kline et al. 1978) to dichloroacetic acid (McKinney et al. 1955). In our work (Kline and Van Duuren 1977), the reactions of TCE with a series of nucleophiles (e.g., 2-mercaptobenzimidazole, *p*-nitrothiophenol, and 3,4-dichlorothiophenol) were examined and the reaction products were characterized. These derivatives were synthesized and characterized with the aim of using some of them as trapping agents, in vivo and in vitro, for epoxide intermediates of unsaturated halogenated hydrocarbons.

On the basis of the above studies, the in vitro covalent binding of [^{14}C]TCE to microsomal proteins and exogenous DNA from mouse and rat liver were determined (Van Duuren and Banerjee 1976; Banerjee and Van Duuren 1978). These studies also indicated that the in vitro covalent binding of TCE to liver microsomal protein is significantly higher for B6C3F1 hybrid mice than for Osborne-Mendel rats. In addition, microsomal protein from male B6C3F1 hybrid mice binds more TCE than does microsomal protein of female mice of the same strain. These findings correlate well with carcinogenicity results of the NCI (Public Health Service 1976a). The irreversible binding of [^{14}C]TCE to mouse liver protein in vivo and in vitro has also been recently reported (Uehleke and Poplawski-Tabarelli 1977).

Purity of TCE Used in Biological Studies

The major industrial process for the manufacture of TCE is by the chlorination and subsequent dehydrochlorination of EDC. For some uses of TCE, stabilizers are added. It has been shown that one of the impurities in TCE is epichlorohydrin (ECH; 1-chloro-2,3-epoxypropane), a known carcinogen (Van Duuren et al. 1974; Van Duuren 1980), which is present at a level of 0.22% in some commercial preparations (Henschler et al. 1977). This prompted the authors to suggest that the carcinogenicity of TCE (Public Health Service 1976a) and mutagenicity of TCE (Greim et al. 1975) are probably due to ECH or other impurities. However, earlier work in this laboratory has shown that some epoxides are not carcinogenic in limited feeding experiments (Van Duuren et al. 1966). The lack of carcinogenicity of epoxides is ascribed to their rapid rate of hydrolysis at pH < 7.0, which is likely to be encountered in the stomach. For this reason, ECH cannot be held responsible for the reported carcinogenicity of commercial grade, >99.0%, TCE (Public Health Service 1976a). On the other hand, the reported mutagenicity of TCE (Henschler et al. 1977) may well be due to ECH.

TCE Carcinogenicity and Metabolic Overload

The chemical structure, known metabolites, possible activated intermediates, and in vitro covalent binding to tissue macromolecules, described above, tend to support the NCI finding of the carcinogenicity of TCE. The reported mutagenicity and cell culture transforming ability of TCE are unclear because small amounts of impurities such as ECH may be responsible for the latter two biological effects.

The negative carcinogenicity findings summarized in this paper and described in detail elsewhere (Van Duuren et al. 1979), as well as the negative findings of C. Maltoni (this volume), suggest that other factors may have played a role in the NCI feeding studies.

It has recently been shown (Watanabe et al. 1977) that metabolic pathways, protein binding, etc., are at times different for low and for high doses. They performed intravenous injection (i.v.) experiments with dioxane at various dosages and found that the metabolism of dioxane to its detoxification product, β-hydroxyethoxyacetic acid, is a saturable process and that the metabolism of dioxane is markedly induced with increasing dose. This phenomenon can be referred to as metabolic overload.

Laib et al. (1979) compared the binding of [^{14}C]VC and [^{14}C]TCE to Wistar rat liver microsomes, NADPH, and yeast RNA. They found notable differences in the extent of protein and RNA covalent binding for the two compounds. These data and other experiments reported in the same paper suggest distinct differences in the liver carcinogenicity of VC and the possible effects of TCE in the liver.

In another study (Pessayre et al. 1979), the activation, inhibition, and destruction of drug-metabolizing enzymes by TCE was explored using in vitro incubation of [^{14}C]TCE in liver microsomal preparations derived from Sprague-Dawley rats. From their study, the authors concluded that TCE is unique in its capability to inhibit, activate, destroy, or simultaneously destroy and induce microsomal enzymes.

Inhalation studies with 1,1-dichloroethylene (also known as vinylidene chloride, VDC) (Andersen et al. 1979) have shown a saturable metabolism similar to that described by Watanabe et al. (1977) for dioxane, which was referred to earlier. Andersen and coworkers stressed once again that in the past little attention has been given to the implications involved when saturable enzymatic activation regulates toxicity.

All the above studies serve only to complicate an interpretation of the conflicting carcinogenicity data, which will probably be aired in the discussion of this paper.

ALLYL CHLORIDE AND 1,4-DICHLOROBUTENE-2

Allyl chloride (3-chloro-1-propene) and 1,4-dichlorobutene-2 merit special attention because they are unsaturated chlorinated hydrocarbons that can conceivably be direct-acting carcinogens, as shown in Figure 3, because of their allylic halide structures.

Allyl chloride was tested in the NCI bioassay program using B6C3F1 mice and Osborne-Mendel rats of both sexes (Public Health Service 1978b). The results suggested that the compound causes papillomas and carcinomas of the forestomach in male and female B6C3F1 mice but not in rats of either

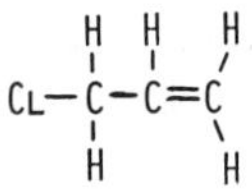

$$\text{Cl}-\underset{\underset{H}{|}}{\overset{\overset{H}{|}}{C}}-\overset{H}{C}=\underset{\diagdown}{\overset{\diagup}{C}}\underset{H}{\overset{H}{}}$$

10

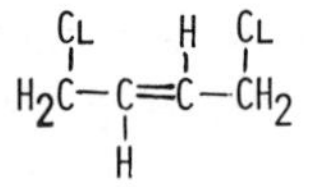

$$\underset{\underset{H}{|}}{\overset{\overset{Cl}{|}}{H_2C}}-\overset{H}{C}=\overset{Cl}{\underset{|}{C}}-CH_2$$

11

$$H_2\underset{\underset{Br}{|}}{C}-\underset{\underset{Br}{|}}{\overset{\overset{H}{|}}{C}}-\underset{\underset{Cl}{|}}{C}H_2$$

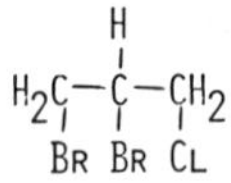

12

Figure 3
Other halogenated hydrocarbons: allyl chloride (*10*); 1,4-dichlorobutene-2 (*11*); and DBCP (*12*).

sex. The conclusion, however, was that the bioassay was inadequate. In our bioassay (Van Duuren et al. 1979), allyl chloride gave a statistically significant incidence of tumors ($P < 0.025$) in initiation-promotion experiments (Table 1) but was not tumorigenic by repeated skin application or subcutaneous injection. We have suggested earlier (Van Duuren 1977) that this compound may be metabolized via the epoxides ECH and glycidaldehyde, both of which are carcinogens. Nevertheless, the carcinogenicity of this compound has yet to be firmly established.

1,4-Dichlorobutene-2 has been shown to induce a low incidence of sarcomas by subcutaneous injection (Van Duuren et al. 1975). It has more recently been shown to induce a statistically significant incidence of malignant nasal tumors in rats by inhalation at a concentration of 5.0 ppm (B. C. McKusick, pers. comm. 1978). The mode of action of 1,4-dichlorobutene-2 remains to be determined.

DIBROMOCHLOROPROPANE

Dibromochloropropane (DBCP; 1,2-dibromo-3-chloropropane) (compound 12, Fig. 3) has been widely used for many years, mainly as a nematocide for many agricultural crops. It has been shown to be carcinogenic in B6C3F1 mice and Osborne-Mendel rats of both sexes, causing statistically significant incidences of squamous carcinoma of the stomach in mice and rats and mammary adenocarcinomas in female rats (Olson et al. 1973; Public Health Service 1978a). In these studies, technical grade DBCP, >90% pure, was used; it contained minor impurities as determined by gas-liquid chromatography (GLC).

The question of the purity of the DBCP used in these studies, as in the case of TCE, again raised the question of the validity of the NCI tests, that is, the role played by impurities. This came about when the Environmental Protection Agency (EPA) became concerned about the environmental dispersion of DBCP in surface and well waters as a result of its use in agriculture.

In our recent carcinogenicity study with DBCP and other halogenated hydrocarbons (Van Duuren et al. 1979), technical grade DBCP (Dow Chemical Company, Fumazone-F, lot no. MO-3196) was redistilled (boiling point 86-89°C/12mm). The technical grade material was shown by gas chromatography (GC) to be 96.2 mole percent pure with 2.8 mole percent allyl chloride and 1.0 mole percent ECH as impurities. The redistilled material, used in our bioassays, was analyzed by GC under conditions similar to those used for the technical grade and was found to have the following mole percent composition: DBCP, 99.88; allyl chloride, 0.06; and ECH, 0.06 (Solomon and Van Duuren, unpubl. data, 1979).

The compound was tested by us in random-bred ICR/Ha Swiss female mice using 30 animals per group. Three dorsal skin tests were performed: a single application of DBCP (69.0 mg/0.2 ml acetone) followed by repeated application three times weekly, of the tumor promoter, PMA, at the same site; DBCP was applied at 35 mg/0.2 ml acetone, three times weekly; and DBCP was applied at two doses, 35.0 mg and 11.7 mg/0.2 ml acetone, three times weekly. Skin application was maintained for the duration of the experiments, which was 427-590 days.

The initiation-promotion experiment resulted in six mice having one papilloma each and no malignant tumors. In the repeated skin application tests, the higher dose (35.0 mg) resulted in one animal with a skin papilloma. However, in both repeated skin application tests, significant incidences of lung and stomach tumors were observed, based on the calculated P values. At the higher dose (35.0 mg), ten of thirty mice had squamous carcinomas of the forestomach and at the lower dose (11.7 mg), five of thirty mice developed carcinomas of the forestomach. In both groups, there were also significantly higher incidences of benign tumors of the lung compared to controls. Full details of these experiments appear in the recent paper by Van Duuren et al. (1979), referred to earlier.

The induction of stomach and lung tumors by DBCP following skin application must be ascribed to skin absorption and also to animal grooming that results in ingestion. Skin absorption studies (Kodama and Dunlap 1956) showed that DBCP is indeed systemically absorbed by rabbit skin. DBCP was recently banned by EPA for most agricultural uses.

SUMMARY

A major outcome of these proceedings has been to draw attention to the rapid growth of information on the pharmacokinetics, toxicology, chemistry, and other aspects of EDC and related compounds. Equally important, these presentations and discussions raise major and difficult questions concerning the available information. This is not only of pure scientific interest, but it is also of great importance to the public at large, federal regulatory agencies, and industry. It is hoped that this discussion has contributed in its own small way to this important area of research.

ACKNOWLEDGMENTS

The author is indebted to his coworkers whose names appear in articles from our laboratory referenced in this paper. The work reported here was supported by National Science Foundation grant PFR76-10656 and U.S. Public Health Service grants ES-01150, ES-00260, and CA-13343.

REFERENCES

Andersen, M. E., J. E. French, M. L. Gargas, R. A. Jones, and L. J. Jenkins, Jr. 1979. Saturable metabolism and the acute toxicity of 1,1-dichloroethylene. *Toxicol. Appl. Pharmacol.* **47**:385.

Banerjee, S. and B. L. Van Duuren. 1978. Covalent binding of the carcinogen trichloroethylene to hepatic microsomal proteins and to exogenous DNA in vitro. *Cancer Res.* **38**:776.

Bonse, G. and D. Henschler. 1976. Chemical reactivity, biotransformation, and toxicity of polychlorinated aliphatic compounds. *CRC Crit. Rev. Toxicol.* **4**:395.

Byington, K. H. and K. C. Leibman. 1965. Metabolism of trichloroethylene in liver microsomes. II Identification of the reaction product as chloral hydrate. *Mol. Pharmacol.* **1**:247.

Clayson, D. B. 1962. Testing chemicals for carcinogenic activity. I. Methods. In *Chemical carcinogenesis*, p. 65. Little, Brown and Co., Boston.

Daniel, J. W. 1963. The metabolism of ^{36}Cl-labelled trichloroethylene and tetrachloroethylene in the rat. *Biochem. Pharmacol.* **12**:795.

Enterline, P. E. 1980. *J. Occup. Med.* (in press).

Fishbein, L. 1979a. Potential halogenated industrial carcinogenic and mutagenic chemicals. I. Halogenated unsaturated hydrocarbons. *Sci. Total Environ.* **11**:111.

―――――. 1979b. Potential halogenated industrial carcinogenic and mutagenic chemicals. II. Halogenated saturated hydrocarbons. *Sci. Total Environ.* **11**:163.

Greim, H., G. Bonse, Z. Radwan, D. Reichert, and D. Henschler. 1975. Mutagenicity in vitro and potential carcinogenicity of chlorinated ethylenes as a function of metabolic oxirane formation. *Biochem. Pharmacol.* **24**:2013.

Henschler, D., E. Eder, T. Neudecker, and M. Metzler. 1977. Carcinogenicity of trichloroethylene: Fact or artifact? *Arch. Toxicol.* **37**:233.

Johnson, A. W., J. G. Buchanan, D. T. Elmore, W. E. Harvey, and J. Walker. 1951. Aldehydes and ketones. In *Chemistry of carbon compounds* (ed. E. H. Rodd), vol. IA, p. 478. Elsevier North-Holland, Inc., Amsterdam.

Kline, S. A. and B. L. Van Duuren. 1977. Reactions of epoxy-1,1,2-trichloroethane with nucleophiles. *J. Heterocycl. Chem.* **14**:455.

Kline, S. A., J. J. Solomon, and B. L. Van Duuren. 1978. Synthesis and reactions of chloroalkene epoxides. *J. Org. Chem.* **43**:3596.

Kodama, J. K. and M. K. Dunlap. 1956. Toxicity of 1,2-dibromo-3-chloropropane. *Fed. Proc.* **15**:448.

Laib, R. J., G. Stöckle, H. M. Bolt, and W. Kunz. 1979. Vinyl chloride and trichloroethylene: Comparison of alkylating effects of metabolites and induction of preneoplastic enzyme deficiencies in rat liver. *J. Cancer Res. Clin. Oncol.* **94**:139.

Leibman, K. C. 1965. Metabolism of trichloroethylene in liver microsomes. I. Characteristics of the reaction. *Mol. Pharmacol.* **1**:239.

Maltoni, C. 1977. Recent findings on the carcinogenicity of chlorinated olefins. *Environ. Health Perspect.* **21**:1.

Maltoni, C. and G. Lefemine. 1975. Carcinogenicity bioassays of vinyl chloride: Current results. *Ann. N.Y. Acad. Sci.* **246**:195.

McKinney, L. L., E. H. Uhing, J. L. White, and J. C. Picken Jr. 1955. Autoxidation of trichloroethylene. *J. Agric. Food Chem.* **3**:413.

Olson, W. A., R. T. Habermann, E. K. Weisburger, J. W. Ward, and J. H. Weisburger. 1973. Brief communication: Induction of stomach cancer in rats and mice by halogenated aliphatic fumigants. *J. Natl. Cancer Inst.* **51**:1993.

Pessayre, D., H. Allemand, J. C. Wandscheer, V. Descatoire, J.-Y. Artigou, and J.-P. Benhamou. 1979. Inhibition, activation, destruction, and induction of drug-metabolizing enzymes by trichloroethylene. *Toxicol. Appl. Pharmacol.* **49**:355.

Powell, J. F. 1945. Trichloroethylene: Absorption, elimination and metabolism. *Brit. J. Ind. Med.* **2**:142.

Price, P. J., C. M. Hassett, and J. I. Mansfield. 1978. Transforming activities of trichloroethylene and proposed industrial alternatives. *In Vitro.* **14**:290.

Public Health Service. 1975. *Memorandum of alert: Carcinogenesis bioassay of trichloroethylene.* National Cancer Institute, Washington, D.C.

——————. 1976a. *Bioassay of trichloroethylene for possible carcinogenicity.* NCI-CG-TR-2. DHEW publication number (NIH) 76-802. Government Printing Office, Washington, D.C.

——————. 1976b. *Report on the carcinogenesis bioassay of chloroform.* National Cancer Institute Technical Report. Government Printing Office, Washington, D.C.

——————. 1977a. *Bioassay of tetrachloroethylene for possible carcinogenicity.* NCI-CG-TR-13. DHEW publication number (NIH) 77-813. Government Printing Office, Washington, D.C.

——————. 1977b. *Epichlorohydrin: Nasal cancer by inhalation in rats.* Memorandum from Center for Disease Control, Government Printing Office, Washington, D.C.

——————. 1978a. *Bioassay of dibromochloropropane for possible carcinogenicity.* NCI-CG-TR-28. DHEW publication number (NIH) 78-828. Government Printing Office, Washington, D.C.

——————. 1978b. *Bioassay of allyl chloride for possible carcinogenicity.* NCI-CG-TR-73. DHEW publication number (NIH) 78-1323. Government Printing Office, Washington, D.C.

Reuber, M. D. 1979. Carcinogenicity of chloroform. *Environ. Health Perspect.* **31**:171.

Uehleke, H. and S. Poplawski-Tabarelli. 1977. Irreversible binding of ^{14}C-labeled trichloroethylene to mice liver constituents in vivo and in vitro. *Arch. Toxicol.* **37**:289.

Van Duuren, B. L. 1975. On the possible mechanism of carcinogenic action of vinyl chloride. *Ann. N.Y. Acad Sci.* **246**:258.

―――――. 1977. Chemical structure reactivity and carcinogenicity of halohydrocarbons. *Environ. Health. Perspect.* **21**:17.

―――――. 1980. Prediction of carcinogenicity based on structure, chemical reactivity and possible metabolic pathways. *J. Environ. Pathol. Toxicol.* (in press).

Van Duuren, B. L. and S. Banerjee. 1976. Covalent interaction of metabolites of the carcinogen trichloroethylene in rat hepatic microsomes. *Cancer Res.* **36**:2419.

Van Duuren, B. L., B. M. Goldschmidt, and I. Seidman. 1975. Carcinogenic activity of di- and trifunctional α-chloro ethers and of 1,4-dichlorobutene-2 in ICR/Ha Swiss mice. *Cancer Res.* **35**:2553.

Van Duuren, B. L., G. Witz, and B. M. Goldschmidt. 1978. Structure-activity relationships of tumor promotors and cocarcinogens and interaction of phorbol myristate acetate and related esters with plasma membranes. In *Carcinogenesis, mechanisms of tumor promotion and cocarcinogenesis* (ed. T. J. Slaga, A. Sivak, and R. K. Boutwell), vol. 2, p. 491. Raven Press, New York.

Van Duuren, B. L., A. Sivak, B. M. Goldschmidt, C. Katz, and S. Melchionne. 1969. Carcinogenicity of halo-ethers. *J. Natl. Cancer Inst.* **43**:481.

Van Duuren, B. L., B. M. Goldschmidt, C. Katz, I. Seidman, and J. S. Paul. 1974. Carcinogenic activity of alkylating agents. *J. Natl. Cancer Inst.* **53**:695.

Van Duuren, B. L., L. Langseth, L. Orris, G. Teebor, N. Nelson, and M. Kuschner. 1966. Carcinogenicity of epoxides, lactones, and peroxy compounds. IV. Tumor response in epithelial and connective tissue in mice and rats. *J. Natl. Cancer Inst.* **37**:825.

Van Duuren, B. L., B. M. Goldschmidt, C. Katz, L. Langseth, G. Mercado, and A. Sivak. 1968. Alpha-haloethers. A new type of alkylating carcinogen. *Arch. Environ. Health* **16**:472.

Van Duuren, B. L., B. M. Goldschmidt, G. Loewengart, A. C. Smith, S. Melchionne, I. Seidman, and D. Roth. 1979. Carcinogenicity of halogenated olefinic and aliphatic hydrocarbons in mice. *J. Natl. Cancer Inst.* **63**:1433.

Van Dyke, R. A. and M. B. Chenoweth. 1965. Metabolism of volatile anesthetics. *Anesthesiology* **26**:348.

Viola, P. L., A. Bigotti, and A. Caputo. 1971. Oncogenic response of rat skin, lungs, and bones to vinyl chloride. *Cancer Res.* **31**:516.

Watanabe, P. G., J. D. Young, and P. J. Gehring. 1977. The importance of nonlinear (dose-dependent) pharmacokinetics in hazard assessment. *J. Envir. Pathol. Toxicol.* **1**:147.

COMMENTS

TER HAAR: I was wondering, Dr. Van Duuren, if you could comment on glycidol. It seemed like it was so similar to the other compounds you were looking at. Do you have any thoughts why that's inactive?

VAN DUUREN: I don't see any reason why it should have any carcinogenic activity. It's very closely related to glycerin in its chemical structure.

TER HAAR: The epoxide doesn't bother you?

VAN DUUREN: The epoxide function? Not it doesn't bother me. We have looked at 50 or more epoxides—monofunctional versus difunctional—the monofunctional epoxides have been always inactive in our bioassay test systems, whereas, the bifunctional epoxides which have molecular flexibility are carcinogenic.

PLOTNICK: You said you were presently testing by skin painting bromoacetaldehyde (2-bromoacetal). Is that correct?

VAN DUUREN: By drinking water exposure in mice.

PLOTNICK: How do you stabilize it, or are you getting it as an acetaldehyde? The material is highly unstable, and I don't think it would exist in the bottle long enough to be painted on the skin. Do you synthesize it yourself?

VAN DUUREN: Yes. It's an extremely time-consuming effort and it is stabilized in solution.

PLOTNICK: Yes. Well, also, skin painting may not show up anything because of the fact that it's going to react almost instantaneously.

VAN DUUREN: That does not bother me at all.

PLOTNICK: It's going to stay right where it is applied.

VAN DUUREN: That does not trouble me at all. Chloromethyl ether and *bis*(chloromethyl) ether are extremely reactive compounds and decompose instantaneously in water. For example, you cannot measure the rate of hydrolysis of chloromethyl ether. By inhalation testing it was a lung carcinogen.

So, whether it's one molecule or 100, it reaches the site. It's immaterial when you're doing it with a potent carcinogen.

INFANTE: In your studies with TCE, were all of these skin applications?

VAN DUUREN: No. First we did skin application. Second, we repeated the skin application in an initiation-promotion sequence—that's a single application of the TCE followed by repeated application of the promoter, which is a useful test system, as I showed you, for allyl chloride, and VDC and DBCP. The third was feeding, and the fourth was s.c. injection. And they were all inactive.

INFANTE: I think there was a second study also, wasn't there, on TCE that shows it's carcinogenic? I wonder if anyone knows about that.

MALTONI: Just a comment to Dr. Van Duuren. National Cancer Institute (NCI 1976) data showed that in one single strain of mice, the usual one used for the NCI tests, extremely massive doses of TCE were shown to produce hepatomas in the mice. Thereafter, there were, I think three kinds of experiments—those by Dr. Van Duuren, one sponsored by industry which was not finished, and our experiments, which had results in line with the results of Dr. Van Duuren. We treated for 52 weeks, five times weekly with .5g/kg body weight. We didn't get any type of carcinogenic effect.

It was found that probably the TCE used by NCI was not pure TCE. And so industry promoted an experiment of mutagenesis which showed negative results on one carcinogenicity test.

We are now conducting what we do believe are the largest studies ever done on TCE—it's the largest one performed in our laboratory at once on rats. We are using Sprague-Dawley rats and our own Swiss mice. A large group of 800 animals were kindly provided to us by NCI. We are testing by inhalation starting by very heavy dose—600 ppm down to 30 ppm on 3500 animals altogether. This study was started in February 1979 and we are confident in 1 year we will have enough time to finalize the results.

On the basis of our present evidence, we agree entirely with what Dr. Van Duuren has said this morning. We didn't have any carcinogenic effect up to this point.

INFANTE: Did NCI use technical-grade TCE?

VAN DUUREN: May I respond to that? Yes, there is ECH and other impurities or additives in technical-grade TCE.

MALTONI: There was 2%.

VAN DUUREN: ECH was not present in the experiments that we performed, but even if we did feed ECH, I would suspect, based on the chemistry of ECH, that it would not pass the stomach, which means that it would not reach the liver as unchanged ECH because it would be rapidly hydrolyzed, especially below pH 7. I showed you on a slide that the rate of hydrolysis is fairly rapid.

INFANTE: Is it a reasonable assumption, then, that the TCE that NCI used, if it was contaminated, isn't the contamination with ECH that might have induced the hepatocellular effects.

VAN DUUREN: The NCI result, in my opinion, has nothing to do with ECH being present as an impurity, and the same applies to DBCP where ECH is also an impurity (which is a potent carcinogen as I described this morning). In the case of TCE, which contained ECH when tested by NCI, ECH cannot account for the observed effect. Now, we may be asked to account for, if I may say so, and you'll excuse me, a false-positive result. I would ascribe it to metabolic overload. And that is, that if one gives any chemical in such high dosages, as were given in the NCI experiments with TCE, so that the normal detoxifying mechanisms are destroyed, then metabolic pathways are altered and the compound causes carcinogenic activity that would not occur at lower doses, as in our experiments. But I don't think that either Dr. Maltoni or myself have to be put in a position of trying to explain a false-positive result in the carcinogenicity assay of the NCI.

ANDERS: You proposed the formation of VC as a metabolite of EDC. We've looked in a preliminary way in our laboratory for VC as a metabolite, and have not been successful in identifying it. I guess we probably think it wouldn't be a metabolite.

MARLOW: With all due respect to your hypothesis, doesn't it bear waiting until Dr. Maltoni's results have been determined from his study of the B6C3F1 mouse?

VAN DUUREN: I will let Dr. Maltoni respond.

MALTONI: As a matter of fact, it seems that it was a suggestion of Dr. Van Duuren that this may be just a result of further strengthening of our already strong opinion that TCE may not be a carcinogenic. But of course after we evaluate all this evidence from 3500 animals, we can say more and with a larger certitude.

References

NCI. 1976. *Carcinogenesis bioassay of trichloroethylene.* Technical report series number 2. DHEW publication number (NIH) 76-802. Government Printing Office, Washington, D.C.

SESSION 3:
Uses of Ethylene Dichloride: Worker Exposure

Human Exposures to Ethylene Dichloride

LOIS SWIRSKY GOLD
Department of Biochemistry
University of California
Berkeley, California 94720

Attempts to regulate human exposures to carcinogenic substances on a rational basis require valid and, where possible, quantitative information about the prevailing sources and levels of exposure. Such data need to be evaluated in conjunction with scientific judgments about the carcinogenic strength of a chemical, its other health effects, statistical estimates of the carcinogenic risk to humans, and the costs and benefits of various regulatory strategies.

This discussion aims at synthesizing and evaluating a large variety of estimates and measurements of human exposures to ethylene dichloride (EDC; 1,2-dichloroethane) and at improving the estimates wherever possible. From a methodological perspective it may be useful to think of EDC as one example of a whole range of chemicals that are potential carcinogens but have yet to be investigated in depth. Many of the data bases and government sources used here for EDC provide bits and pieces of important information about a large number of substances. In general, quantitative information about human exposures is difficult to obtain and estimates are subject to considerable uncertainty.

Three areas of exposure are discussed: production and consumption patterns, occupational exposures (where a relatively small subgroup of the total population potentially has relatively great exposure), and the much larger general population (which may be exposed to relatively lower levels in air, drinking water, food, and consumer products). The largest exposure to EDC is likely to be via inhalation, because of its volatility; however, absorption through the skin or through ingestion of drinking water and food can also occur.

Currently, federal regulation of exposures to EDC is based upon toxic effects other than carcinogenicity.

PRODUCTION AND CONSUMPTION PATTERNS

With an annual production of 11 billion pounds, EDC is the largest volume synthetic organic chemical manufactured in the United States. Worldwide capacity production is 51 billion pounds. This enormous volume is due primarily to demand for vinyl chloride (VC), for which EDC is a feedstock (Fig. 1).

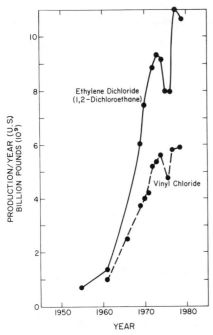

Figure 1
Production of EDC and VC. Data from *Chemical and Engineering News* (1960-1979.) Data for 1955 from SRI International (1979).

EDC was the first chlorinated hydrocarbon known and was initially produced in 1795 (Hardie 1964). Commercial production in the U.S. has been reported since 1922 (U.S. Tariff Commission 1923); however, rapid growth did not begin until the 1960s, with increases of about 10% per year. (The production decline in the mid-1970s was due to a general slowdown of the U.S. economy.) Projections for the next few years are for an annual growth rate of 5%.

Most EDC is used captively by the 11 companies that produce it in 17 plants. Only about 15% is sold on the open market, and the bulk of this amount is used by 10 additional plants for the manufacture of other chemicals (SRI International 1979). EDC production facilities are geographically clustered on the Gulf Coast; only two plants—one in Calvert, Kentucky, and one near Los Angeles—lie outside this region.

Ninety-eight percent of all EDC is used in the chemical industry for the production of VC, 1,1,1-trichloroethane, ethyleneamines, vinylidene chloride (VDC; 1,1-dichloroethene), perchloroethylene (PCE; tetrachloroethylene), and trichloroethylene (TCE). Much of the remainder (196 million pounds) is used as a lead scavenger in gasoline (Table 1). Miscellaneous uses of EDC (lb/yr) are pesticides (2 million); textile and equipment cleaning (3 million); extracting oil from seeds, processing animal fats, pharmaceuticals (2 million); production

Table 1
Estimated U.S. Consumption Pattern for EDC, 1977

	Pounds (10^6)	% Total consumption	Projected annual growth rate 1977-1982 (%)
VC	9,460	85	5 to 8
1,1,1-Trichloroethane	473	4	4 to 5
Ethyleneamines	299	2	-2 to -4
VDC	213	2	5 to 7
PCE	191	2	0 to 2
TCE	205	2	-2 to 3
Lead scavenger	196	2	-15
Miscellaneous	11	<1	—
Total	11,048	100	5 to 6

Adapted from data in *Chemical economics handbook,* SRI International (1979).

of polysulfide elastomers (1 million); and other uses (1 million) (SRI International 1979; E. Fry, pers. comm.).

An overview of the amount of EDC estimated to have been dispersed in commercial products and emitted during manufacture in 1977 is presented in Figure 2. EDC is not known to occur in nature (Johns 1976). The amount

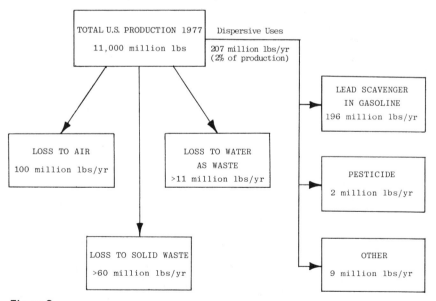

Figure 2
Estimated environmental release of EDC during 1977. Estimates are based upon information contained in Auerbach Associates (1978); Storm (1978); Drury and Hammons (1979); and SRI International (1979); E. Fry (pers. comm.); P. Williams (pers. comm.).

released is estimated to be more than 378 million pounds annually. Large amounts are emitted during the manufacture of EDC and its chemical end products: approximately 100 million pounds as air emissions, more than 11 million pounds in wastewater, and more than 60 million pounds in waste tar, which is primarily a remainder from the production of VC. An estimated 207 million pounds of EDC were dispersed in commercial products; of this, 196 million were used in leaded gasoline. Although relatively small in volume compared to total production, the 2 million pounds used in pesticides and the 9 million pounds present in other products may result in exposures of varying magnitude in many occupational and nonoccupational settings.

WORKER EXPOSURES

The most recent estimates from the National Occupational Hazard Survey 1972-1974 conducted by the National Institute for Occupational Safety and Health (NIOSH) indicate that between 60,000 and 1.6 million workers are potentially exposed to EDC (Table 2). (In 1977 NIOSH reported 1.9 million exposed; however since that time more complete information on product contents has reduced the estimate to 1.6 million [NIOSH 1977; NIOSH 1979b].) The lower estimate is based upon observed exposures to the chemical in its normal state or as an ingredient in a trade name product. The higher estimate is possible because it includes potential exposures to similar products, which NIOSH thinks might contain the chemical and whose contents have yet to be ascertained. A midrange estimate of 1.3 million is obtained by projections based on similar products for only those 34 industries where actual products containing EDC have already been identified (Table 2).

Although production data show that only a small proportion of all EDC is dispersed outside the chemical industry, these occupational estimates indicate that a large number of workers are potentially exposed across a broad spectrum of industrial classifications. Between 35,000 and 1.5 million workers outside of the chemical industry are estimated to be in occupations where EDC is used (Table 2). Further evidence for this widespread use is provided in Tables 3 and 4, which list the kinds of products identified as containing the chemical in 1972-1974 and the industries recently inspected by NIOSH and the Occupational Safety and Health Administration (OSHA) for exposures to EDC.

The current OSHA standard for employee exposure to EDC is 50 ppm (200 mg/m^3) as an 8-hour time-weighted average (TWA), with an acceptable ceiling concentration of 100 ppm for 5 minutes in any 3-hour period. Concentrations should never exceed 200 ppm (OSHA 1978). No data are available about the prevailing levels of exposure on a TWA basis; however, information from the OSHA Management Information System datafile indicates that of the 28 companies identified as having been inspected for EDC between 1972 and 1979, three had exposure levels exceeding the standards. In one chemical plant the ceiling or peak limit was exceeded; in a footwear plant and a plastic

Table 2
Estimates of Numbers of Workers Potentially Exposed to EDC, 1972-1974

Standard industrial classification	Estimate based on actual and trade name products[a]	Estimate based on similar as well as actual and trade name products[a]
Agriculture services and hunting	100	7,700
Oil and gas extraction	100	2,100
General building contractors	600	5,500
Heavy construction contractors	600	5,800
Ordnance and accessories	<100	1,500
Food and kindred products	2,000	28,200
Tobacco manufactures	700	1,000
Apparel and other textile products	300	4,300
Lumber and wood products	<100	3,700
Furniture and fixtures	100	11,000
Paper and allied products	700	30,400
Printing and publishing	100	100,700
Chemicals and allied products	25,400	74,100
Petroleum and coal products	200	8,000
Rubber and plastics products, NEC[b]	1,400	24,700
Leather and leather products	100	3,600
Stone, clay, and glass products	200	34,400
Primary metal industries	800	50,200
Fabricated metal products	800	52,200
Machinery, except electrical	3,200	101,000
Electrical equipment and supplies	1,700	60,000
Transportation equipment	1,000	56,200
Instruments and related products	400	21,000
Miscellaneous manufacturing industries	500	13,000
Local and interurban passenger transit	1,900	7,500
Trucking and warehousing	800	15,600
Transportation by air	1,000	38,700
Electric, gas, and sanitary services	100	17,400
Wholesale trade	2,200	114,900
Retail general merchandise	200	18,000
Automotive dealers and service stations	4,300	142,000
Personal services	1,600	12,300
Miscellaneous business services	<100	27,900
Medical and other health services	8,000	177,100
Total	60,600	1,268,000
Total including 328,300 workers potentially exposed in 26 other industries		1,596,300

Data from National Occupational Hazard Survey Datafile (NIOSH 1979b).
[a] Numbers rounded to nearest hundred.
[b] Not elsewhere classified.

Table 3
Types of Products Containing EDC, Which Were Observed in the Workplace, 1972-1974

Acrylic cement	masking compound
Tack primer	pesticide
Solvent	automobile antiknock additive
Deletion fluid	seam sealer
Paint	carburetor and parts cleaner
Disinfectant	

Data from National Occupational Hazard Survey, Trade Name Products resolved for 1,2-dichloroethane as of December 1979 (NIOSH 1979b).

EDC was known to be an ingredient in these products in 1972-1974. Since that time, products may have been reformulated to remove the chemical, or the whole product may have been removed.

Table 4
Industries Inspected for Exposures to EDC

OSHA inspections, 1972-1979
    Chemical and chemical preparations[a]
    Rubber and plastic footwear[b]
    Miscellaneous plastic products[c]
    Macaroni and spaghetti
    Electrical compounds
    Shipbuilding and repair
    Miscellaneous manufacturing
    Bags, except textile
    Surgical and medical instruments
    Vinyl resins
    Commercial printing
    Refuse system
    Grain
    Oil field
    Aircraft parts
    Agricultural chemicals and pesticides
NIOSH inspections, 1973-1977
    Steel tool and engineering
    Grain elevators
    Police scientific laboratory

Data from OSHA Management Information System, 1977 and 1979 (OSHA 1979); NIOSH Health Hazard Evaluations Search (NIOSH 1979a).
    [a] In one plant the acceptable ceiling or peak concentration was exceeded.
    [b] In one plant the 50-ppm standard was exceeded by 1.1-2.0 times, according to the OSHA severity rating codes.
    [c] In one plant the 50-ppm standard was exceeded by 2.1-3.0 times, according to the OSHA severity rating codes.

products plant the 50-ppm standard was exceeded. All 3 establishments were cited. Exposures in the other 25 companies and in the 3 companies inspected by NIOSH were all within the allowable limits (Table 4; NIOSH 1979a; OSHA 1979).

The current 50-ppm standard is based upon toxic effects other than cancer. In 1976 NIOSH recommended an exposure limit of 5 ppm on the basis of reports that workers chronically exposed to 10-15 ppm experienced nervous system and liver effects. Nursing mothers were also warned not to work with the chemical because a study indicated that it has been found in the milk of exposed women (NIOSH 1976). Following the positive National Cancer Institute (NCI) cancer bioassay, NIOSH recommended that EDC be handled as a carcinogen in the workplace and that the standard be revised downward to 1 ppm (NIOSH 1978a,b,c). These recomendations have not become government policy.

A detailed risk assessment on EDC is beyond the scope of this discussion, but it does seem that a reevaluation of the current OSHA standard is called for. Ideally, such an evaluation should include a more complete understanding of the apparently divergent results of the animal feeding and inhalation studies (Hooper et al., this volume).

From the NCI bioassay, our project on carcinogenic potency at the University of California at Berkeley, has estimated that for rats a daily dose of about 10 mg EDC/kg body weight for a lifetime will induce tumors in half of otherwise tumor-free animals. For mice the corresponding daily dose is about 70 mg/kg body weight (Ames et al., this volume; Hooper et al., this volume).

These estimates may be compared to the levels that workers currently are permitted to inhale. (The amount that workers actually do breathe in will vary considerably and may be much lower.) If a worker who weighed 70 kg breathed in 7 m^3 of air containing 50 ppm (200 mg/m^3) EDC daily, he would be exposed to 20 mg/kg each day. If we assume all of this is absorbed, the rate (kg body weight) is close to that which we estimate will induce tumors in 50% of rats and mice. The similarity of these levels suggests the wisdom of reevaluating the current threshold limit value on the basis of a detailed risk assessment.

Quantitative information on the actual levels to which workers are exposed is scarce; such data obviously would be useful in any evaluation of hazard from toxic substances. Currently the OSHA Management Information System reports the severity rating of inspections that exceed OSHA standards; it would be useful to report the levels monitored even when these are within the allowable limits. If future NIOSH occupational hazard surveys were to include data on the length of time or type of exposures workers in various occupations have to specific substances, this too would be an important source for regular use in assessing hazards of chemicals that are candidates for regulatory action.

LEADED GASOLINE

The largest dispersive use of EDC is as a lead scavenger in gasoline, for which 196 million pounds were consumed in 1977. This amount is expected to decline in the future as unleaded gasoline captures the fuel market; however, leaded gas will continue to fuel some trucks and older cars. In addition to gas station attendant and other worker exposures from this use, about 30 million Americans are exposed to EDC while filling their tanks in self-service gas stations. These exposures are estimated at 1.5 ppb (6 $\mu g/m^3$) for 2.2 hours per year—a TWA of only 0.0004 ppb (Suta 1979). Most EDC is destroyed in the engine during combustion. A small amount evaporates from gas tanks and from refueling; this results in exposures estimated at 0.01–0.03 ppb to about 14 million people who live near gas stations or in urban areas (Suta 1979).

FOOD-RELATED PRODUCTS

EDC is potentially available for human consumption if it is present as a residue in finished products from the following uses: solvent to extract spice oleoresins, pesticide, grain fumigant, solvent to clean grain-mill machinery, and adhesive coating in food packaging.

Federal regulations allow 30 ppm as a residue in spice oleoresins (U.S. Code of Federal Regulations). Actual residues have been found in 11 of the 17 common spices examined, at levels of 2–23 ppb (Page and Kennedy 1975). Since only small quantities of spices are consumed, ingestion from this source is not expected to be very large.

EDC is registered for agricultural use in a variety of formulations and is used commercially for postharvest fumigation, bark treatment, soil fumigation, and spraying of agricultural premises. In 1977, 84 such pesticide products were registered with EPA by 52 different companies; together they contained 2,190,261 pounds of EDC (E. Fry, pers. comm.). A list of these products is presented in Table 5 (pages 218-219) along with the concentration of EDC in each one (Drury and Hammons 1979; EPA 1979a).

Since 1956 EDC has been exempted from the requirement of a tolerance for residues when used as a postharvest fumigant on barley, corn, oats, popcorn, rice, rye, wheat, and sorghum. This exemption was based on baking tests, which showed no residues in bread cooked from fumigated flour; sensitivity of the method was 2 ppm (Monro 1969). More recently bread residues were measured usually below 0.05 ppm (Wit et al. 1969). Another recent study was unable to find residues in fumigated cereal samples and concluded that further research was needed to explain this result. The author noted that it was possible that either the EDC was only partially sorbed by the wheat because of rapid volatilization or the EDC was changed to a non-EDC residue (Berck 1974).

In 1971 the Joint Meeting of the FAO Working Party of Experts and the WHO Expert Committee on Pesticide residues evaluated EDC. It was noted that little information was available on residues in food reaching the consumer, and guidelines were suggested for residue limits of 10 ppm in milled cereal products and 0.1 ppm in cooked cereal products and bread (WHO 1972). Since these guidelines were suggested, the NCI bioassay has been done and more sensitive methods to detect residues have become available.

It seems advisable to review the policy of an exemption from a tolerance for fumigated grains. As a grain fumigant, EDC is commonly mixed with carbon tetrachloride to reduce flammability. It is often combined with ethylene dibromide (EDB; 1,2-dibromoethane) which is a considerably more potent carcinogen than EDC. Both EDB and carbon tetrachloride are on the RPAR list of EPA. Thus, subsitution by some other less toxic fumigant would be desirable.

Pesticide residues may also result from spraying home-grown fruits and vegetables. Currently, the consumer can purchase EDC in pesticides for home garden use to control tree borers and as a solution for a general insecticide. In northern California it is possible to purchase a product containing EDC that is recommended as a diluted spray for food crops like strawberries and cabbage within 1 day of harvest. These foods are eaten raw and hence any residues would not be removed by cooking.

CONSUMER PRODUCTS

A 1979 industry profile prepared for the Consumer Product Safety Commission reports that consumer products do not contain significant amounts of EDC (Winslow and Barr 1979). Manufacturers of some products that once contained EDC—certain fumigants and some solvent cements for acrylic plastics—have voluntarily removed EDC from their products following the positive NCI bioassay. Fur and garment fumigants containing the chemical are not sold as consumer products.

Although the industry profile reports that California is considering legislation to ban the use of EDC in home-use pesticides, I could not confirm this legislation from state agencies.

Since industrial products that were observed to contain EDC in the NIOSH Occupational Hazard Survey might also be consumer products, the product names from the list in Table 3 were sent to the Consumer Product Safety Commission. A hazard evaluation by the Commission has just determined that only one product is currently sold to consumers; this is a cleaning solvent containing 50 ppm EDC (P. Preuss, pers. comm.). The fact that EDC has been found in domestic sewage (Versar, Inc. 1975) suggests that it may be or has been used in consumer products.

Table 5
Pesticide Products Containing EDC, 1977

Product name	% EDC	Product name	% EDC
Big F "LGF" Liquid Gas Fumigant	75.0	914 Weevil Killer and Grain Conditioner	63.1[a]
Farmrite Mushroom Spray	75.0	Cooke Kill-Bore	50.0
Grain Fumigant	75.0	Destruxol Borer-Sol	50.0
Hydrochlor GF Liquid Gas Fumigant	75.0	Ferti-Lome Tree Borer Killer	50.0
Parson Lethogas Fumigant	73.5[a]	Hacienda Borer Solution	50.0
Best 4 Servis Brand 75-25 Standard Fumigant	70.3	Navlet's Borer Solution	50.0
Bug Devil Fumigant	70.3	Staffel's Boraway	50.0
Fumisol[b]	70.3	Chemform Brand Bore-Kill	35.0
Gas-O-Cide	70.3	Okay Mole and Gopher Fumigant	30.0
Riverdale Fumigant	70.3	Tri-X Garment Fumigant	30.0[a]
Standard 75-25 Fumigant	70.3	Brayton EB-5 Grain Fumigant	29.2[a]
Brayton Flour Equipment Fumigant for Bakeries	70.2	De-Pester Weevil Kill	29.2[a]
		Formula 635 (FC-2) Grain Fumigant	29.2[a]
Brayton 75-25 Grain Fumigant	70.2	Grainfume MB	29.2[a]
Cardinalfume	70.2	T-H Vault Fumigant	29.2[a]
De-Pester Fumigant No. 1	70.2	Volcan Formula 635 (FC-2) Grain Fumigant	29.2[a]
Diamond 75-25 Grain Fumigant	70.2		
Hill's Hilcofume 75	70.2	Dowfume EB-5 Effective Grain Fumigant	29.0[a]
Maxkill 75-25	70.2	T&C Fruit & Vegetable Insecticide and Miticide	28.0
Spray-Trol Brand Insecticide Fumi-Trol	70.2		
		Infuco 50-50 Spot Fumigant	26.84[a]
Stephenson Chemicals Stored Grain Fumigant	70.2	Selig's Selcofume	25.0
Vulcan Formula 72 Grain Fumigant	70.2	Cooke Bug Shot Law Special Spray Concentrate	20.0

Product	%	Product	%
Westofume Fumigant	70.2	Dowfume EB-15 Inhibited	20.0[a]
Zep-O-Fume Grain Mill Fumigant	70.2	J-Fume-20	20.0[a]
Diweevil	70.0	Max Spot Kill Machinery Fumigant	20.0[a]
Dowfume 75	70.0	Crest 15 Grain Fumigant	19.6[a]
Excelcide Excelfume	70.0	(FC-13) Mill Machinery Fumigant[b]	19.6[a]
Fume-O-Death Gas No. 3	70.0	Solig's Grain Fumigant No. 15	19.6[a]
Hydrochlor Fumigant	70.0	Spot Fumigant	19.6[a]
Infuco Fumigant 75	70.0	Dyna Fume	12.0[a]
J-Fume-75	70.0	Iso-Fume	11.4[a]
Pearson's Funigrain P-75	67.49	49'er Gold Strike Bonanza Plant Spray	10.25
Coop New Activate Weevil Killer Fumigant	66.0[a]	KLX	10.25
Pioneer Brand Grain Fumigant	66.0[a]	Agway Serafume	10.0[a]
Dowfume F	65.0[a]	Formula MU-39	10.0
FC-7 Grain Fumigant	64.7[a]	Max Kill Spot-59 Spot Fumigant for Mills and Milling Machinery	10.0[a]
De-Pester Grain Conditioner and Weevil Killer	64.6[a]	Serfume	10.0[a]
(FC-4) SX Grain Storage Fumigant	64.6[a]	Dowfume EB-59	9.0[a]
Selig's Grain Storage Fumigant	64.6[a]	Waco-50	9.0[a]
Sure Death Brand Millfume No. 2	64.6[a]	Leitte Spotfume 60	8.5[a]
Patterson's Weevil Killer	63.1[a]	Koppersol	3.0
T-H Grain Fumigant No. 7 Weevil Killer and Grain Conditioner	63.1[a]	Sirotta's Sircofume Liquid Fumigating Gas	1.0

Data from Office of Pesticide Programs, Product Label File, 1979 (EPA 1979a).
[a] Also contains EDB.
[b] This name is registered twice. It is produced by two different companies.

SOLID AND LIQUID WASTES FROM MANUFACTURING PLANTS

The environmental fate of waste products from the EDC-VC industry is of concern because of the large and still increasing production volume and because the waste tars are mutagenic (Jensen et al. 1975; Rannug and Ramel 1977). Information about waste disposal practices and their polluting potential is scarce.

In California, waste water and waste tar from EDC-VC production is disposed of in drums into Class-1 landfill sites; waste disposal manifests reported to the Department of Health Services indicate the composition of these wastes (California Administrative Code; Storm 1978). On the basis of a survey of these manifests, it is estimated that 9.6 million pounds of waste tar containing 25-75% EDC were disposed of in California in 1979 from one plant. Another 2 million gallons of wastewater containing 2% EDC, or about 400,000 pounds, were also buried (California Department of Health Services 1979). (The survey of hazardous waste manifests did not identify the producer or the Class-1 disposal sites.) Because waste volume and composition vary with production processes, it is difficult to assess total national volume. Projections from these California data to the U.S. give estimates of about 11 million pounds annually as EDC waste in water and 60 million pounds of EDC in more than 230 million pounds of waste tar from production of EDC and VC.

Outside California, some waste tar is disposed of by incineration; some is buried in disposal sites and recent air sampling of a landfill site in New Jersey detected EDC in amounts ranging from trace levels to a high of 14 ppb (57 $\mu g/m^3$) at a point 200 yards downwind of the dump (Pelizzari 1979). In 1975, one report indicated that wastes were disposed of as effluent to a river (Brown et al. 1975). In Europe, where large quantities of EDC tar were at one time dumped into the North Sea, waste disposal is now by incineration at sea.

In view of the large volume of waste and its hazardous composition, it would seem advisable to address several aspects of the disposal issue. What are the by-products of incineration and how much is removed? Would incineration be an economically feasible disposal method nationally? To what extent does EDC evaporate from disposal sites? Does EDC from landfills contaminate surface, ground, or drinking waters? Is waste tar or waste water currently being dumped into any river waters? What is the toxicity and fate of the large variety of chlorinated by-products in the EDC tar?

OCCURRENCE IN WATER

Chronic low-dose exposures to people may result from occurrence of EDC in drinking water, in the ppb range. Although these amounts are relatively small, there has been concern for health effects because of a lack of information about synergistic interactions and alterations of metabolic pathways that may result from chronic ingestion.

The National Organics Reconnaissance Survey (NORS) in 1975 reported EDC in the drinking water supplies of 26 of the 80 cities studied, at levels of 0-6 µg/liter (Symons et al. 1975). In a study of surface waters near heavily industrialized sites across the U.S., concentrations at usual levels of 1-2 ppb were found in 53 of 204 samples. Some concentrations were considerably higher, however, and one sample from the Delaware River Basin contained 90 ppb (Ewing et al. 1977). Other water monitoring has detected EDC in river waters in Europe at 0.7 µg/liter (Eurocop-cost 1976) and in tap water in Japan at 0.9 µg/liter (Fujii 1977).

The Office of Drinking Water at EPA has not yet proposed regulation of EDC but may do so under the Safe Drinking Water Act. The Office of Water Planning and Standards of EPA has drafted a proposal on ambient water quality criteria for chlorinated ethanes, including EDC. That office is considering an interim target risk level in the range of 1 in 100,000 to 1 in 10 million. Their calculations, based on the NCI data for mammary adenocarcinomas in female rats and using a one-hit model and a low bioaccumulation factor, indicate that a concentration of EDC in water to keep the cancer risk below 1 in 100,000 is 7 µg/liter. This calculation assumes consumption of 2 liters of water daily and a small amount of contaminated fish (EPA 1979b). A 70-kg person who consumed 2 liters of water every day containing 7 ppb (7 µg/liter) EDC, would be getting 0.2 µg (0.0002 mg)/kg body weight each day.

Whether or not EDC is produced as a result of the chlorination process is under discussion at EPA. Charcoal filtration is known to be 90-100% effective in eliminating EDC from drinking water.

EMISSIONS TO AIR

Attempts to measure EDC in atmospheric air have not detected it even in the ppb and ppt ranges (Grimsrud and Rasmussen 1975; Singh et al. 1977). Estimates of the half-life of EDC in the atmosphere range from weeks to months (Pearson and McConnell 1975). Exposures do result, however, for those people who live near EDC production plants. A recent EPA study monitored ambient air near production plants (PEDCo Environmental, Inc. 1979), and the data were subsequently used to estimate average exposures to the local populations. In the monitoring study, ambient EDC levels varied considerably and depended on such factors as plant production process and rate, emission control technologies used by the manufacturer, meteorological conditions, characteristics of the EDC point sources such as stack height, and location of the monitoring sites (Drury and Hammons 1979).

Dispersion modeling based on the living patterns of the nearby populations provides an estimate that 12.5 million people are exposed to average annual EDC concentrations of 0.01-10 ppb (Suta 1979; Kellam and Dusetzina, this volume; Table 6). A 70-kg person breathing in daily 20 m^3 of air containing 10 ppb (40 $µg/m^3$) EDC would be getting an exposure of 11 µg (0.011 mg)/kg body

Table 6
Estimated Human Population Exposures to Atmospheric EDC Emitted by Producers

Annual average atmospheric EDC concentration (ppb)	Number of people exposed
10.0	1,700
6.00-10.00	3,300
3.00- 5.99	28,000
1.00- 2.99	280,000
0.60- 0.99	400,000
0.30- 0.59	1,500,000
0.10- 0.29	4,300,000
0.060-0.099	1,900,000[a]
0.030-0.059	3,500,000[a]
0.010-0.029	550,000[a]
Total	12,500,000

Data from Suta (1979).
[a] These are underestimates because the dispersion modeling results were not extrapolated beyond 30 km from each EDC production facility.

weight each day. It is worth pointing out that an equivalent ppb in water would result in exposures about 40 times smaller: 10 ppb in water is 10 μg/liter and a 70-kg person drinks about 2 liters a day. This would result in an exposure of 0.3 μg (0.0003 mg)/kg body weight each day.

Most manufacturing losses of EDC to the environment occur as air emissions. The amount released is about half as large when EDC is produced by direct chlorination of ethylene than by the oxychlorination method. Currently, nearly 50% of all U.S. production is accomplished by the oxychlorination method (Drury and Hammons 1979). Some EDC is also released from manufacturing plants as a result of storage and distribution of the chemical.

Current control mechanisms for these emissions include scrubbers and condensers. Condensers are far more effective than scrubbers but have substantially higher capital and operating costs (Pervier et al. 1974). Control technologies also are available for emissions from storage tanks, and these are reportedly under study by several manufacturers (Schwartz et al. 1974).

EPA is currently considering regulation of EDC as an air pollutant and is in the process of assessing the risk to humans based on the NCI cancer bioassay.

ACKNOWLEDGMENTS

I am indebted to B. N. Ames and K. Hooper for helpful criticisms and to Elizabeth Higgins for bibliographical assistance. This work was supported by Department of Energy contract EY76-S-03-0034 PA156, by a California Policy Seminar grant to B. N. A., and by National Institute of Environmental Health Sciences Center grant ES-01896.

REFERENCES

Auerbach Associates, Inc. 1978. *Miscellaneous and small volume consumption of ethylene dichloride. EPA-68-01-3899 and AAI-2431-104-TN-1. EPA.* Auerbach Associates, Inc., Philadelphia, Pennsylvania.

Berck, B. 1974. Fumigant residues of carbon tetrachloride, ethylene dichloride, and ethylene dibromide in wheat, flour, bran, middlings, and bread. *J. Agric. Food Chem.* 22:977.

Brown, S. L., S. Y. Chan, J. L. Jones, D. H. Liu, K. E. McCaleb, T. Mill, K. N. Sapios, and D. E. Schendel. 1975. *Research program on hazard priority ranking of manufactured chemicals. Phase II. Final Report to the National Science Foundation.* SRI International, Menlo Park, California.

California Administrative Code. Title 22, Division 4. Chapter 30: Minimum standards for management of hazardous and extremely hazardous wastes. California Department of Health Services.

California Department of Health Services. 1979. Source of survey of waste disposal manifests: Paul Williams

Chemical and Engineering News. 1950-1979. "Top-Fifty Chemical Products."

Drury, J. S. and A. S. Hammons. 1979. *Investigations of selected environmental pollutants: 1,2-Dichloroethane.* EPA 560/2-78-006. EPA Oak Ridge National Laboratory, Oak Ridge, Tennessee.

EPA (Environmental Protection Agency). 1979a. Office of Pesticide Programs. Product Label File Printout from Information on Microfiche, 3rd Edition, July 1978.

_____. 1979b. "Chlorinated Ethanes. Ambient Water Quality Criteria." Draft, Criteria and Standards Division, Office of Water Planning and Standards. EPA. Washington, D.C.

Eurocop-cost. 1976. *A comprehensive list of polluting substances which have been identified in various fresh waters, effluent-discharges, aquatic animals, and plants, and bottom sediments,* 2nd ed., p. 41. EUCO/MDU/73/76, Commission of the European Communities. Compiled by Water Research Center (Stevenage Laboratory) Stevenage, Hertfordshire, England.

Ewing, B. B., E. S. K. Chian, J. C. Cook, C. A. Evaris, P. K. Hopke, and E. G. Perkins. 1977. Monitoring to detect previously unrecognized pollutants in surface waters. EPA 560/6-77-015, p. 63-64, 73.

Fujii, T. 1977. Direct aqueous injection gas chromatography-mass spectrometry for analysis of organohalides in water at concentrations below the parts per billion level. *J. Chromatogr.* 139:297.

Grimsrud, E. P. and R. A. Rasmussen. 1975. Survey and analysis of halocarbons in the atmosphere by gas chromatography-mass spectrometry. *Atmos. Environ.* 9:1014.

Hardie, D. W. F. 1964. Vinyl chloride. In *Kirk-Othmer encyclopedia of chemical technology,* 2nd ed., vol. 5, p. 171. John Wiley & Sons, Inc., New York.

Jensen, S., R. Lange, G. Berge, K. H. Palmork, and L. Renberg. 1975. On the chemistry of EDC-tar and its biological significance in the sea. *Proc. R. Soc. Lond. B Biol. Sci.* 189:333

Johns, R. 1976. *Air pollution assessment of ethylene dichloride.* MTR 7164. The Mitre Corporation, McLean, Virginia.

Monro, H. A. U., ed. 1969. Manual of fumigation for insect control. *FAO Agric. Stud.* 79.

NIOSH (National Institute for Occupational Safety and Health). 1976. *Criteria for a recommended standard ... occupational exposure to ethylene dichloride (1,2-dichloroethane).* DHEW (NIOSH) publication number 76-139. Government Printing Office, Washington, D.C.

———. 1977. *National Occupational Hazard Survey.* Vol. 3. *Survey analysis and supplemental tables.* DHEW, NIOSH Division of Surveillance, Hazard Evaluations and Field Studies. Cincinnati, Ohio.

———. 1978a. *Revised recommended standard ... occupational exposure to ethylene dichloride (1,2-dichloroethane).* DHEW (NIOSH) publication number 78-211. Government Printing Office, Washington, D.C.

———. 1978b. Ethylene dichloride. *Current Intelligence Bulletin,* No. 25. DHEW.

———. 1978c. Chloroethanes: Review of toxicity. *Current Intelligence Bulletin,* No. 27. DHEW.

———. 1979a. Health Hazard Evaluations Search, as of December 1979.

———. 1979b. National Occupational Hazard Survey Data File as of December 1979. Source of printout: Joseph A. Seta.

OSHA (Occupational Safety and Health Administration). 1978. *General industry: OSHA safety and health standards.* OSHA 2206. Revised November 7, 1978.

———. 1979. Management Information System Datafile 1977 and 1979. Source of Data Analysis: William Cloe.

Page, B. D. and B. P. C. Kennedy. 1975. Determination of methylene chloride, ethylene dichloride, and trichloroethylene as solvent residues in spice oleoresins, using vacuum distillation and electron-capture gas chromatography. *J. Assoc. Offic. Anal. Chem.* 58:1062.

Pearson, C. and G. McConnell. 1975. Chlorinated C_1 and C_2 hydrocarbons in the marine environment. *Proc. R. Soc. London B Biol. Ser.* 189:305.

PEDCo Environmental, Inc. 1979. *Monitoring of ambient levels of ethylene dichloride (EDC) in the vicinity of EDC production and user facilities.* EPA 600/4-79-029. EPA, Research Triangle Park, North Carolina.

Pelizzari, E. D. 1979. *Ambient air carcinogenic vapors: Improved sampling and analytic techniques and field studies.* EPA 600-2-79-081. EPA, Research Triangle Park, North Carolina.

Pervier, J. W., R. C. Barley, D. E. Field, B. M. Friedman, R. B. Morris, and W. A. Schwartz. 1974. *Survey reports on atmospheric emissions from the petrochemical industry,* vol. 2. EPA-450/3-73-005-b. EPA, Research Triangle Park, North Carolina.

Rannug, U. and C. Ramel. 1977. Mutagenicity of waste products from vinyl chloride industries. *J. Toxicol. Environ. Health* 2:1019.

Schwartz, W. A., F. G. Higgins, Jr., J. A. Lee, R. Newirth, and J. W. Pervier. 1974. *Engineering and cost study of air pollution control for the petrochemical industry,* vol. 3. *Ethylene dichloride manufacture by oxychlorination.* EPA-450/3-73-006-c. EPA, Research Triangle Park, North Carolina.

Singh, H. B., L. J. Salas, and L. A. Cavanagh. 1977. Distribution, sources and sinks of atmospheric halogenated compounds. *J. Air Pollut. Control Assoc.* 27:332.

SRI International. 1979. *Chemical economics handbook.* Menlo Park, California.
Storm, D. 1978. *Handbook of industrial waste compositions in California.* California Department of Health Services, Hazardous Materials Management Section. Berkeley, California.
Suta, B. 1979. *Assessment of human exposures to atmospheric ethylene dichloride.* Final Report to EPA Office of Air Quality Planning and Standards. SRI International, Menlo Park, California.
Symons, J. M., T. A. Bellar, J. K. Carswell, J. DeMarco, K. L. Kropp, G. G. Robeck, D. R. Seeger, C. J. Slocum, B. L. Smith, and A. A. Stevens. 1975. National organics reconnaissance survey for halogenated organics. *J. Am. Water Works Assoc.* **67**:634.
U.S. *U.S. Code of Federal Regulations.* Title 21, Sections 172, 173, 175.
U.S. Tariff Commission. 1923. Census of dyes and other synthetic organic chemicals 1922. *Tariff Information Series,* no. 31. Washington, D.C.
Versar, Inc. 1975. *Identification of organic compounds in effluents from industrial sources.* Final Report to EPA Office of Toxic Substances, Washington, D.C. EPA 560/3-75-002.
Winslow, R. A. and H. W. Barr, Jr. 1979. *Final report on 1,2-dichloroethane in consumer products.* U.S. Consumer Product Safety Commission. Battelle, Columbus, Ohio.
Wit, S. L., A. F. H. Besemer, H. A. Das, W. Goedkoop, F. E. Loosjes, and E. R. Meppelink. 1969. *Results of an investigation on the regression of three fumigants (carbon tetrachloride, ethylene dibromide and ethylene dichloride) in wheats during processing to bread.* Report No. 36/69. National Institute of Public Health, Bilthoven, Netherlands.
WHO (World Health Organization). 1972. 1971 Evaluations of some pesticide residues in food. In *World Health Organization, Pesticide Residues Report No. 2,* p. 276. Geneva, Switzerland.

Production, Uses, and Environmental Fate of Ethylene Dichloride and Ethylene Dibromide

LAWRENCE FISHBEIN
Department of Health, Education, and Welfare
Food and Drug Administration
National Center for Toxicological Research
Jefferson, Arkansas 72079

In recent years there has been recognized concern over both the environmental and toxicological effects of a spectrum of halogenated hydrocarbons, primarily the organochlorine pesticides and related derivatives and congeners, e.g., DDT, heptachlor, endrin, dieldrin, Mirex, Kepone, PCBs (polychlorinated biphenyls), PBBs (polybrominated biphenyls) dibromochloropropane (DBCP; 1,2-dibromo-3-chloropropane), toxaphene, and the chlorofluorocarbons. This concern has now been extended to practically all the major commercial halogenated hydrocarbons, many of which have extensive utility as solvents, fumigants, aerosol propellants, degreasing agents, dry-cleaning fluids, refrigerants, flame retardants, synthetic feedstuffs, cutting fluids, and intermediates in the production of textiles, plastics, and other chemicals. Hence, the halogenated aliphatic hydrocarbons represent one of the most important categories of industrial chemicals from a consideration of volume, use categories, environmental and toxicological effects, and populations potentially at risk.

The major objective of this overview is to present the salient features of the production, uses, and environmental fate of ethylene dichloride (EDC; 1,2-dichloroethane) and ethylene dibromide (EDB; 1,2-dibromoethane) that may have a bearing in assessing the magnitude of exposure of these two extremely important industrial chemicals.

EDC

Production and Utility

EDC is produced industrially primarily by the vapor or liquid phase reaction of ethylene and chlorine in the presence of a catalyst. In an illustrative manufacturing process (Fig. 1), chlorine gas is bubbled through a tank of EDB and the mixed vapors are then reacted with a stream of ethylene via passage into a chlorinating tower maintained at 40–50°C. The subsequent gas from the chlorination stage is passed through a partical condenser maintained at a temperature of 85–130°C to condense only the EDB, which is returned to the process.

From Ethylene and Chlorine

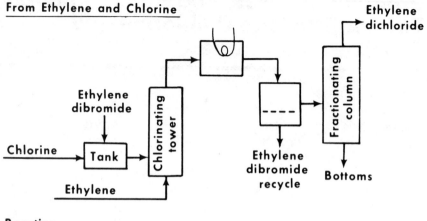

Reaction

$$CH_2=CH_2 + Cl_2 \rightarrow ClCH_2CH_2Cl$$

96 to 98% yield

Figure 1
Production of EDC. (Reprinted, with permission, from Lowenheim and Moran 1975a.)

Gaseous EDC is subsequently condensed and purified by fractional distillation to yield a refined product in yields of about 96-98% (Hardie 1964; Lowenheim and Moran 1975a).

The above reaction can yield a mixture of EDC, 1,1-dichloroethane, and 1,1,2-trichloroethane. The production of the desired addition product is enhanced by controlling the temperature of the reaction and by the use of specific catalysts, e.g., ethyl bromide, aluminum, ferric, copper, or antimony chloride (Hardie 1964; Lowenheim and Moran 1975a).

Significant amounts of ethylene chloride are also produced by an oxychlorination process involving the reaction of ethylene, hydrogen chloride, and air in a fluid-bed catalytic process. The catalyst is a mixture of copper and other chlorides (Lowenheim and Moran 1975a).

Until recently, a considerable amount of EDC was obtained as a by-product of ethylene chlorohydrin manufacture for the production of ethylene oxide. However, this process is no longer used in the United States (Hardie 1964; Lowenheim and Moran 1975a). Additionally, EDC has also been obtained to some extent from the direct chlorination of ethane to ethyl chloride, and as a by-product of trichloroethylene synthesis (Faith et al. 1965).

Commercial EDC is usually stabilized by addition of small amounts of alkyl amines (about 0.1% by weight) (Hardie 1964). It is made by 11 producers in the U.S. with an estimated 1978 production of 10.5 billion pounds. (The worldwide capacity for EDC production is approximately 51 billion pounds). Table 1 lists the producers, sites, and estimated annual capacity for 1979

Table 1
Estimated Annual Capacity for Production of EDC in the U.S. in 1979

Producer	Site	Annual capacity (millions of pounds)
Continental Oil Co.	Lake Charles, Louisiana	1,155
Diamond Shamrock Corp.	Deer Park, Louisiana	320
	LaPorte, Texas	1,585
Dow Chemical Co., USA	Freeport, Texas	1,600
	Oyster Creek, Texas	1,100
	Plaguemines, Louisiana	2,100
Ethyl Corp.	Baton Rouge, Louisiana	700
	Pasadena, Texas	260
B. F. Goodrich Co.	Calvert City, Kentucky	1,000
ICI Americas Inc.	Baton Rouge, Louisiana	695
PPG Industries, Inc.	Lake Charles, Louisiana	1,200
Shell Chemical Co.	Deer Park, Texas	1,400
	Norco, Louisiana	1,200
Stauffer Chem. Co.	Carson, California	340
Union Carbide Corp.	Taft, Louisiana	150
	Texas City, Texas	150
Vulcan Materials Co.	Gersimer, Louisiana	300
Total		15,255

Data from SRI International (1979).
Capacities are flexible depending on finishing capacities for VC and chlorinated solvents. PPG Industries, Inc., shut down 835 million pound/year facility at Guayanilla, Puerto Rico, in November, 1978.

production, which totaled approximately 15 billion pounds at 16 sites predominantly in Louisiana and Texas (SRI International 1979). EDC is currently the 15th-largest-volume U.S. commercial chemical (Chemical and Engineering News 1978b).

U.S. production of EDC increased from about 510 million pounds in 1955 to almost 8 billion pounds in 1972 (NIOSH 1976). This 16-fold increase was mostly due to increased vinyl chloride (VC) production, for which EDC is one of the basic raw materials.

It is estimated that approximately 83% of EDC never leaves the producing company and at least 90% of the chemical is converted immediately to VC monomer. Methyl chloroform (1,1,1-trichloroethane), trichloroethylene (TCE; 1,1,2-trichloroethene), ethyleamines, and perchloroethylene (PCE; tetrachloroethylene) account for 2-3% each of EDC use.

EDC has been used in large amounts as a lead scavenger agent in gasoline (over 10^8 kg in 1970) (SRI International 1977). According to the National Institute for Occupational Safety and Health (NIOSH), 36 U.S. firms put EDC at levels as high as 20% by weight into mixtures with EDB for use as a gasoline additive (Toxic Materials News 1978). There are at least six different formulations of antiknock fuel additives containing EDC available on the market (NIOSH 1976). On the basis of the best available information, the 1977 consumption of EDC for minor uses was estimated at about 11 million pounds, or slightly less than 0.1% of the total annual U.S. production. EDC is also used in various solvent applications and as a component in fumigants (with EDB) for grain, upholstery, and carpets (SRI International 1972). A mixture of 25% EDC and 75% carbon tetrachloride is generally used for grain fumigation in the U.S. At least 29 chemical companies manufacture at least 45 fumigant-insecticides that contain EDC (Billings 1974). As an intermediate insecticidal fumigant, it is primarily active against the peach tree bores, Japanese beetle, and root-rot nematodes.

Occurrence and Environmental Fate

The U.S. Environmental Protection Agency (EPA) estimated that about 163 million pounds of EDC entered the environment in 1974 from its use applications in the U.S. alone (Toxic Materials News 1978). Best estimates of current emissions indicate that approximately 11,000-44,000 metric tons of EDC are emitted annually from production and process facilities (EPA 1979). Losses of EDC to the environment arise principally from vapors released during primary production or during end-product manufacture, during dispersive use applications, and as a contaminant in waste water and in waste solids, principally as EDC tars. EDC tar is a complex mixture consisting chiefly of chlorinated aliphatic hydrocarbons (e.g., approximately 33% EDC, 1,1,2-trichloroethane, and about 0.06% VC monomer [Jensen et al. 1975]). The composition of EDC tar varies not only from factory to factory but also from time to time within the same factory. EDC in EDC tars from VC monomer production is estimated at about 60 million pounds/year. In the U.S., disposal of EDC tar is usually accomplished by burial in landfills or by incineration. Because of its volatility, there is a probability that EDC in buried wastes may eventually leak into the atmosphere.

Dispersive uses of EDC generally are considered to result in the release of all EDC to the environment. Use of EDC as a solvent or fumigant is a relatively minor source of emission, with such uses estimated to release 5000 metric tons. Auto emissions and use of EDC as an intermediate in the synthesis of other organic compounds were estimated to release about 4000 metric tons annually, and use as a grain fumigant may account for annual emissions of an additional 500 metric tons (EPA 1979).

On the basis of reaction rates with hydroxyl free radicals that constitute the principal removal mechanism concerning ambient air levels, EDC has an approximate tropospheric lifetime of 0.75-1.3 years (EPA 1979). Photooxidative reactions involving EDC are believed to result in the formation of monochloroacetylchloride, HCl, formylchloride, and monochloroacetic acid. Ambient air measurements made to date indicate that levels are in the ppb range.

Until recently, relatively few data were available regarding ambient air concentrations of EDC. However, recent measurements made at sites in three geographical study areas central to EDC production and user facilities in the U.S. indicated EDC concentrations as high as 745 $\mu g/m^3$ (186 ppb) (PEDCo Environmental, Inc. 1979). In Calvert City, Kentucky, ambient concentrations ranged from < 0.5 to 59.9 $\mu g/m^3$ (<0.123 to 14.8 ppb). Levels recorded in the New Orleans study area ranged from < 0.5 to 169 $\mu g/m^3$ (< 0.123 to 42 ppb) (PEDCo Environmental, Inc. 1979). Concentrations of EDC in air distant from point sources appears to be very low, probably in the ppt range.

Elkins (1959) reported the following average concentrations (ppm) of EDC in American industries: rubber cementing, 85-110; leather finishing, 65; fabric spreading, 125; drum filling, 35; and metal cleaning, 180. EDC concentrations ranging from 0.4 to 5 ppm were found in a wool-scouring plant in Massachusetts (NIOSH 1976). EDC has also been reported in an aircraft factory at levels ranging from less than 5 ppm to 52 ppm, where EDC was the solvent for glue used to make rubber tanks (Kozik 1957).

The EPA National Organics Reconnaissance Survey (NORS) of 80 water supplies indicated that 13% of the finished water contained EDC at a mean concentration of 1 μg/liter (EPA 1975a). Of the 80 water supplies surveyed in 1974, 26 contained EDC at less than 0.2-31 μg/liter (EPA 1975a). A concentration of 8 μg/liter has been reported in New Orleans finished water. The EPA Region-V survey of 83 water supplies concluded that EDC is not produced during chlorination of water. It should be noted however, that a recent EPA report concluded that EDC occurs with somewhat greater frequency in municipal finished water supplies apparently as a result of chlorine treatment (VERSAR 1975). EDC is difficult to degrade biologically; 1 part of EDC is soluble in about 120 parts of water (about 9000 ppm) (NAS 1977). In a compilation of pollutants found in water, Shackelford and Keith (1976) identified EDC as having been found in industrial effluents, finished drinking water, and river waters. The concentrations of EDC in surface waters distant from point sources are about 1 ppb, but some samples may contain a 100-fold greater amount. Although EDC is relatively stable in water, the relatively high vapor pressure of EDC can result in its rapid volatization from aqueous effluents.

In regard to bioaccumulation there is no current evidence to suggest that EDC bioaccumulates in the marine environment or in other biota, although the uptake of EDC by fish and oysters has been observed.

The use of EDC in fumigant mixtures (in admixture with EDB or carbon tetrachloride) for disinfecting fruits, vegetables, foodgrains, tobacco, seeds, seedbeds, mills, and warehouses, suggests the possibility that their residues per se or the residues of their respective hydrolytic products (ethylene chlorohydrin or bromohydrin) may be present in fumigated materials (Berck 1974; Fishbein

1976). Residues of EDC ranging from 23 to 43 ppm have been found in various portions of sacks of wheat several weeks after fumigation (Wit et al. 1969).

Residues of EDC (ranging from 3 to 23 $\mu g/g$) have also been detected in spice oleoresins for which EDC is employed as an extractant during preparation (Page and Kennedy 1975). Spice oleoresins identified as containing residues of EDC include black pepper, capsicum, celery, cinnamon, ginger, mace, marjoram, paprika, rosemary, sage, thyme and turmeric. There was a wide variability in EDC concentrations from similar products from each manufacture.

Potential Populations at Risk

Baier (1978) estimated that a total of about 2 million workers in the U.S. may receive some exposure to EDC, with perhaps as many as 200,000 receiving a substantial exposure primarily during its use as a solvent in textile cleaning and metal degreasing, in certain acrylic type adhesives and rubber cements, and as a component in fumigants (Baier 1978). However, NIOSH estimated in 1976 that 18,000 people are potentially exposed to EDC in their working environment (NIOSH 1976).

Formulators of insecticide mixtures and agricultural workers involved in the fumigation of a variety of crops are potentially exposed to EDC in their occupation. Additionally, formulators of antiknock compounds containing tetraethyl and tetramethyl lead might be exposed to EDC, which is a constituent of antiknock mixtures. This category of worker exposure is decreasing with decreasing use of tetraethyl and tetramethyl lead additives for gasoline in the U.S.

In 1975 NIOSH (Chemical Register Reporter 1977) recommended a reduction in the still current Occupational Safety and Health Administration (OSHA) standard of 50 ppm (202 mg/m^3) (8-hr time-weighted average [TWA]) to 5 ppm (20.2 mg/m^3) (TWA for up to a 10-hr workday, 40-hr workweek) (Chemical Register Reporter 1978). In September, 1978, NIOSH recommended a further reduction of the previously recommended standard from 5 ppm to 1 ppm (4.05 mg/m^3). This recommendation was based on the National Cancer Institute (NCI) findings regarding the carcinogenicity of EDC.

EDB

Production and Utility

EDB (also known as *sym*-dibromoethane, ethylene bromide) is prepared commercially by the liquid-phase bromination of ethylene. In an illustrative manufacturing process (Fig. 2), ethylene and bromine are reacted at 35-85°C at atmospheric pressure with EDB as the solvent in a glass packed-column vessel. After cooling, the reaction mixture is neutralized to remove free acid and the product purified by distillation (Lowenheim and Moran 1975b).

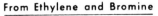

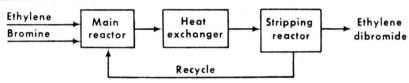

Reaction

$$CH_2{=}CH_2 + Br_2 \rightarrow BrCH_2CH_2Br$$

97% yield on ethylene
99% yield on bromine

Figure 2
Production of EDB. (Reprinted, with permission, from Lowenheim and Moran 1975b.)

Natural bromide-containing brines are treated with chlorine to release elemental bromine via anionic replacement. It is important to note that reaction between gaseous ethylene and liquid bromine, which can contain traces of chlorine, yields in addition to EDB, side products that can include very small amounts of vinyl bromide, ethyl bromide, and ethyl chlorobromide. However, if the reaction temperatures and pressures are carefully controlled, the purity of the EDB can reach 99.95% (NIOSH 1977b). EDB can also be commercially produced by the hydrobromination of acetylene, although this process is of current minor value (Olmstead 1972).

The U.S. production of EDB has increased from an estimated 64 million pounds in 1940 to a peak of 332 million pounds in 1974 (SRI International 1975). This fivefold increase can be primarily related to the increased consumption of gasoline containing EDB as an additive, which has always been its largest single use (NIOSH 1977b).

Although the U.S. production of EDB was 244 million pounds, down 32% from the peak 332 million pounds in 1974, it remains the leading bromine-based chemical. The outlook is for further sharp declines in demand, possibly by as much as 75% by 1980 because of phase-down of lead additives in gasoline and possible restriction or elimination of use in pesticides (e.g., DBCP) (Chemical Engineering News 1978a).

Approximately 85-90% of EDB production during the last few years has been as a constituent of antiknock mixtures containing tetraethyl lead (SRI International 1975; Johns 1976). Here, it is used as a scavenger to convert lead oxides in automobile and other engines to lead halides, which escape with engine exhaust. Approximately 213 million pounds of EDB were formulated into tetraethyl lead antiknock mixes in the U.S. in 1974. The concentration of EDB in these formulations varies. Motor fuel antiknock mixes can contain approximately 18% by weight of EDB (2.8 g/liter) and aviation gasoline antiknock mixes can contain about 36% by weight EDB (IARC 1977). Although aviation gasoline requires pure EDB, mixtures of EDC and EDB can be used in automobiles (Lowenheim and Moran 1975b).

Other areas of application (approximately 10-15%) include its use as a fumigant-insecticide or nematocide, a synthetic intermediate in the production of dyes and pharmaceuticals, and a specialty solvent for resins, gums, and waxes (Johns 1976).

In the U.S., EDB is registered for use as a soil fumigant on a variety of vegetable, fruit, and grain crops and has been used for disinfecting fruits, vegetables, grains, tobacco, and seeds in storage (EPA 1970). An estimated 5 million pounds were used in the U.S. in 1975 as a fumigant (mostly on agricultural crops). Although over 100 pesticides registered with the EPA contain EDB (Billings 1974; NIOSH 1977b), the compound is not available as an over-the-counter product (Johns 1976).

Previous uses of EDB now extremely limited or nonexistent include its use in fire extinguishers or as an industrial solvent.

EDB is produced in the U.S. by four companies at the following sites: Dow Chemical Company, USA, and Ethyl Corporation (Magnolia, Arkansas), Great Lakes Chemical Corporation (El Dorado, Arkansas), and PPG Industrial Inc. (Beaumont, Texas) (SRI International 1979). The production of EDB is the largest single use of bromine, and as such, the locations of manufacturing plants have usually been near the major sources of bromine, for example, Arkansas (NIOSH 1977b).

It should also be noted that EDB is currently produced in the United Kingdom, Benelux, France, Spain, Italy, and Switzerland, with production estimated to be about 10-66 million pounds per year. Another estimate of West European production of EDB for 1974 was about 90 million pounds (IARC 1977). In 1975, two Japanese companies produced about 3.0 million pounds, of which approximately 1.5 million pounds were imported (Muto 1976).

Occurrence and Environmental Fate

The chief sources of EDB emissions are from automotive sources via evaporation from the fuel tank and carburetor of cars operated on leaded fuel. Emissions from these sources have been estimated to range from 2 to 25 mg/day for 1972-1974 model-year cars in the U.S. (EPA 1975b).

Very limited and preliminary air monitoring data for EDB show air concentration values of 0.07-0.11 $\mu g/m^3$ (about 0.01 ppb) in the vicinity of gasoline stations along traffic arteries in three major U.S. cities, 0.2-1.7 $\mu g/m^3$ (about 0.1 ppb) at an oil refinery, and 90-115 $\mu g/m^3$ (10-15 ppb) at EDB U.S. manufacturing sites, suggesting that it is present in ambient air at very low concentrations (EPA 1975b).

It should be noted that the increased use of unleaded gasoline should result in lower ambient air levels of EDB from its major sources of emissions (EPA 1975b; Toxic Materials News 1976).

EDB has also been found in concentrations of 96 $\mu g/m^3$, up to a mile away from a U.S. Department of Agriculture fumigation center (Toxic Materials News 1976). Air concentrations of EDB have been measured in and around citrus fumigation centers in Florida. Average air levels to which site personnel were exposed ranged from 370 to 3100 $\mu g/m^3$ inside the facilities and from about 0.1 to 29 $\mu g/m^3$ outside (Pesticide Chemical News 1977).

Concentrations of EDB on the order of 1 ppb have been found in samples from streams of water on industrial sites. Limited information suggests that EDB degrades at moderate rates in both water and soil (EPA 1975b).

The use of EDC and EDB in fumigant mixtures for disinfecting fruits, vegetables, foodgrains, tobacco, seeds, seedbeds, mills, and warehouses suggests that their residues per se or their respective hydrolytic products (e.g., ethylene chlorohydrin or bromohydrin) may be present in fumigated materials (Olomucki and Bondi 1955; Wit et al. 1969; Berck 1974; Fishbein 1976). For example, when apples were fumigated with 12 or 24 mg/liter EDB at 13°C for 4 hours and stored at this temperature, the residues after 1 day were 36 and 75 mg/kg, which decreased after 6 days to 1.2 and 1.6 mg/kg (Dumas 1973).

Although materials such as EDC and EDB are volatile, and their actual occurrence in processed or cooked foods can possibly be considered negligible, more significant exposure is considered more likely among agricultural workers or among those fumigating grain and crops in storage facilities and the field than among consumers of the food products (Olson et al. 1973).

Potential Populations at Risk

NIOSH estimates that approximately 9000 employees (primarily manufacturing, formulating, fumigating) are potentially exposed to EDB in the workplace (Table 2); another 650,000 gasoline station attendants are exposed. No estimate can be currently given to the number of motorists who are potentially exposed to EDB during "self-serve" operations at the gas pump.

The Criteria and Evaluation Division of EPA has recently made a preliminary estimate concerning the exposure of professional applicators involved with EDB soil fumigation applications. Those individuals applying EDB for 30-40 days/year would receive a total annual inhalation dose of 3-40 mg/kg and farmer-applicators applying EDB for 7-10 days/year would receive a total annual inhalation dose of 0.7-10 mg/(kg · year) (EPA 1977).

The current OSHA environmental standard for EDB is 20 ppm (150 mg/m^3) for an 8-hour time-weighted exposure in a 40-hour workweek, a 30-ppm (230 mg/m^3) acceptable ceiling, and a 50-ppm (380 mg/m^3) maximum peak (5 min). NIOSH has recommended a 1-mg/m^3 ceiling (0.13 ppm) (15 min) environmental exposure limit (Pesticide Chemical News 1977; NIOSH 1977a).

Table 2
Occupations with Potential Exposures to EDB

Antiknock compound makers	lead scavenger makers
Cabbage growers	motor fuel workers
Celluloid makers	nematode controllers
Corn growers	oil processors
Drug makers	organic chemical synthesizers
EDB workers	resin makers
Fat processors	seed corn maggot controllers
Fire extinguisher makers	soil fumigators
Fruit fumigators	termite controllers
Fumigant workers	tetraethyl lead makers
Gasoline blenders	waterproofing makers
Grain elevator workers	waxmakers
Grain fumigators	wood insect controllers
Gum processors	wool reclaimers

Data from NIOSH (1977b).

REFERENCES

Baier, E. J. 1978. Statement on ethylene dichloride before the Subcommittee on Oversight and Investigations, House Committee on Interstate and Foreign Commerce, Jan. 23. Washington, D.C.

Berck, B. 1974. Fumigant residues of carbon tetrachloride, ethylene dichloride and ethylene dibromide in wheat, flour, bran, middlings and bread. *J. Agric. Food Chem.* **22**:977.

Billings, S. C. 1974. *Pesticides handbook-entoma,* 25th ed. Entomological Society of America, College Park, Maryland.

Chemical and Engineering News. 1978b. Ethylene dichloride cancer issue heats up. *Chemical and Engineering News,* September 25, p. 6.

───────. 1978a. Debate renewed over ethylene halides. *Chemical and Engineering News,* November 20, p. 7.

Chemical Register Reporter. 1977. Ethylene dibromide: Institute recommends ceiling limit one milligram, engineering controls. *Chemical Register Reporter* **1**:7.

───────. 1978. Current standards for EDC are inadequate, NIOSH bulletin says. *Chemical Register Reporter* **2**(6):197.

Dumas, T. 1973. Inorganic and organic bromide residues in foodstuffs fumigated with methyl bromide and ethylene dibromide at low temperatures. *J. Agric. Food Chem.* **21**:433.

Elkins, H. B. 1959. *The chemistry of industrial toxicology,* 2nd ed. John Wiley & Sons, New York.

EPA (Environmental Protection Agency). 1970. *U.S. environmental compendium of registered pesticides,* pp. III-E-0.1-III-E-9.5. Government Printing Office, Washington, D.C.

———. 1975a. *Draft report for congress: Preliminary assessment of suspected carcinogens in drinking water.* Office of Toxic Substances, Government Printing Office, Washington, D.C., October 17.

———. 1975b. *Sampling and analysis of selected toxic substances.* Government Printing Office, Washington, D.C.

———. 1977. EPA notice of rebuttable presumption against registration and continued registration of pesticide products containing ethylene dibromide. *Federal Register* 42:63134, Dec. 14; *Chemical Register Reporter* 1:1436.

———. 1979. *Ethylene dichloride* (June 22). Environmental Protection Agency, Environmental Criteria and Assessment Office, Research Triangle Park, North Carolina.

Faith, W. L., D. B. Keyes, and R. L. Clark. 1965. *Industrial chemicals*, 3rd ed., p. 368, 757, 805. John Wiley & Sons, Inc., New York.

Fishbein, L. 1976. Potential hazards of fumigant residues. *Environ. Health Perspect.* **14**:39.

Hardie, D. W. F. 1964. Chlorocarbons and chlorohydrocarbons. In *Kirk-Othmer encyclopedia of chemical technology*, p. 149. John Wiley & Sons, Inc., New York.

IARC (International Agency for Research on Cancer). 1977. Ethylene dibromide. *IARC Monogr.* **15**:195.

Jensen, S., R. Lange, G. Berg, K. H. Plark, and K. Renberg. 1975. On the chemistry of EDC-tar and its biological significance. *Proc. R. Soc. Lond. B Sci.* **189**:333.

Johns, R. 1976. *Air pollution assessment of ethylene dibromide*, p. 33. Environmental Protection Agency, Office of Toxic Substances. Government Printing Office, Washington, D.C.

Kozik, I. 1957. Problems of occupational hygiene in the aviation industry. *Gig. Tr. Prof. Zabol.* **I**:31.

Lowenhein, L. A. and M. K. Moran. 1975a. Ethylene dichloride. In *Faith, Keyes and Clark's industrial chemicals*, 4th ed., p. 389. John Wiley & Sons, Inc., New York.

———. 1975b. Ethylene dibromide. In *Faith, Keyes, and Clark's industrial chemicals*, 4th ed., p. 389. John Wiley & Sons, Inc., New York.

Muto, T. 1976. *Noyaku Yoran*, p. 35. Nippon Plant Boeki Association, Tokyo.

NAS (National Academy of Sciences). 1977. *Drinking water and health*, p. 723. National Academy of Sciences, Washington, D.C.

NIOSH (National Institute for Occupational Safety and Health). 1976. *Criteria for a recommended standard: Occupational exposure to ethylene dichloride (1,2-dichloroethane)*, p. 16. Government Printing Office, Washington, D.C.

———. 1977a. Summary of NIOSH recommendations for occupational health standards, October, 1977. *Chemical Register Reporter* **1**(40):1376.

———. 1977b. *Criteria for a recommended standard: Occupational exposure to ethylene dibromide*, p. 25. Government Printing Office, Washington, D.C.

Olmstead, E. V. 1972. Dibromoethane. In *Encyclopedia of occupational health and safety*, p. 384. International Labour Office, Geneva.

Olomucki, E. and A. Bondi. 1955. Ethylene dibromide fumigation of cereals. I. Sorption of ethylene dibromide by grain. *J. Sci. Food Agric.* 6:592.

Olson, W. A., R. T. Haberman, E. K. Weisburger, J. M. Ward, and J. H. Weisburger. 1973. Induction of stomach cancer in rats and mice by halogenated aliphatic fumigants. *J. Natl. Cancer Inst.* 51:1993.

OSHA (Occupational Safety and Health Administration). 1976. Air Contaminants. *U.S. Code of Federal Regulations,* Title 29, part 1919.1000 (e), p. 30.

Page, B. D. and B. P. C. Kennedy. 1975. Determination of methylene chloride, ethylene dichloride and trichloroethylene as solvent residues in spice oleoresins, using vacuum distillation and electron capture gas chromatography. *J. Assoc. Off. Anal. Chem.* 40:206.

PEDCo Environmental Inc. 1979. *Monitoring of ambient levels of ethylene dichloride (EDC) in the vicinity of EDC production and user facilities.* Environmental Protection Agency, Contract No. 68-02-2722, January.

Pesticide Chemical News. 1977. EDB presents close to 100% cancer risk for citrus fumigators, EPA finds. *Pesticide Chemical News* 5(46):3.

Shackelford, W. M. and L. H. Keith. 1976. *Frequency of organic compounds identified in water.* EPA Report No. 600/4-76-062. Environmental Research Laboratory, Office of Research and Development, Environmental Protection Agency, Athens, Georgia, December.

SRI International. 1972. *Chemical economics handbook.* Stanford Research Institute, Menlo Park, California.

———. 1975. Ethylene dibromide—salient statistics. In *Chemical economics handbook,* p. 650. Stanford Research Institute, Menlo Park, California.

———. 1977. *A study of industrial data on candidate chemicals for testing.* Prepared for U.S. Environmental Protection Agency. EPA-560/5-77-06. Menlo Park, California, August.

———. 1979. *Directory of chemical producers, USA,* p. 598. Stanford Research Institute, Menlo Park, California.

Stenger, V. A. and G. J. Atchison. 1964. Bromine compounds. In *Kirk-Othmer encyclopedia of chemical technology,* p. 771. John Wiley & Sons, Inc., New York.

Toxic Materials News. 1976. Ethylene dibromide "ubiquitous" in air, EPA report says. *Toxic Materials News* 3:12.

———. 1978. NCI finds ethylene dichloride to be carcinogenic. *Toxic Materials News* 5:284.

VERSAR, Inc. 1975. *Identification of organic compounds in effluents from industrial sources.* Report for EPA: EPA-560/3-75-002, April.

Wit, S. L., A. F. H. Besemer, H. A. Das, W. Goedkoop, F. E. Loosjes, and E. R. Meppelink. 1969. *Results of an investigation on the regression of three fumigants (carbon tetrachloride, ethylene dibromide and ethylene dichloride) in wheats during processing to bread.* Report No. 36/69. National Institute of Public Health, Bilthoven, Netherlands.

Use and Air Quality Impact of Ethylene Dichloride and Ethylene Dibromide Scavengers in Leaded Gasoline

EMMETT S. JACOBS
Petroleum Laboratory
E. I. du Pont de Nemours & Company, Inc.
Wilmington, Delaware 19898

Soon after the discovery of the antiknock activity of tetraethyl lead by Midgley and Boyd in 1921 (Boyd 1950), it was found that tetraethyl lead (TEL) alone could not be used in gasoline. It left solid deposits of lead oxides and lead sulfates, which accumulated on the combustion chamber walls, piston top, exhaust valves, and spark plugs and caused significant deterioration in engine performance. However, Midgley, who had worked for years to discover the antiknock activity of TEL, quickly determined that the trouble could be corrected by adding an organic compound of bromine, or bromine and chlorine, to the gasoline along with TEL (Boyd 1950). Thus followed the first commercial sale of gasoline with TEL to a motorist in Dayton, Ohio, in February 1923. Also in 1923, the first commercial production of TEL was started by E. I. du Pont de Nemours & Company for the Ethyl Gasoline Corporation.

Ever since 1923 all alkyl lead compounds used in gasoline to suppress engine knock have contained ethylene dichloride (EDC; 1,2-dichloroethane) and ethylene dibromide (EDB; 1,2-dibromoethane), as halide scavengers, to reduce the accumulation of inorganic lead deposits in engines. This paper will provide a brief review of the history of halide scavengers use, their mechanism of action, current compositions, and past and future estimates of their use in gasoline in the United States.

As part of the effort to evaluate exposure due to use of EDC and EDB in gasoline, this paper also presents information on measurement of these compounds in occupational and ambient air. These data include workplace exposure in manufacture of alkyl lead antiknock mixes, concentration in vehicle exhaust, gasoline service station attendant exposure, and ambient air levels in cities near gasoline stations and heavily trafficked roads.

EXPERIMENTAL

Monitoring EDB and EDC in Air

Sampling in the antiknock blending plant was done by drawing a known volume of air through a charcoal tube to trap the EDC and EDB vapors present. The vapor trapped on the charcoal was desorbed with carbon disulfide and an aliquot of each desorbed sample was analyzed by gas chromatography using an electron capture detector. This procedure is essentially the same as the National Institute for Occupational Safety and Health (NIOSH) analytical procedure (NIOSH 1978a).

Personal samples were collected in the breathing zone of individual employees by attaching the charcoal sample tube to the shirt collar or lapel of each employee's coat. Plastic tubing was used to connect the charcoal tube to the personal sampling pump on the employee's belt. The air-flow rate of this sampler system was checked before and after sampling.

Service Station Monitoring for EDC and EDB

A gasoline service station exposure test to EDC and EDB was conducted by Du Pont at its Petroleum Laboratory fuel handling facilities, located on the Du Pont Chambers Works Plant, Deepwater New Jersey. A diagram of the fuel handling area is shown in Figure 1. The fueling operation consisted of refueling

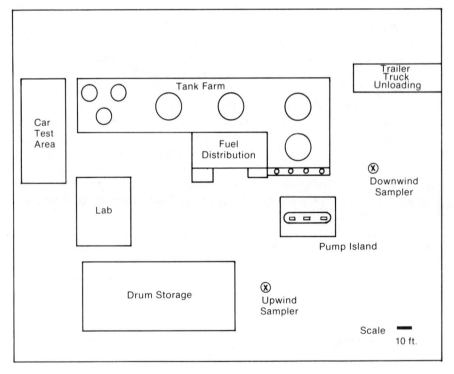

Figure 1
Du Pont Petroleum Laboratory fuel-handling area

88 privately owned automobiles with 1577 gallons of gasoline containing 1.61 g lead/gallon. The fueling operation was performed by two operators who wore personal air monitors during the 1.8-hour operation. This operation is equivalent to a gas station handling 250,000–300,000 gallons of gasoline a month, which is a very large operation when compared to the average of about 50,000 gallons per month for all gas stations in the United States. The test was conducted in July, with ambient temperatures of 89°F and a 4-mph wind from the southwest. Thus, the test conditions provided for maximum exposure to vapors of EDB from gasoline.

Personal air monitors sampled the air at the breathing level of each service station operator by drawing a known volume of air through a charcoal absorption tube. The two operators stayed next to the vehicle during complete refueling of each car. In addition to the personal air monitoring samples, a continuous air sample was also collected on charcoal at a point 55 feet upwind and 35 feet downwind of the fuel pump island (Fig. 1). The charcoal from each of the personal and the continuous air samplers was desorbed with carbon disulfide and a portion was analyzed by gas chromatography using an electron capture detector (NIOSH 1978a).

In addition to the charcoal air samplers, samples of ambient air were collected in previously evacuated glass containers and the air was analyzed directly for EDB by gas chromatography using an electron capture detector. One sample was taken at the fuel pump island, four samples were taken near vehicle fuel tanks during fueling, and one each at the upwind and downwind sampler sites.

EDC and EDB in Vehicle Exhaust

The sample collection system shown in Figure 2 was used to obtain samples of vehicle exhaust. The test vehicles, a 1966, 307 C.I.D. Chevrolet and a 1974,

Sample Collection

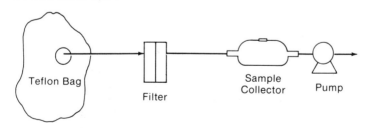

Figure 2
Sample collection method for measurements of scavenger in vehicle exhaust. Analysis is by gas chromatography with electron capture detector.

350 C.I.D. Chevrolet, were driven on a Clayton Dynamometer at idle, 30 mph steady cruise, and on the Federal Test Procedure used for vehicle emission tests (EPA 1972). For each driving condition, the exhaust was collected in a plastic sampling bag using the constant volume sampler system employed with the Federal Test Procedure. A portion of this diluted exhaust was sucked through a filter and into a glass sample bulb. A sample from the glass bulb was analyzed by gas chromatography using an electron capture detector.

The cars were tuned to manufacturer's specifications and, in the case of the 1974 model, the exhaust emissions of hydrocarbons, carbon monoxide, and nitrogen oxides met the 1974 Federal vehicle emission standards. Both vehicles were operated on gasoline containing tetraethyl lead motor mix at a concentration of 2.0 g lead/gallon, 0.95 g EDC/gallon, and 0.91 g EDB/gallon.

DISCUSSION

Halide Scavengers in Gasoline

One of the important factors limiting the power and efficiency of the internal combustion engine is the occurrence of knock. Knock is identified as an audible pinging sound in gasoline-powered internal combustion engines. The sound is caused by high-frequency vibration of the engine resulting from uneven burning, explosive combustion, or detonation of the unburned fuel and air mixture ahead of the advancing flame front in a cylinder. Broadly speaking, the occurrence of knock is a technical barrier to the efficient operation of engines. The two methods used to overcome this limitation are better design of engines and preventive use of chemicals. Considerable progress in engine design has been achieved and undoubtedly more progress will be made in the future, but the major method for overcoming knock limitations of engine performance and efficiency has been by chemical means.

Chemical means include refinery processing to increase the antiknock quality of the hydrocarbon fuel. This is expensive since it uses some fuel to operate the refining processes as well as loses fuel because of less than 100% yield. The most effective means to improve fuel antiknock quality is through the use of lead antiknock agents such as TEL. Tetraalkyl lead compounds suppress knock by decomposing during combustion into lead oxides, which deactivate the chain reactions responsible for the detonation. Lead antiknocks are less expensive than refinery processing and, more importantly, use of lead antiknock agents in gasoline saves energy loss in refining and allows greater fuel economies and performance in automobiles. The use of lead antiknocks in gasoline can provide an overall saving of 6-12% in total gasoline used in the U.S. (Cantwell et al. 1978).

Although almost all of the combustion products from lead antiknock compounds are removed from the combustion chamber with the exhaust gases, a small amount remains on the combustion chamber surfaces in the form of inorganic lead salts. In addition to lead salts, the combustion chamber deposits contain carbonaceous materials resulting from incomplete combustion of the fuel and lubricating oil, and minor amounts of metallic and nonmetallic compounds introduced into the combustion chamber with the air, fuel, and oil, and produced as a result of engine wear. Although deposits vary in composition with fuel and engine operation, a typical deposit consists of 5-10% by weight of carbonaceous material, the remainder being a complex mixture of lead salts, as shown in Table 1 (Newby and Dumont 1953). The uncontrolled buildup of these deposits on the combustion chamber surfaces can cause significant effects on engine operation.

Deposits on combustion chamber walls and piston tops increase engine octane requirements by occupying space, which increases the effective compression ratio of the engine. Such deposits also insulate the fuel-air mixture from the cooler combustion chamber walls and raise the mixture temperature, which also increases the engine octane requirement.

The formation of deposits on spark plugs can lead to misfiring by providing a low-resistance electrical path to ground; worse, deposits can bridge the spark plug gap and prevent sparking.

Deposits on exhaust valves can insulate the valve from the cooler cylinder head, which can cause overheating and burning. The deposits may flake away from small areas, resulting in valve overheating and burning and leakage of the fuel-air mixture. Burned exhaust valves reduce power output.

Composition of Halide Scavengers

The initial work by Midgley identified that organic compounds of bromine and chlorine when added to leaded gasoline would correct or control the deposit problems (Boyd 1950). This is based on the fact that lead halides are more

Table 1
Lead Compounds Formed in Engine Combustion Chamber Deposits

Compound	Melting point (°C)
$PbCl_2$	496
$PbBr_2$	370
$PbSO_4$	1170
PbO	888
$(PbBr_2)_x (PbCl_2)_y$	370-496
$PbO \cdot PbCl_2$	524
$2PbO \cdot PbCl_2$	693
$PbO \cdot PbBr_2$	497 (Decomposed)
$2PbO \cdot PbBr_2$	709
$PbO \cdot PbSO_4$	975
$2Pbo \cdot PbSO_4$	895

volatile than lead oxide or lead sulfate (see Table 1). Investigators at Du Pont (Newby and Dumont 1953) have shown that the organohalides added along with lead antiknock compounds form halogen acids, which help remove lead deposits by reacting with lead oxide after it has been deposited on combustion chamber surfaces, i.e., halide scavenging takes place on the deposit surface and not in the gas phase. Typical reactions in the combustion chamber illustrating this mechanism of deposit scavenging are shown in Table 2.

Over the years, considerable study of the lead scavenging problem has resulted in several modifications of Midgley's original formulation, as shown in Table 3. Today all leaded gasolines use EDC and EDB as scavengers added in amounts proportional to the amount of lead present. All Du Pont antiknock mixes are listed in Table 4. The names used by other suppliers are slightly different, but the compositions are the same. Conventional motor-mix formulations contain 1 mole EDC and .5 mole EDB/mole tetraalkyl lead. Thus, an antiknock is said to contain one theory of halogen scavenger when it contains an amount sufficient to convert all of the lead present to lead chloride ($PbCl_2$) or lead bromide ($PbBr_2$). All current antiknock mixes contain 1.0 theory of chlorine as EDC and 0.5 theory of bromine as EDB. The one exception is TEL Aviation Mix, which contains only one theory of halogen as EDB (1 mole EDB/mole TEL). The antiknock mixtures are formulated such that each contains 39.39 weight percent of lead calculated as lead metal. The balance of the antiknock mixture consists of small quantities of dye to aid in leak detection, antioxidant for stability, and inert diluent, such as kerosene, to assist in standardizing mixture strength.

Antiknock mixes are shipped worldwide in a variety of containers including portable tanks, drums, tank trucks, rail tank car, and as bulk cargo

Table 2

Lead Antiknock and Scavenger Reactions in Engine Combustion Chamber

Gas phase — before and during combustion
  $(Ethyl)_4\,Pb \rightarrow PbO$ (active antiknock)
  $EDB \rightarrow HBr$
  $EDC \rightarrow HCl$
  $Sulfur \rightarrow SO_2$

Solid phase — after combustion
  $PbO + HBr \rightarrow PbBr_2 + H_2O$
  $PbO + 2\,HCl \rightarrow PbCl_2 + H_2O$
  $PbO + HBr + HCl \rightarrow PbClBr + H_2O$
  $PbBr_2 + n\,PbO \rightarrow PbBr_2 \cdot n\,PbO$
  $PbCl_2 + n\,PbO \rightarrow PbCl_2 \cdot n\,PbO$

  $PbO + O + SO_2 \rightarrow PbSO_4$

Table 3
History of Scavenger Use in the U.S.

Marketing period	Halogens (theories)	
	bromine[a]	chlorine[b]
1926-1928	1.5	0.1
1928-1929	1.15	0.1
1929-1930	1.0	0 (aviation mix)
1930-1933	0.85	0.3
1933-1934	0.75	0.4
1934-1942	0.70	0.45
1942-present	0.50	1.0 (motor mix)

[a] As EDB.
[b] As EDC.

in specially designed ships and barges dedicated to that service. Within the U.S., most are transported by railroad tank car and shipped as Department of Transportation class-B poisons.

Use of Halide Scavengers in the U.S.

The use of EDC as halide scavengers is directly tied to the use of lead antiknock compounds in gasoline. In the U.S., the Environmental Protection Agency (EPA) has issued two separate regulations that will eventually eliminate the use of lead antiknock compounds and halide scavengers in gasoline.

The first regulation, issued in 1973 (EPA 1973a), required that unleaded gasoline be made available for use in cars using catalytic exhaust emission control devices. Since 1975, all new cars made in the U.S. use exhaust catalytic converters to meet federally mandated automotive exhaust emission standards; the same is true for virtually all foreign cars sold in the United States since 1979. Thus, all new cars require unleaded gasoline. With time, all the older cars on the road that could use leaded gasoline will be scrapped and disappear from the gasoline market. Thus, the market for leaded gasoline will decline and the volume of unleaded gasoline will increase. The use of leaded gasoline in passenger cars will disappear around 1990 but some leaded gasoline will still be required for gasoline-powered trucks.

In December 1973, EPA issued the second lead regulation requiring a gradual reduction of the lead content in the total gasoline pool (EPA 1973b). The lead phase-down regulation currently in effect restricts the lead content of the total gasoline pool to an average of 0.5 g lead/gallon. However, refiners may obtain a waiver from the EPA permitting production of gasoline with an average lead content of 0.8 g/gallon until October 1, 1980, if they adhere to certain unleaded gasoline production volume requirements (EPA 1979).

Table 4
Composition of Du Pont Antiknocks (Weight %)

	TEL[a]		TML[b]	Tetramix®			PM-10	PM-25	PM-50	PM-75
	aviation	motor		25	50	75				
Total lead alkyls				58.82	56.16	53.49				
TEL	61.41	61.49	—				55.34	46.11	30.75	15.37
TML	—	—	50.82				5.08	12.71	25.41	38.12
EDB	35.68	17.86	17.86	17.86	17.86	17.86	17.86	17.86	17.86	17.86
EDC	—	18.81	18.81	18.81	18.81	18.81	18.81	18.81	18.81	18.81
Solvents, antioxidant, dye, and inerts	2.91	1.84	12.51	4.51	7.17	9.84	2.91	4.51	7.17	9.84

[a] Tetraethyl lead.
[b] Tetramethyl lead.

As a consequence of these EPA regulations, the use of lead and associated halogen scavengers (EDC and EDB) in gasoline has declined; the peak use occurred in the 1969-1970 period. As shown in Table 5, the use of EDC and EDB as lead scavengers has decreased from around 100,000 metric tons in 1970 to about 60,000 metric tons in 1979. Because of a further decline in use of leaded gasoline in the U.S., as required by EPA regulations and use of unleaded gasoline in all new cars, by 1985 the use of EDC and EDB in leaded gasoline will decline to 23,000 and 22,000 metric tons, about 38% of today's use. Overall, the use of EDC and EDB as lead scavengers in 1985 will decline by 80% from the peak use in 1970.

The data in Table 5 are based on Du Pont information on average lead levels in gasoline and annual gasoline volume used in the U.S. as reported by the Department of Energy (DOE 1979). Future estimates of use are made with Du Pont's ESCON econometric and demographic model for projecting gasoline demand (Cantwell et al. 1978).

At present, the use of EDB as a scavenger in leaded gasoline accounts for about 90% of the total U.S. use in 1976 (SRI International 1979). Because of the restrictions on leaded gasoline noted above, and unless other uses are found, this material may essentially disappear from the market.

EDC used in leaded gasoline accounted for less than 3% of the total EDC used in the U.S. in 1970 and only 1.4% of the total in 1978 (SRI International 1979). Thus, its use as a lead scavenger in gasoline has been and will continue to be very small in comparison to total EDC use in the U.S.

Table 5
U.S. Consumption of EDC and EDB as Lead Scavengers

Year	Metric tons (10^3)		
	lead[a]	EDC	EDB
1970	216	103	98
1971	206	98	93
1972	212	101	96
1973	204	98	93
1974	176	84	80
1975	166	79	75
1976	171	82	78
1977	161	77	73
1978	151	72	68
1979[b]	126	60	57
1980[b]	78	37	35
1985[b]	49	23	22

[a] Lead metal used as antiknock agent.
[b] Estimated by Du Pont.

Workplace Exposure to EDC and EDB

EDC and EDB exposures were measured at a Du Pont plant where lead antiknock blends are prepared. The facilities for the production of the antiknock blends using EDC and EDB have been in operation since 1923. The halide scavengers are not made by Du Pont; instead they are brought in by rail tank car and unloaded into storage tanks for later use in blending to make antiknock mixes.

The antiknock blends are made using batch-type operations in which measured amounts of EDC, EDB, and other antiknock blend constituents are mixed in tanks to obtain a homogeneous mixture. These blends are physical mixtures of the constituents rather than chemical reactions of them. The finished blend or antiknock mixes are transferred by pumps to storage tanks for shipment by tank car, truck, or drums.

Blending is a closed-system operation in which ingredients and products are automatically transported from one point to another by piping and pump systems. Pumps are activated and the process is controlled from a control room. The only manual operations that are conducted include connecting and disconnecting hoses while loading and unloading tank cars, taking quality control samples, and processing and loading drums. Personnel involved in drum processing and loading and tank car loading and unloading are required to use rubber gloves and a full facepiece respirator with organic vapor canister. Quality control sampling is conducted under hood exhaust ventilation, so only rubber gloves are required for handling the sampling bottles.

The results of monitoring worker exposure to EDC and EDB in the antiknock blending and storage area are given in Tables 6 and 7. The job descriptions are the same as reported in the recent NIOSH report on EDB (NIOSH 1978b). In addition to these long-time personal samplings, some short-time additional monitoring of specific tasks was conducted for EDB. These results are included in Table 8.

NIOSH has recommended a 1-ppm standard as a time-weighted average (TWA) exposure to EDC for a 10-hour workshift. They further recommend a

Table 6
Personal Air Levels of EDC in Antiknock Blending Plant

Job description	8-hr TWA, EDC concentration (ppm)	
	range	median
Tank sampler[a]	0.001-2.2	0.5
EDC storage building	—	0.05
Blender operator	0.003-0.18	0.023

[a] Full-face respirator worn during this task.

Table 7
Personal Air Levels of EDB in Antiknock Blending Plant

Job description	Number of samples	8-hour TWA EDB concentration (ppm)	
		range	median
Blend operator	6	0.001-0.009	0.006
Relief operator	2	0.0005-0.007	0.004
Reactor operator	2	0.001-0.003	0.002
Bulk operator	2	0.001-0.008	0.004
Lab technician	4	0.0001-0.0005	0.0004
Drum loader	4	0.008-0.018	0.014
Drum processor	3	0.012-0.036	0.016
Raw material handler	2	0.027-0.082	0.054

Table 8
Short-Term Air Levels of EDB in Antiknock Blending Plant

Task[a]	Site	Sampling time	EDB concentrations (ppm)
Quality control sample	EDB tank car	13 min, 10 sec	0.7
Loading tank car	tank car	7 min	0.14

[a] Respirator worn during these tasks.

ceiling concentration of 2 ppm over a 15-minute sampling period. The present federal standard is 50 ppm measured as a TWA with a ceiling of 100 ppm, and a maximal allowable peak above the ceiling of 200 ppm for no more than 5 minutes during any 3-hour period (NIOSH 1978b).

Existing engineering controls at the plant surveyed were found to be adequate in maintaining EDC vapor concentrations within the current federal standard of 50 ppm and the NIOSH recommended standard of 1 ppm.

Existing engineering controls were also adequate in maintaining EDB vapor concentration within the current 20 ppm 8-hour TWA federal standard. However, additional protective measures in the form of respiratory protection are used for performance of selected tasks, such as quality control sampling and loading or unloading tank cars. At present, the concentration of EDB vapors in air during these short-time, special tasks, exceeds the recent NIOSH recommended 0.13-ppm ceiling for a 15-minute sampling period (NIOSH 1978a). The presence of EDB vapors in the air in blending plants surveyed did not exceed an 8-hour TWA concentration of 0.13 ppm.

EDB in Vehicle Exhaust

The results of measurements of EDB in vehicle exhaust, given in Table 9, show the concentration of EDB in the raw, undiluted exhaust from vehicles using leaded gasoline ranged from 5 to 19 ppb. These data are from duplicate runs for each vehicle for each driving condition. EDC was not detected, primarily because the detection level for EDC by the electron capture detector was much higher, 150 ppb, than the 1-ppb detection limit for EDB.

Based on studies of the ratio of carbon monoxide concentrations in exhaust and in ambient air, the exhaust from cars usually is diluted from 100- to 1000-fold in normal driving. Thus, based on these levels of EDB in exhaust, the concentrations of EDB in air alongside roads owing to vehicle exhaust emissions may range from 0.005 to 0.19 ppb.

These results concerning EDB in vehicle exhaust are very similar to recent data reported for cars in England (Leinster et al. 1978). Leinster et al. reported EDB concentrations in the exhaust of a 1.6-liter Ford, driven on the European Test Driving Cycle, at idle and steady speeds of 10, 30, 40, and 50 mph were 15, 9, 10, 8, 5, and 0-3 ppb respectively. The smaller concentrations seen in the exhaust of the Ford and the 1966 Chevrolet as compared to the 1974 Chevrolet are probably related to much lower fuel economy (only 12.2 mpg) observed with the larger 1974 vehicle.

Gas Station Attendant Exposure to EDB

An obvious potential exposure to halide scavengers used in leaded gasoline could occur during fueling of automobiles. This could occur due to vapor emissions from the gas tank as well as evaporation from gasoline spills. To determine the extent of this potential exposure, Du Pont conducted personal air monitoring during vehicle refueling at its Petroleum Laboratory fuel handling facilities at

Table 9
Concentration of Halide Scavengers in Vehicle Exhaust

Vehicle	Driving cycle	Scavenger in exhaust (ppb)[a]	
		EDB	EDC
1966 Chevrolet	FTP[b]	7.2	n.d.[c]
	idle	6.0	n.d.
	30 mph	5.0	n.d.
1974 Vehicle	FTP[b]	19	n.d.
	idle	16	n.d.
	30 mph	15	n.d.

Gasoline: 2.0 g Pb/gal of TEL, Motor Mix.
[a] Detection limit for EDB is 1 ppb, for EDC, 150 ppb.
[b] Federal Test Procedure.
[c] Not detected.

the Du Pont Chambers Works Plant in July 1975. The test conditions, described earlier, provided for maximum exposure to EDB vapors from gasoline during refueling of vehicles.

The results of analyses of air samples collected during this test are given in Table 10. These data show the average exposure of the gas station attendants to EDB during refueling was only 0.24 ppb. Measurements of the vapors at the car fuel tank filler pipe showed a maximum instantaneous EDB concentration of 13.8 ppb, with an average of four samples of 1.3 ppb. This represents the maximum for a short-term exposure. The concentration of EDB in air at the fuel pump island was not much different than the values measured at the upwind and the downwind sites. Overall, the very low air levels measured in this test indicate the potential for gas station attendant exposure to EDB while refueling cars is very low and is much less than any current or proposed air standard for EDB.

Calculated Air Levels of EDC and EDB from Gasoline

In addition to the exposure to the halide scavengers EDC and EDB during fueling operations, there is also potential air exposure due to evaporative emissions of gasoline from carburetors and open storage containers and to vapors from a spill of leaded gasoline. It is difficult to conduct tests to evaluate these possibilities, since conditions of potential exposure vary widely. However, calculation of the maximum potential air levels from these evaporative emissions conditions can be performed using known vapor pressures of EDC and EDB.

The results of calculated air levels of EDB and EDC from leaded gasoline because of vapors leaked from a tank and concentrations owing to leaded gasoline spill are given in Table 11. In each case, it is assumed that the gasoline hydrocarbon vapor concentration is 500 ppm. This, of course, is an arbitrary choice. The gasoline vapor concentration can be higher or lower depending on

Table 10
Gas Station Attendant Exposure to EDB During Vehicle Refueling

Sample	Concentration of EDB in air (ppb)
Upwind background	<0.1
Downwind background	<0.1
Fuel pump island	0.13
Near vehicle fuel pipe during refueling	1.3[a]
	13.7[b]
Personal air sampler	0.28
	0.20

[a] Average.
[b] Maximum.

Table 11
Air Levels of EDC and EDB from Leaded Gasoline Vapors and Spills

Component	Concentration in gasoline (g/gal)	Calculated air concentration (ppm)[a]	
		equilibrium vapor	gasoline spill
Lead (as TEL)	3.18	0.00022	0.262
EDC	1.52	0.054	0.262
EDB	1.44	0.0036	0.132

Gasoline characteristics: vapor pressure, 6 psi at 20°C; molecular weight, 95; specific gravity, 0.73; temperature, 20°C.

[a] Assumes gasoline hydrocarbon vapor concentration of 500 ppm.

the extent of vapor leakage from tanks or volume of gasoline spilled. The gasoline hydrocarbon vapor concentration of 500 ppm, however, is about the maximum value reported around bulk gasoline storage and transfer terminals.

The calculated air levels, shown in Table 11, represent maximum values for these situations. Overall, these EDC and EDB air levels are less than the recommended NIOSH standards of 0.13 ppm for EDB and 1.0 ppm for EDC.

Measured Ambient Air Levels of EDB and EDC

The U.S. EPA has reported the results of several studies on the measurement of EDC and EDB in the ambient atmosphere of cities as well as urban and rural area (Going and Lang 1975; Going and Spigarelli 1976). Presumably, sources of these air levels are emissions from gas stations dispensing leaded gasoline and evaporative emissions from cars using leaded gasoline.

Sampling was done by drawing a known volume of air through charcoal tubes to trap the vapors of EDC and EDB. The vapors trapped on the charcoal were desorbed with hexane and an aliquot of each desorbed sample was analyzed by gas chromatography using an electron capture detector.

A summary of the results reported by EPA is presented in Table 12. Overall, these results indicate the air levels are very low, ranging from 0.006 to 0.45 ppb in worse-case conditions near gas stations and heavy traffic in cities. These ambient air levels are consistent with measurements at service stations shown above and the concentrations to be expected in urban air based on evaporative and exhaust emissions. The air levels are very low and are about 100 to 10,000 times less than the most stringent air level recommended by NIOSH.

Table 12
Ambient Air Levels of EDC and EDB Reported by EPA

Location	Traffic (vehicles/day)	Air concentration (ppb) EDC	EDB
Near gas stations			
Phoenix, Arizona	38,000	–	0.32
Phoenix, Arizona	36,000	0.008	0.006
Los Angeles, California	46,000	–	0.11
Los Angeles, California	53,000	0.013	0.01
Camden, New Jersey	13,000	–	0.45
Seattle, Washington	32,000	0.01	0.08
Heavy traffic			
Phoenix, Arizona	95,000	–	0.36
Los Angeles, California	144,000	–	0.13
Light traffic – suburban			
Kansas City, Missouri	100	–	0.06
Maryville, Missouri	–	–	0.07

SUMMARY

EDC and EDB have been used as a source of halogen scavenger of leaded combustion chamber deposits since the initial introduction of lead antiknock agents in 1923. Although other materials have been studied, bromine and chlorine remain the only effective scavengers of lead oxide deposits from spark ignition engines. Lead deposits accumulate on spark plugs and exhaust valves and can cause significant loss of engine performance if they are not removed. Lead antiknock agents are used to increase the octane quality of gasoline and thus provide more efficient engine operation.

The use of EDC and EDB as lead scavengers in gasoline is directly tied to the use of lead antiknocks in gasoline. In the U.S., EPA regulations on the use of lead in gasoline and the use of unleaded gasoline for all post-1975 automobiles will eventually eliminate the use of lead and halide scavengers in gasoline. The use of halide scavengers in leaded gasoline in 1979 was only 60% of the 100,000 metric tons used in 1970. This use will continue to decline until only 20% as much lead scavengers will be used in 1985 as were used in 1970.

Concentration of EDB vapors measured in the exhaust from vehicles using leaded gasoline are very low, ranging from 5 to 19 ppb in undiluted automobile exhaust. The concentration did not vary greatly with mode of vehicle operation. On the basis of a 100- to 1000-fold dilution of exhaust, which occurs in urban air (using lead and CO measurements), this would indicate potential EDB air levels from auto exhaust of 0.005 to 0.019 ppb.

The maximum air exposure of two service station attendants to EDB during almost continuous refueling of cars with leaded gasoline was measured as 0.20 and 0.28 ppb. Ambient air levels measured by EPA in heavy trafficked areas near gasoline service stations in three cities showed that EDC concentrations varied from 0.008 to 0.013 ppb. EDB ambient air concentrations varied widely (0.006–0.45 ppb) in six different cities.

Potential workplace air exposure to EDC and EDB in an antiknock blending plant varied with the work activity. The 8-hour average exposure to EDC ranged from 0.023 to 0.5 ppm, whereas EDB 8-hour average exposure ranged from 0.002 to 0.05 ppm.

Overall, it would appear that ambient air levels of EDC and EDB owing to use of leaded gasoline are very low and do not exceed present or proposed air standards.

REFERENCES

Boyd, T. A. 1950. Pathfinding in fuels and engines. *SAE Quarterly Trans.* **4**:182.
Cantwell, E. N., E. N. Castellano, and J. M. Pierrard. 1978. "Projections of motor vehicle fuel demand and emissions." SAE Paper No. 780933 presented at International SAE Fuels and Lubricants Meeting, Toronto, Canada.
DOE (Department of Energy). 1979. *Monthly Energy Review – September 1979.* DOE/EIA-0035/9 (79), Government Printing Office, Washington, D.C.
EPA (Environmental Protection Agency). 1972. *Federal Register* **37** (221, part II):24250.
⸻. 1973a. *Federal Register* **38**(6):1254.
⸻. 1973b. *Federal Register* **38**(234):23734.
⸻. 1979. *Federal Register* **44**(178):53144.
Going, J. E. and S. Lang. 1975. *Sampling and analysis of selected toxic substances task II–ethylene dibromide.* EPA report No. 560/6-75-001, U.S. Environmental Protection Agency, Office of Toxic Substances, Washington, D.C. (September).
Going, J. E. and J. L. Spigarelli. 1976. *Sampling and analysis of selected toxic substances task IV–ethylene dibromide.* EPA report No. 560/6-76-021, U.S. Environmental Protection Agency, Office of Toxic Substances, Washington, D.C.
Leinster, P., R. Perry, and R. J. Young. 1978. Ethylene dibromide in urban air. *Atmos. Environ.* **12**:2383.
Midgley, T., Jr. and T. A. Boyd. 1922. Chemical control of gaseous detonation, with particular reference to internal combustion engines. *J. End. Eng. Chem.* **14**:894.
NIOSH (National Institute for Occupational Safety and Health). 1978a. An industry wide industrial hygiene study of ethylene dibromide. DHEW publication number 79-112. Cincinnati, Ohio.
⸻. 1978b. *Revised recommended standard-occupatioanl exposure to ethylene dichloride.* DHEW publication number 78-211. Cincinnati, Ohio.

Newby, W. E. and L. F. Dumont. 1953. Mechanism of combustion chamber deposit formation with leaded fuels. *J. End. Eng. Chem.* **55**:1336.

SRI International. 1979. Product reviews on antiknock mixes, ethylene dichloride and ethylene dibromide. In *Chemical economics handbook.* Stanford Research Institute, Menlo Park.

Medical Aspects of Ethylene Dichloride in the Workplace

MAURICE N. JOHNSON
B. F. Goodrich Company
Akron, Ohio 44318

The whole problem of occupational health examinations and what we can and cannot reasonably expect them to accomplish is an important one, and this meeting provides an excellent framework for its examination. The acute effects of ethylene dichloride (EDC; 1,2-dichloroethane) are well documented in the literature and are only seen extremely rarely in the chemical industry today. The classical story is a very heavy exposure to EDC followed, sometimes after several hours delay, by severe nausea, vomiting, and general signs of toxicity up to and including coma. The principal underlying pathology is severe damage to the liver and kidneys, and treatment is generally supportive, i.e., the attending physician tries to maintain proper fluid and electrolyte balance, assure adequate oxygenation of the blood, and attempts to prevent complicating infections. There is no specific antidote and the goal in treatment is simply to maintain basic body functions while the normal excretion, detoxification, and repair mechanisms take place.

Our concerns here are with the long-term effects of chronic exposure. This leads us to the general problem of regular periodic medical examinations for employees exposed to EDC and, for that matter, to a host of other chemicals in the working environment.

THE VALUE OF PERIODIC EXAMINATIONS

Let me begin by saying first of all that I personally believe strongly that periodic medical examinations are valuable and should be done regularly on all employees with potential exposure to hazardous working environments. But at the same time, I think it is extremely important that we recognize what routine examinations can and cannot do.

In the broadest sense as they apply to the general public, periodic medical examinations are of value, however we must recognize that they are limited in what they can accomplish. Although they can be invaluable in finding some important conditions, they are virtually useless in finding others. For an example, let me name several conditions and indicate what value a routine examination may have in their detection.

First, consider cancer of the head of the pancreas. This is not a common condition but neither is it rare among persons of middle and advancing years. It is extremely difficult to diagnose even if one is suspicious that it may be present. Many times this disease escapes detection even when the most sophisticated diagnostic techniques are used to try to find it. It is apparent then that the ordinary routine examination is of no value in early detection of this problem.

Another example is chronic glaucoma, which is by far the most common cause of blindness in the United States today. This is a disease that can be discovered by simply measuring the pressure in the anterior chamber of the eye. It is a simple and painless procedure, which I am sure most of you have experienced, and it is of enormous value because if done regularly it can detect the disturbed physiology long before there are any clinical symptoms. In most cases, the condition can be controlled with simple medication and the patient is spared a very serious problem.

In between these two extremes there are a number of conditions that can be detected at an asymptomatic stage, but about which there is no uniform agreement as to the value of early detection. Examples of these would be the finding of lung cancer and of adult diabetes. In each case it is clear that the disease can be found before it is symptomatic, but there is wide disagreement by experts as to whether early discovery and intervention really changes the outcome of the disease process.

Thus I think it is very important that we always keep in mind, as we do periodic examinations, these variations in what the examination can and cannot accomplish and then correctly report the results to the patient or employee.

When we come to the more specific problems of routine examinations for chemical exposures, we again find that there are some situations in which our examinations can be quite specific and meaningful and others in which their value is extremely limited. Two examples with which I have had some personal experience are organophosphorous insecticides and vinyl chloride (VC) exposure. In the case of organophosphorous insecticides, the effect is well understood and measurable. One need merely determine the level of cholinesterase in the blood and one can say with considerable confidence that the employee has or has not absorbed an excessive amount of this chemical. There is no need to look for early evidence of the clinical signs of cholinesterase depression such as pupillary constriction, sweating, muscle weakness, etc.

Perhaps at the other end of the scale is the examination of employees for the long-term effects of VC exposure. Here the target disease is angiosarcoma of the liver, and we are immediately thrown into the problem of early detection of cancer, a disease process that we do not fundamentally understand and for which we have virtually no ability to detect the altered physiology until there is evidence of the tumor itself. Regrettably, the evidence of tumor cannot be found, in most cases, until it is too late for effective therapy.

Faced with this problem in the VC case, the Occupational Safety and Health Administration prescribed a series of tests that would form the basis of the periodic physical examinations. I do not disagree particularly with these tests and, indeed, they are the very ones we were doing before the regulation was issued. But I think we must recognize that we do them simply because we feel that we must do something rather than because we have much confidence that they enable us to detect disease at a stage when it is either reversible or readily treatable.

In view of what I have said so far, I think it is obvious that I believe that the control of hazardous exposures in the workplace cannot and should not be based on periodic medical examinations of the employee population with the goal of identifying those with adverse effects and then reducing their exposures or treating their disease. In my view, the periodic medical examination should serve rather as a general health maintenance tool which on occasion will identify employees who, for any of a variety of medical reasons, should not be permitted the chemical exposures that are tolerated with no problem by the healthy workers.

THE NEED TO COLLECT HEALTH DATA

We all recognize that there are a great many chemicals in use in industry today and only a small percentage of these have been adequately studied in animals. Human epidemiologic data is available for relatively few of these chemicals. One important aspect of periodic health examinations is that if they are done broadly and records are well kept so that the information in them is retrievable, they may enable us to learn more about the effects of chemical exposures in the workplace. To do this most effectively, they need to be part of a general health surveillance system so that for each worker we build over the years a record, not only of the findings on periodic medical examinations, but also a record of illnesses, hospitalizations, dispensary visits, and any other health information available up to and including the death certificate. Obviously this is a long, tedious, and expensive process, but I know of no way other than collecting this information along with a history of the employee's chemical exposures during his working lifetime, to gain new information about the actual effect of exposures to chemicals on the worker. Also, I believe that the protection of the chemical worker's health will be achieved by limiting exposures to levels that past experience and animal studies suggest are safe. If we were to rely on medical examinations to play a major role in the ongoing determination of safe levels, I think we would be relying on far too crude a tool.

In summary, I believe we should do medical examinations on chemically exposed workers, keep good records in a form that will be useful in the future, communicate the results to the workers in a realistic way, and educate our employees to protect their health by limiting their exposures and not relying too much on their annual check-up.

COMMENTS

WARD: In the case of workers exposed to vinyl chloride (VC), the ones that developed liver disease and tumors are one group, perhaps of the high exposure. What about the other people being followed—the ones exposed to lower levels and those who didn't get sick.

JACOBS: Yes, they are being followed regularly. But, again, it's just simply building this sort of data base. The law prescribes a certain series of tests each year, and that data keeps building up. It's unsatisfactory in that, if we are looking at this target problem of angiosarcoma of the liver, you just don't find a tumor of the liver until it's quite large. By that time, they are not amenable to surgical treatment. Chemotherapy is the only hope, and it's a rather poor one. I only know of one surviving angiosarcoma that has gone 3 years past diagnosis.

HINDERER: I have a sort of statement-question. Every now and then you'll hear someone make a comment about how "there's very little being done in the way of diagnostic testing or test development." I'm personally aware of hundreds of thousands of dollars that have been spent in this area, particularly in the VC area; my guess is that maybe we're really talking about millions of dollars.

The real question is where do we go now? I think we're really looking for creative new techniques to follow up, and right now we're at a loss for what else to do. It's somewhat disturbing.

Right now you're saying the answer is to put our emphasis on stopping exposure. But I think we still have that potential for people developing health problems, and you can't leave that go.

FABRICANT: I would like to ask a question relevant to that. Have you done any cytogenetics on some of the workers for unscheduled DNA synthesis or sister chromatid exchange (SCE)? To my knowledge, these have not been done. Is that possible?

JOHNSON: At the moment we are frustrated in this area. We did a rather big SCE study with another chemical compound and it got contaminated with fungus, so we are starting over again on that one. I think a number of those tests are being done now.

HINDERER: That isn't really a diagnostic test, though, in the sense of early detection of a disease. I think that's what I was looking at primarily. Certainly, we can pick up situations that are suggestive of an event, but we can't pick up something where we can intervene and perhaps prevent the final occurrence.

FABRICANT: I think that findings of increased frequency of structural aberrations in chromosomes, for example, or increased frequency of SCE above the control value for a worker would indicate that he or she should be moved out of that area into another area.

JOHNSON: If you presume that the person is unusually sensitive, which I think probably wouldn't be right, then that is the remedy. But also the answer probably would be that you have too high an exposure level in that area, and you should reduce the exposures with an engineering solution.

FABRICANT: Well, I think there could be genetic sensitivity. We know that certain diseases, for example ataxia-telangiectasia or xeroderma pigmentosa are repair deficient. Perhaps heterozygotes for certain diseases, do have increased frequencies of certain kinds of chromosomal aberrations.

JOHNSON: Well, yes, but then another problem occurs if you start trying to move people around because of biological peculiarities, some of which are not so peculiar, including basic sex.

KARY: With regard to VC, I believe that the University of Louisville group either has or will be doing some cytogenetics work on a worker population.

GOLD: To what extent are people who work with vinyl chloride also exposed to EDC?

JOHNSON: I am sure there are some people who overlap, particularly in the people that work around the units that crack EDC to vinyl. The people with the high exposures to VC have virtually no exposure to EDC, because they are working in the polymerization unit itself, which is a separate operation.

GOLD: The polymerization workers may not be exposed to EDC; but operators in an EDC-VC production plant may work in both units over a period of years. Maintenance workers who move around the plant might also be exposed to both substances.

JOHNSON: Yes, they might bid from one job to another. Also, a mechanic, who is going around repairing leaks or whatever, could cross.

MALTONI: I think that probably the time is coming for trying to better evaluate what we want to get with medical control, because we have to realize that medical control is an expression that is used too often. We also know that in many cases what you call medical surveillance is just an obsolete exercise just to do something, and it doesn't produce anything at all.

Also, when medical surveillance reaches some point technically, we have to analyze what it produces. We have some tools that we know are productive at three different levels: diagnosis of tumor, a sort of a common consensus of the precursor to have some broad evaluation of risk, and identification in a qualitative way of whether there is some risk.

When our tools do produce this type of effect, then do we find out if there is more or less precursor immediate to us, or do we find out that there are some more chromosomal aberrations that are indications of risk?

If we use early diagnosis, it could be that it brings no real success. For example, a lot of effort was put in the past at the time of the great onset of bladder tumor by aromatic amines and then by lung tumors by the chromium compounds or by uranium ore or by anything else we know. And so, occupational bladder tumors and lung tumors are the best examples of the possibility of an early diagnosis.

But despite the early diagnosis, the people still die the same way. It's just that time of survival from detection is prolonged. If you're correct by the factor of early diagnosis of the early stage of the tumor, you do not get any benefit whatsoever of life-style. We have to try to make judgments to spend less, or to spend in the same way, but to spend better, because a lot of medical surveillance is absolutely obsolete.

Medical surveillance should be used to collect indications of the medical risk in histopathology used as an epidemiological service rather than a public health service. It would be very good for epidemiology; it will not be good for the care of sick people.

Do you know one single type of tumor, one single type of environmental occupational tumor, in which medical intervention at the therapeutic level has improved the fate of patients one single bit?

JOHNSON: It's very disappointing because in the case of cancer of the cervix, the exfoliative cytology has been a great help. But in the bladder . . . .

MALTONI: We have to keep in mind that tumors of lung, bladder, and liver are all multifoci tumors. If in the case of bladder cancer, complete obliteration of the bladder may bring some benefit, but such a mutilation occurs. And, you still have the risk of having a tumor in the pelvis.

We have to keep in mind that early diagnosis in the case of occupational and environmental tumors has been a fantasy. We have been extremely frustrated after 10 years of continual screening of every cancer in the bladder of people exposed to hematic amines. Early diagnosis does not seem to benefit. Moreover, you have to deal with 80 or 90% of all types of tools used in medical surveillance, and it doesn't give us any information. We are going to repeat these tests since vinyl chloride is the best example we have.

JOHNSON: True. I hope you will make that speech many times in many places because it needs to be said.

MALTONI: It doesn't save one single life. We haven't collected enough epidemiological data in doing this type of medical surveillance on VC workers on both sides of the ocean.

JOHNSON: Well, I don't think we have done any good in the field of epidemiology yet, concerning this area. We must perform these tests in a strict scientific manner with a very selective group and then do a bioassay in 5 or 10 years.

HOOPER: Is there some agreement on a correct protocol for collecting worker exposure and worker health data that would insure its usefulness in future epidemiological studies?

JOHNSON: Sort of a prospective epidemiology?

HOOPER: Yes.

JOHNSON: Yes, I think there's a general understanding of what needs to go in.

HOOPER: But I have heard it said that exposure and health ill-effects are still drastically underreported for workers. Dr. Ephram Kahn, in the State of California Health Department, for example, estimates that farmworker work-related illness is 100 times greater than that reported, based on interviews with doctors and workers.

Human Exposure to Ethylene Dichloride: Potential for Regulation via EPA's Proposed Airborne Carcinogen Policy

ROBERT G. KELLAM AND MICHAEL G. DUSETZINA
Office of Air Quality Planning and Standards
Environmental Protection Agency
Research Triangle Park, North Carolina 27711

Ethylene dichloride (EDC; 1,2-dichloroethane) is one of a number of chemical and radioactive substances currently under consideration by the Environmental Protection Agency (EPA) for regulation as a potential airborne carcinogen. The procedures for identifying such substances, assessing the associated health risks, and determining the appropriate level of control are the subjects of EPA's proposed airborne carcinogen policy, released for public comment in October, 1979 (EPA 1979a).

Under the proposed policy, EPA will list as hazardous air pollutants those substances that, in the judgment of the administrator, pose a significant risk of cancer as a result of emissions into the ambient air from stationary sources. This listing, under section 112 of the Clean Air Act, signals EPA's intent to develop national emission standards for new and existing sources of such pollutants, which will protect public health with an ample margin of safety.

The operation of the proposed policy, as it would apply to EDC, is briefly described in the following sections along with the results of preliminary assessment by EPA of human exposure to EDC.

METHODOLOGY

Figure 1 outlines the operation of EPA's proposed airborne carcinogen policy. The procedures are divided into three phases: identification, assessment, and regulation.

Identification

Substances that are possible candidates for regulation as airborne carcinogens are identified through surveys of available information on production, use, toxic effects, sampled or modeled ambient concentrations, and other physical or chemical properties. The sources of this information include EPA programs, the programs of other federal agencies, the scientific literature, and the submission

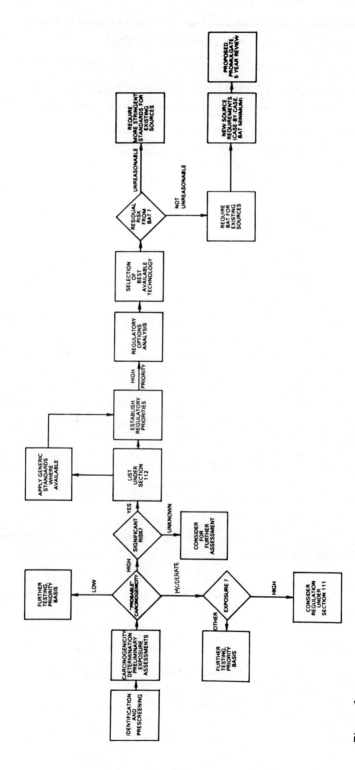

Figure 1
Policies and procedures for identifying, assessing, and regulating substances posing a risk of cancer in ambient air

of data by external groups. Although other toxic properties or routes of exposure may lead to regulation by EPA through the air or other media programs, evidence of carcinogenicity and potential for human exposure via the ambient air establish the basis for further assessment as airborne carcinogens.

In 1976 over 600 chemicals were identified as potential candidates for regulation. On the basis of available health effects data, this list was reduced to 140. Approximately 40 of these, including EDC, have been identified as candidates for further assessment.

Assessment

The decision to list a substance as an airborne carcinogen under section 112 is based on the results of two preliminary assessments: a determination of the quality of the evidence of human carcinogenicity and an evaluation of the extent of human exposure via the ambient air. The carcinogenicity determination is performed by the EPA Carcinogen Assessment Group (CAG) under the auspices of interim guidelines published by the agency in May 1976 (EPA 1976) and more recently by the report in July 1979 of the Risk Assessment Work Group of the Interagency Regulatory Liaison Group (IRLG 1979) on the identification of carcinogens and the estimation of carcinogenic risks.

Assessment of carcinogenicity by CAG is a weight of evidence judgment on the quantity and quality of available scientific information. For a particular substance, the evidence may be regarded as best, substantial, suggestive, ancillary, or inadequate. Generally, best evidence would require the presence of definitive human epidemiology supported by animal testing. Weak epidemiological studies or well-conducted animal tests could lead to a judgment of substantial evidence. Suggestive evidence may be provided by the Ames *Salmonella* test or other short-term bioassay procedures. Structural or functional similarities to known carcinogens could be the basis for concluding that ancillary evidence exists.

In addition to a qualitative evaluation of the evidence of carcinogenicity, the CAG, where possible, provides a quantitative estimate of the lifetime risk of cancer to an individual exposed to a unit air concentration. This unit risk is combined with information obtained from the exposure assessment to develop a quantitative assessment of risks used in the establishment of regulatory priorities and in the determination of the risks remaining after the application of controls to certain categories of emission sources.

Under the proposed policy, substances found to present best or substantial evidence of carcinogenicity may be considered (for regulatory purposes) for designation as high-probability carcinogens. High-probability carcinogens found to be emitted into the air in such a way that significant human exposure results are candidates for listing as hazardous air pollutants.

The preliminary exposure assessment focuses on the number of sources and source categories that emit the candidate substance, the nature of these sources, the magnitude and distribution of air emissions, the number of people exposed, and the concentrations to which they are exposed. Figure 2 outlines the inputs to this assessment.

Emitting sources are identified by category and, where relevant, plant location. For each source or model source (depending on the availability of information), production, consumption, and emission levels, and the efficiency of present emission controls are estimated. To determine ambient concentrations in the vicinity of emitting sources where actual data are lacking, computerized dispersion models are employed. Depending on the number of plants and the nature of emissions, the models may include consideration of meteorological conditions and terrain features.

The modeling results characterize emissions in terms of ambient ground-level concentration rings radiating outward from specific sources. Population exposure is calculated by overlaying census data.

EDC Preliminary Exposure Assessment

For EDC, four major source categories have been identified: EDC-producing facilities, production facilities of other chemicals in which EDC is an intermediate, gasoline service stations marketing EDC-containing fuels, and automobiles emitting EDC through evaporation from gasoline (Table 1). As indicated, the major emission source category appears to be the primary production of EDC.

It is important to note that the two major industrial source categories that emit EDC are not generally separate and discrete facilities. All the sources that synthesize EDC utilize the chemical to some extent as an intermediate in the production of other synthetic organics. For the purpose of characterizing source category emissions, however, the synthetic and intermediate processes have been separated.

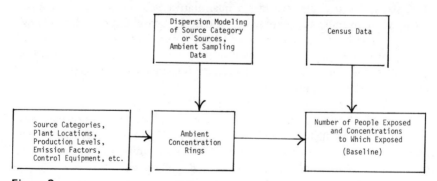

Figure 2
Conduct of exposure assessment

Table 1
Source Characteristics (1977)

Source categories	Number of plants	Total production (10³ metric tons)	Total consumption (10³ metric tons)	Emissions (1000 metric tons/yr)	% Total emissions
EDC production	18	5200	–	43.9	83.3
Production: EDC intermediate product	28	–	4928	2.5	4.7
Gasoline service stations	184,000	–	89	0.1	0.2
Automobile evaporative emissions[a,b]	–	–	–	1.2	2.3
Other	–	–	5	5.0	9.5
Total		5200[c]	5022[c]	52.7[d]	100

[a] EDC content in leaded gasoline 0.02% by volume or 1.1 g EDC/gallon by weight.
[b] EDC evaporative emission factor 0.017, EDC/gas.
[c] Difference accounted for by export.
[d] Transportation and waste disposal were not included.

Ambient Concentrations

Estimated ambient concentrations of EDC in the vicinity of emitting production sources have been calculated using rough-cut Gaussian plume techniques. The model predicts ground-level, 1-hour concentrations when assumptions regarding meteorological conditions (wind speed, 4 m/sec and neutral D stability conditions) and release height are made. Concentrations are calculated at various radii from the sources and adjusted to annual average omnidirectional concentrations.

Evaporation of EDC from leaded gasoline during fueling of automobiles constitutes a third major source of exposure. Atmospheric concentrations of EDC at and in the vicinity of service stations have been estimated by proportional scaling of monitored benzene levels and through the use of the Single Source (CRSTER) model (EPA 1977).

Although EDC is to a large extent destroyed in the combustion process, some evaporative emissions occur from automobile fuel tanks and carburetors. Emission factors have been derived from empirical data obtained in studies of ethylene dibromide (EDB; 1,2-dibromoethane) from these sources. Ambient concentrations have been estimated through the use of an area-wide dispersion model developed by Gifford and Hanna (1973).

Dispersion models have been used to estimate ambient concentrations owing to a general lack of monitoring data. Of the data that are available, the agreement with modeled results does not appear to be unreasonable (Table 2).

Table 2
Comparison of EDC Monitoring and Modeling Atmospheric Concentrations in the Vicinity of Producing Facilities (ppb)

Distance (km)	Monitoring average concentrations[a]			Three-location average	Three-location modeling average[a]
	Calvert City	Lake Charles	New Orleans		
0.7–1.0	2.0	36.7	6.3	15.0	13.5
1.1–1.5	n.d.[b]	6.0	1.4	3.7	9.1
1.6–2.0	1.0	7.0	1.6	3.1	6.5
2.1–3.0	2.1	1.1	0.7	1.3	4.3
3.1–4.0	0.9	n.d.	0.3	0.6	2.1
4.1–5.0	n.d.	n.d.	n.d.	n.d.	1.6
5.1–6.0	n.d.	n.d.	0.9	0.9	1.2
14.0	n.d.	n.d	0.6	0.6	0.4

[a]Data are the average 24-hr concentrations over 10–13 days for monitoring and estimated annual averages for modeling.
[b]No monitoring data were collected.

Population Exposure

For the production and intermediate use facilities, the magnitude of population exposure to EDC was calculated by overlaying the ambient concentration rings developed from the dispersion modeling of individual plants with census data available in 1-km^2 grids across the United States (Table 3). To obtain a better basis for comparison, midpoint concentrations can be multiplied by the population in each ring (Table 4).

Table 3
Summary of Estimated Population Exposures to Atmospheric EDC from Specific Emission Sources (10^3)

Annual average EDC concentration (ppb)	EDC	Production facilities	Gasoline service stations	Automobile emissions	Automobile refueling
> 10	2				
6.0–10.00	3				
3.00–5.99	28				
1.00–2.99	280				
0.60–0.99	400	1			
0.30–0.59	1500	0.4			
0.10–0.29	4300	51			
0.060–0.099	1900	83			
0.030–0.059	3500	509			
0.010–0.029	550	1977	1000	13,000	
< 0.01					30,000
Total	12,500	2621	1000	13,000	30,000

Table 4
Exposure to Atmospheric EDC Emissions in Person-ppb/yr (10^3)

Annual average EDC concentration (ppb)	EDC	Production facilities	Gasoline service stations	Automobile emissions	Automobile refueling
> 10	20				
6.0–10.00	24				
3.00–5.99	126				
1.00–2.99	560				
0.60–0.99	318	1			
0.30–0.59	668	2			
0.10–0.29	839	10			
0.060–0.099	151	7			
0.030–0.059	156	23			
0.010–0.029	11	39	20	254	
< 0.01					300
Total	2873	82	20	254	300

REGULATION

The assessments of carcinogenicity and human exposure for potential airborne carcinogens will form the basis under the proposed policy for the listing of these substances as hazardous air pollutants (Fig. 1). The EDC assessment documents are being drafted and will be publicly available prior to review by the EPA Science Advisory Board. Following that review, EPA will decide whether or not EDC should be listed under section 112 as presenting a significant carcinogenic risk to public health.

Where applicable, generic standards will be proposed concurrent with the listing of an airborne carcinogen to expedite reductions in emissions that can be achieved through good housekeeping practices in the manufacturing, handling, or use of hazardous materials. At present, the generic standards, published as an advance notice of proposed rulemaking (EPA 1979b) at the time of the proposal of the policy, apply only to the synthetic organic chemical manufacturing industry. The standards would require that applicable emission sources conduct routine monitoring and maintenance programs to detect and repair leaks and undertake other measures to reduce fugitive emissions of listed pollutants.

EPA anticipates that several substances will be listed in the near future as airborne carcinogens. For each substance there may be several source categories presenting significant risks. To insure that the agency's resources are allocated in such a way that the most important or tractable problems are addressed first, regulatory priorities will be established. The factors considered in the priority assignment include the magnitude of the risks, both the risk to most-exposed individuals and the projected incidence of excess cancers; the ease of expeditious standards development and implementation; and the feasibility of significant improvements in control.

High-priority source categories will undergo a regulatory options analysis. This analysis will identify a number of technologically feasible control alternatives, which may range from no further control to a complete ban on emissions. On the basis of assessments of environmental, economic, and energy impacts, one of the control options identified will be designated as the best available technology (BAT) for the control of emissions from the sources in the category. This level of control will be that technology, which in the judgment of the administrator, is the most advanced level of control adequately demonstrated, considering economic, energy, and environmental impacts.

The control level designated BAT may be different for new and existing facilities in a category. For practical purposes, this level of control for new sources will, as a minimum, be equivalent to that which would be selected as the basis for a New Source Performance Standard (NSPS) under section 111. The requirement of BAT for new sources would consider economic feasibility and would not preclude new construction.

The selection of BAT for existing sources may require consideration of the technological problems associated with retrofit and related differences in the economic, energy, and environmental impacts. In practice, BAT for existing sources would consider economic feasibility and would not exceed the most advanced level of technology that at least most members of an industry could afford without plant closures.

Following the identification of BAT for existing sources, the quantitative risk assessment will be used to determine the risks remaining after the application of BAT to the source category. If the residual risks are not judged by the administrator to be unreasonable, further controls would not be required. If, however, there is a finding of unreasonable residual risk, a more stringent alternative would be required.

The determination of unreasonable residual risk will be based primarily on the protection of public health. To the extent possible, quantitative or qualitative estimates of various factors will be made for pusposes of comparison. Among these are

1. the range of health risks to the most exposed individuals;
2. the range of total expected cancer incidence in the existing and future exposed populations through the anticipated operating life of existing sources;
3. readily identifiable benefits of the substance of activity;
4. the economic impacts of requiring additional control measures;
5. the distribution of the benefits of the activity versus the risks it presents;
6. other possible health and environmental effects resulting from the increased use of substitutes.

For new sources, the policy proposes special requirements to encourage the consideration of potential health risks in siting decisions. EPA is particularly interested in receiving comments on this procedure or on possible alternative means to achieve the same objective.

Regulations proposed under the policy will be subject to public reviews and comment. Standards will be reviewed for possible modification at no more than 5-year intervals.

SUMMARY

As one element in the implementation of a proposed policy for the identification, assessment, and regulation of airborne carcinogens, EPA has performed a preliminary assessment of human exposure to EDC in the ambient air. Because of a general lack of ambient monitoring data, the assessment relies heavily on disspersion modeling techniques in the estimation of ambient EDC concentrations.

The exposure assessment and an assessment of carcinogenicity for a suspect airborne carcinogen form the principal basis for determining if the substance should be listed by EPA as a hazardous air pollutant. Following listing, significant source categories would be required, as a minimum, to install BAT to reduce emissions. If the residual risk remaining after the application of BAT is judged unreasonable, more stringent standards would be required.

REFERENCES

Environmental Protection Agency (EPA). 1976. Health risk and economic impact assessments for suspected carcinogens, interim procedures and guidelines. *Federal Register* **44**:24102.

_____. 1977. *User's manual for single-source (CRSTER) model.* EPA-450/2-77-013, Environmental Protection Agency, Washington, D.C.

_____. 1979a. Proposed policy and procedures for identifying, assessing, and regulating airborne substances posing a risk of cancer. *Federal Register* **44**: 58642.

_____. 1979b. National emission standards for hazardous air pollutants; advance notice of proposed generic standards. *Federal Register* **44**:58662.

Gifford, F. A. and S. R. Hanna. 1973. Technical note: Monitoring urban air pollution. In *Atmospheric Environment,* vol. 7. Pergamon Press, New York.

IRLG (Interagency Regulatory Liaison Group). 1979. Scientific bases for identification of potential carcinogens and estimation of risks. Report by the Work Group on Risk Assessment of the Interagency Regulatory Liaison Group. *Federal Register* **44**:39858.

United States. *Clean Air Act.* National Emission Standards for Hazardous Air Pollutants, section 112.

COMMENTS

TER HAAR: Is EDC listed by the EPA in the hazardous group because of the NCI studies; and, if so, are there any number of studies that could be done in the rest of this century that would take it off the list?

KELLAM: Okay. No, it has not been listed. It's one of the chemicals that we have under assessment.

TER HAAR: You are considering it?

KELLAM: Yes. The answer is how did it come into consideration—I would say that the NCI study had a lot to do with it.

TER HAAR: Well, related to that, from a policy point of view—suppose you decide, based on the NCI study, to list it—are there any experiments that could ever be done that could take it off that list, once you decide that the NCI study is the reason for listing it?

KELLAM: The real reason for listing is, as we've set it out, weight of evidence judgment based on the data. I would think that at some point if there was a preponderance, in attempts to repeat the NCI study—as Dr. Van Duuren and Dr. Maltoni are now doing—that if the preponderance of the studies are negative that that would be an important factor. Although certainly negative evidence is considered, I don't think we'd ever go so far as to say it was noncarcinogenic. If it doesn't cause a threshold of significant risk, then we probably would not list it.

REITZ: Is there any point in the rule-making process at which the actual quantitative carcinogenic risk estimation, which you say does play an integral role, would be exposed for public comment? These risk estimations are kind of buried, and it's very hard to find out where the final number actually comes from.

KELLAM: At present, when the Carcinogen Assessment Group makes their determination, they are looking first at the quality of the evidence, the qualitative assessment.

The second thing that they attempt to do is to develop the numbers, the unit risk, the lifetime individual risk to a particular concentration. That unit risk is what we later use in conjunction with our exposure study to do the quantitative risk assessment.

The unit risk and how that number is determined will be subject to public comment before that Assessment Group advises the Science Advisory Board.

I would expect in the case of EDC over the next few months, as we get closer to Science Advisory Board (SAB) review, that those documents, both the CAG assessment and the exposure assessment, will be released for about three months for public comment before we go to the SAB.

GOLD: Does the Carcinogen Assessment Group assess risk for water quality and for air as well? Does the group use a threshold, one-hit model?

KELLAM: My understanding of the water quality criteria documents is that they've established target risk goals, postregulatory goals. If the technology they can apply to reduce concentrations in water results in a risk that is no greater than 10^{-5} or 10^{-7}, that is as far as they will go.

I don't know exactly how they calculate that risk in water.

SESSION 4:
Related Chemicals

Dietary Disulfiram Enhancement of the Toxicity of Ethylene Dibromide

HARRY B. PLOTNICK, WALTER W. WEIGEL, DONALD E. RICHARDS,
KENNETH L. CHEEVER, AND CHOUDARI KOMMINENI
Division of Biomedical and Behavioral Science
National Institute for Occupational Safety and Health
Cincinnati, Ohio 45226

Ethylene dibromide (EDB; 1,2-dibromoethane) is used both as a fumigant and as an additive in leaded gasoline. The current U.S. occupational standard for EDB is 20 ppm (Department of Labor 1979). A study of the toxicity of EDB in rats at the 20-ppm level has recently been completed by the Midwest Research Institute, Kansas City, Missouri (Midwest Research Institute 1979). In that study, male and female Sprague-Dawley rats were exposed to EDB or to filtered air (controls) 7 hours/day, 5 days/week for 18 months to simulate occupational exposure conditions. At the time that the inhalation study was planned, it was believed that a portion of an administered dose of EDB underwent oxidative dehalogenation to 2-bromoethanol followed by oxidation to bromoacetaldehyde. This belief has since been substantiated by investigators at the Southern Research Institute (Hill et al. 1978). Accordingly, some of the animals were placed on a diet containing 0.05% disulfiram (tetraethylthiuram disulfide), an inhibitor of aldehyde dehydrogenase used in the management of alcoholism in man. It was believed that disulfiram would block the further oxidation of the bromoacetaldehyde formed, resulting in an increase in the toxicity of inhaled EDB. Four groups of animals were employed in the inhalation study, each group consisting of 48 male and 48 female rats. The group designations were:

1. Group I. Control-control, which was filtered-air exposed and fed on a standard rat diet;
2. Group II. Control-disulfiram, which was filtered-air exposed and fed on a diet containing 0.05% disulfiram;
3. Group III. EDB-control, which was 20-ppm-EDB exposed and fed a standard rat diet; and
4. Group IV. EDB-disulfiram, which was 20-ppm-EDB exposed and fed a diet containing 0.05% disulfiram.

The histopathologic findings from this study appear in Table 1. The data in the table indicate that EDB, under these exposure conditions, is a carcinogen in the

Table 1
Selected Histopathologic Findings in Rats Exposed to 20 ppm EDB or to Filtered Air with and without Disulfiram in the Diet for 18 Months

	Control-control		Control-disulfiram		EDB-control		EDB-disulfiram	
	male	female	male	female	male	female	male	female
Liver								
Hepatocellular carcinoma	0	0	1	0	2	3	36[a]	32[a]
Hemangiosarcoma	0	0	0	0	1	0	3	4[b]
Spleen								
Hemangiosarcoma	0	0	0	0	10[c]	6[c]	15[a]	10[a]
Atrophy	0	0	0	0	6[c]	0	30[a]	19[a]
Mesentery or omentum								
Hemangiosarcoma	1	0	0	0	0	0	11[a]	8[a]
Kidney								
Renal tumor	0	0	0	0	3	1	17[a]	7[a]
Hemangiosarcoma	1	0	0	0	0	0	0	2
Testes								
Atrophy	4	—	1	—	2	—	43[a,c]	—

Data from Midwest Research Institute (1979).
[a]Significantly different from both EDB-control and control-disulfiram groups ($p < 0.05$).
[b]Significantly different from control-disulfiram group ($p < 0.05$).
[c]Significantly different from control-control group ($p < 0.05$).

rat. Most striking, however, is the disulfiram-induced enhancement of the carcinogenic and other toxic effects of EDB. Particularly noteworthy is the high incidence of hepatocellular carcinoma in both males and females in the EDB-disulfiram group and the high incidence of testicular atrophy in the males in this group.

The present study was undertaken to determine whether dietary disulfiram modifies the tissue distribution and excretion of orally-administered EDB, utilizing ^{14}C-labeled material, in an attempt to explain the interaction noted in the inhalation study. Special attention was given to the organs affected in the chronic inhalation study. In addition, liver nuclei were isolated for the quantitation of radioactivity to determine whether the disulfiram diet was associated with a preferential distribution of radioactivity to that organelle.

MATERIALS AND METHODS

Male Sprague-Dawley rats, weighing 100–125 g, were purchased from Charles River Breeding Laboratories, Wilmington, Massachusetts. Animals were housed two per cage for a 1-week acclimation period. Disulfiram was obtained from the Sigma Chemical Co., St. Louis, Missouri. [U-^{14}C]EDB was purchased from New England Nuclear, Boston, Massachusetts. Analysis of the radioisotope by the supplier prior to shipment showed this compound to have a radiochemical purity of 98% and a specific activity of 9.35 mCi/mmole. EDB (98%) was obtained from MCB Manufacturing Chemists, Norwood, Ohio. The treatment solution was prepared by mixing the [U-^{14}C]EDB with sufficient unlabeled EDB in corn oil (MCB Manufacturing Chemists, Norwood, Ohio) to yield a concentration of 3.75 mg EDB/g of treatment solution with a specific activity of 6800 dpm/μg of EDB. Upon initiation of the experiment, 24 rats were weighed, randomly assigned to two experimental groups of 12 animals each, and placed in individual stainless steel metabolism cages designed for the separate collection of urine and feces. One group received ground Purina Rodent Laboratory Chow, while the second group received ground Purina Rodent Laboratory Chow containing 0.05% disulfiram. All animals were permitted free access to tap water and their respective diets. After 12 days on these diets, the rats were fasted overnight and subsequently given single 4-g/kg (15 mg EDB/kg) doses of the treatment solution by oral intubation. The rats were returned to their metabolism cages, and urine and feces were collected from each rat at 24-hour intervals from the time of treatment with EDB until the animal was killed. At 24 and 48 hours after administration of the treatment solution, six rats from each experimental group were killed by exsanguination by cardiac puncture with a heparinized syringe following anesthetization by 100-mg/kg i.p. injections of sodium pentobarbital (Nembutal Sodium, Abbott Laboratories, North Chicago, Illinois). In addition to blood, the liver, kidneys, spleen, testes, brain, and suprarenal fat were removed at autopsy for analysis of ^{14}C activity. Plasma was obtained by centrifugation of an aliquot of the heparinized blood. Livers taken at autopsy were immediately placed in ice-cold 0.25 M sucrose in TKM buffer solution (0.05 M *tris*-[hydroxymethyl]-aminomethane, 0.025 M KCl, and 0.005 M $MgCl_2$, adjusted to pH 7.5 at 20°C). The livers were individually homogenized in two volumes of 0.25 M sucrose in TKM buffer, and the nuclei from 9.0 ml of the homogenate (equivalent to 3 g of whole liver) were isolated according to the method of Blobel and Potter (1966). The nuclear pellets obtained were resuspended in 36 ml of TKM buffer and subsequently resedimented by centrifugation at 5000g for 10 minutes. These washed nuclear pellets were then analyzed quantitatively for ^{14}C. A 2-ml aliquot of each liver homogenate was taken as the liver sample to be analyzed for ^{14}C as part of the tissue distribution studies. All biological samples, including urine and feces, were prepared and analyzed for ^{14}C activity by the method of Weigel, Plotnick, and Conner (1978) using a Beckman LS 8100 Liquid Scintillation System. The significance of differences between groups was determined by Student's t-test at a probability level of 0.05.

RESULTS

At both 24 and 48 hours after compound administration, tissue concentrations of ^{14}C in the liver, kidneys, spleen, testes, and brain were significantly higher in the rats receiving the disulfiram diet. Urinary excretion of radioactivity during the first 24 hours after isotope administration was significantly depressed in the disulfiram diet group. A comparison of tissue levels at 24 and 48 hours indicates that the rate of clearance of ^{14}C from liver, kidneys, spleen, testes, and brain was appreciably lower in the disulfiram group. This was particularly true for clearance from the testes. While the levels (μg/g) in the testes of the animals in the control group at 24 and 48 hours were significantly different ($p < 0.001$), a comparison of the corresponding levels in the testes of the disulfiram group at the two time intervals indicated that they did not differ significantly ($p > 0.2$). There were no significant differences between the groups with respect to levels of ^{14}C in the fat and whole blood at either time interval studied. The tissue distribution data, expressed both as tissue concentrations and as a percentage of the administered dose, appear in Tables 2 and 3. The levels of radioactivity in the washed liver nuclei obtained from the disulfiram-treated animals were significantly higher than those of the controls at both 24 and 48 hours. These data appear in Table 4.

DISCUSSION

The study performed by Midwest Research Institute established that the addition of disulfiram to the diet of rats enhanced the toxicity of inhaled EDB. The study reported here demonstrates that addition of disulfiram to the diet of male rats results in significant increases in the tissue levels of subsequently administered [^{14}C]EDB in the organs studied. This increase in tissue levels was evident

Table 2
Effect of Dietary Disulfiram upon the Distribution of ^{14}C in Selected Tissues and Body Fluids of Male Rats 24 Hours after a Single Oral Dose of 15 mg/kg [U-^{14}C]EDB

Tissue	Tissue concentrations[a]		Percent of dose[b]	
	control diet	disulfiram diet	control diet	disulfiram diet
Liver	4.68 ± 0.24	6.52 ± 0.39[f]	1.79 ± 0.07	2.46 ± 0.16[f]
Kidneys	3.32 ± 0.42	6.82 ± 1.37[f]	0.21 ± 0.02	0.45 ± 0.09[f]
Spleen	1.00 ± 0.03	1.56 ± 0.14[f]	0.02 ± <0.01	0.02 ± <0.01
Testes	0.49 ± 0.05	0.88 ± 0.10[f]	0.04 ± <0.01	0.07 ± 0.01[f]
Brain	0.41 ± 0.04	0.64 ± 0.08[f]	0.02 ± <0.01	0.03 ± <0.01[f]
Fat[c]	0.35 ± 0.04	0.48 ± 0.07	0.15 ± 0.02	0.20 ± 0.02
Blood[d]	0.90 ± 0.05	0.95 ± 0.09	0.59 ± 0.03	0.59 ± 0.05
Plasma	0.46 ± 0.04	0.56 ± 0.05	—	—
Urine[e]	—	—	72.38 ± 0.98	64.86 ± 1.94[f]
Feces[e]	—	—	1.65 ± 0.28	1.60 ± 0.37

[a] Values represent mean concentrations in μg/g or μg/ml (expressed as parent compound ± S.E.M. of duplicate determinations on 6 animals.
[b] Values represent the mean percentage of the administered radioactive dose ± S.E.M. of duplicate determinations on 6 animals.
[c] 6% Of body weight (Donaldson 1924).
[d] 9% Of body weight (Donaldson 1924).
[e] $n = 12$ (Includes 24-hr samples obtained from rats killed 48 hr after compound administration).
[f] Significantly different from control values ($p < 0.05$).

Table 3
Effect of Dietary Disulfiram upon the Distribution of ^{14}C in Selected Tissues and Body Fluids of Male Rats 48 Hours after a Single Oral Dose of 15 mg/kg [U-^{14}C]EDB

Tissue	Tissue concentrations[a]		Percent of dose[b]	
	control diet	disulfiram diet	control diet	disulfiram diet
Liver	2.87 ± 0.33	5.23 ± 0.38[f]	1.10 ± 0.12	1.74 ± 0.10[f]
Kidneys	1.06 ± 0.16	4.31 ± 0.40[f]	0.08 ± 0.01	0.27 ± 0.03[f]
Spleen	0.66 ± 0.03	1.29 ± 0.12[f]	0.01 ± <0.01	0.02 ± 0.00[f]
Testes	0.19 ± 0.02	0.72 ± 0.08[f]	0.01 ± 0.01	0.06 ± 0.01[f]
Brain	0.17 ± 0.02	0.50 ± 0.03[f]	0.01 ± <0.01	0.03 ± <0.01[f]
Fat[c]	0.44 ± 0.06	0.53 ± 0.05	0.20 ± 0.03	0.23 ± 0.02
Blood[d]	0.64 ± 0.07	0.81 ± 0.05	0.43 ± 0.04	0.53 ± 0.03
Plasma	0.22 ± 0.02	0.39 ± 0.03[f]	—	—
Urine[e]	—	—	73.54 ± 2.80	66.95 ± 2.48
Feces[e]	—	—	2.42 ± 0.54	1.56 ± 0.45

[a] Values represent mean concentrations in μg/g or μg/ml (expressed as parent compound) ± S.E.M. of duplicate determinations on 5 animals.
[b] Values represent the mean percentage of the administered radioactive dose ± S.E.M. of duplicate determinations on 5 animals.
[c] 6% Of body weight (Donaldson 1924).
[d] 9% Of body weight (Donaldson 1924).
[e] Cumulative 48-hr excretion.
[f] Significantly different from control values ($p < 0.05$).

Table 4
Effect of Dietary Disulfiram upon the ^{14}C Content of Liver Nuclei Isolated 24 or 48 Hours after Administration of a Single Oral Dose of 15 mg/kg [U-^{14}C]EDB

Time interval	Control	Disulfiram
24 Hours	687 ± 82[a]	1773 ± 314[b]
48 Hours	460 ± 42	1534 ± 197[c]

[a] Results are expressed as dpm/pellet (mean ± S.E.M.) of duplicate determinations on 6 animals per group at 24 hr and 5 animals per group at 48 hr.
[b] Significantly different from control values ($p < 0.01$).
[c] Significantly different from control values ($p < 0.001$).

at both 24 and 48 hours following administration of the halogenated hydrocarbon and probably accounts for the interaction noted. Particularly noteworthy is the increase in the amount of radioactivity in washed nuclei isolated from the livers of rats in the disulfiram group. At 24 and 48 hours there are 1.4-fold and 1.8-fold increases, respectively, in the levels of radioactivity in the livers of animals in the disulfiram group as compared with those of the controls. Corresponding ratios obtained for the nuclear pellets are 2.6 and 3.3, respectively, indicating a nonuniform, preferential distribution to this organelle. This increased nuclear uptake of ^{14}C in the liver may account for the high percentage of hepatocellular carcinomas found in the inhalation study in the group exposed to EDB while receiving disulfiram in the diet when compared with those exposed to EDB and receiving a control diet (70% vs 5%). Slower clearance of ^{14}C from the testes in the disulfiram group may explain the significantly greater incidence of testicular atrophy observed in the rats receiving the combined exposure in the inhalation study when compared with those exposed to the halogenated hydrocarbon alone (90% vs 4%).

The mechanism of the enhancement of carcinogenicity by disulfiram is presently unclear. Researchers at the Southern Research Institute (Hill et al. 1978) have identified bromoacetaldehyde as an intermediate of EDB metabolism in the rat. Based upon this observation, one could speculate that disulfiram, a known inhibitor of aldehyde dehydrogenase, blocks the further oxidation of the bromoacetaldehyde formed, resulting in increased tissue levels of this intermediate. While little is apparently known about the toxicity and biological reactivity of bromoacetaldehyde, such information is available on another α-haloaldehyde, chloroacetaldehyde. In the Ames mutagen assay, employing tester strain TA100 without activation, chloroacetaldehyde was found to be 746 times more active as a mutagen, per μmole, than vinyl chloride (McCann et al. 1975). In addition, it is known that chloroacetaldehyde reacts nonenzymatically with the nucleic acid bases adenine and cytosine to form so-called etheno derivatives (Secrist et al. 1972). Such a reaction, if it occurs in vivo, could produce significant alterations in nucleic acids. It is also possible that disulfiram merely inhibits the excretion of EDB metabolites and that the increased tissue levels reflect this impairment of excretion. This alternate hypothesis is supported by a recent study which suggests that disulfiram interferes with the excretion of barbital, a hypnotic agent that does not undergo any significant biotransformation, resulting in an increased sleeping time in rats (Sharkawi and Cianflone 1978). Studies of the biochemical mechanism of this interaction are now in progress in our laboratory.

REFERENCES

Blobel, G. and V. R. Potter. 1966. Nuclei from rat liver: Isolation method that combines purity with high yield. *Science* **154**:1662.

Department of Labor. 1979. *Code of federal regulations,* 29 CFR 1910.1000.

Donaldson, H. H. 1924. *The rat data and reference tables,* p. 181. The Wistar Institute of Anatomy and Biology, Philadelphia.

Hill, D. L., T. Shih, T. P. Johnston, and R. F. Struck. 1978. Macromolecular binding and metabolism of the carcinogen 1,2-dibromoethane. *Cancer Res.* **38**:2438.

McCann, J., V. Simmon, D. Streitweiser, and B. N. Ames. 1975. Mutagenicity of chloroacetaldehyde, a possible metabolic product of 1,2-dichloroethane (ethylene dichloride), chloroethanol (ethylene chlorohydrin), vinyl chloride, and cyclophosphamide. *Proc. Natl. Acad. Sci. USA* **72**:3190.

Midwest Research Institute. 1979. *Study of carcinogenicity and toxicity of inhaled 1,2-dibromoethane in rats treated with disulfiram.* Contract #210-76-0131. National Institute for Occupational Safety and Health, Cincinnati.

Secrist, J. A., III, J. R. Barrio, N. J. Leonard, and G. Weber. 1972. Fluorescent modification of adenosine-containing coenzymes. Biological activities and spectroscopic properties. *Biochemistry* **11**:3499.

Sharkawi, M. and D. Cianflone. 1978. Disulfiram enhances pharmacological activity of barbital and impairs its urinary elimination. *Science* **201**:543.

Weigel, W. W., H. B. Plotnick, and W. L. Conner. 1978. Tissue distribution and excretion of ^{14}C-epichlorohydrin in male and female rats. *Res. Commun. Chem. Pathol. Pharmacol.* **20**:275.

Evidence for the Carcinogenicity of Selected Halogenated Hydrocarbons Including Ethylene Dichloride

PETER F. INFANTE AND PATRICIA B. MARLOW
Office of Carcinogen Identification and Classification
Occupational Safety and Health Administration
U.S. Department of Labor
Washington, D.C. 22010

In the past four decades, there has been a tremendous proliferation of man-made chemicals into the work environment. The potential for many of these chemicals to cause adverse effects on health extends from the primary production segment of industry through fabrication or reformulation processes to the end product usage of these chemicals or of substances containing these chemicals. Of concern are the facts that a majority of these substances never have been tested adequately for potential toxic effects and that the minority that have been tested and found to manifest various degrees of toxicity are usually not posted as such in the workplace nor labeled adequately on consumer products. As a result, workers and consumers find themselves in the untenable position of not being able to avoid unnecessary risks and of not being able to choose intelligently those risks that may have less of an adverse impact on their health status. One of the most striking examples of this uninformed consent to risk can be seen through the toxicity of vinyl chloride (VC; chloroethene). Although VC toxicity was first demonstrated in 1930 (Patty et al.) and carcinogenicity was first demonstrated by 1970 (Viola et al. 1971), in 1973 industrial workers were still being exposed to levels of VC that were 100 times higher than levels now shown to induce cancer in experimental animals. At the same time, cosmetologists and consumers were unknowingly being exposed to high levels of VC because this carcinogenic substance was used as a propellant in some hairsprays. Unfortunately, it was not until late 1973 with the public announcement of rare liver cancer among workers exposed to VC in the United States that action was initiated to reduce substantially exposure to VC in the occupational setting. The Occupational Safety and Health Administration (OSHA) promulgated a final standard for VC on the basis of carcinogenicity in October 1974. It was not until March 1978 that the Consumer Product Safety Commission (CPSC) promulgated a final rule banning VC as an aerosol propellant. The observation of VC inducing cancer in experimental animals and the subsequent confirmation of the carcinogenic effects in humans stimulated investigators to conduct experimental bioassay and epidemiologic studies on the carcinogenicity of chemicals related to VC. Table 1 lists selected substances related to VC by structural similarity or by industrial use, contamination, or formulation. These substances are shown by estimated annual U.S. production, number of workers exposed, and OSHA standards in parts per million (ppm) of air averaged over an 8-hour working day, i.e., time-weighted average (TWA). The chemical in the greatest volume of production is ethylene dichloride (EDC; 1,2-dichloroethane), which is used in the synthesis of VC, followed by VC and the halogenated solvents and extractants, perchloroethylene (PCE; tetrachloroethylene), and carbon tetrachloride (CCl_4). Vinyl bromide (VB; bromoethene), which was introduced into commerce sometime around 1959–1961, is in the lowest volume of production.

The estimated number of workers exposed is derived from data acquired during the National Occupational Hazard Survey (NOHS) of 1972–1973 conducted by the National Institute for Occupational Safety and Health (NIOSH). The number of workers exposed ranges from those with direct occupational exposure to those with a potential for indirect exposure. The current OSHA standards for these substances as expressed in TWAs ranges from a high of 100 ppm for trichloroethylene (TCE; 1,1,2-trichloroethene) and PCE to 1 ppm for VC. As shown in Table 1, there is no occupational standard for either vinylidene chloride (VDC; 1,1-dichloroethene) or VB. Table 2 lists these same substances by major industrial uses and major occupational exposure industries.

As may be seen in Figure 1, some of the halogenated hydrocarbons of concern are straight-chain saturated and unsaturated compounds. Most of these substances are two-carbon structures, connected by single or double bonds with halogens substituted on the carbon atoms to varying degrees. In two cases, the substances have chlorine atoms bonded to a single carbon atom.

Table 1
Estimates of U.S. Annual Production, Numbers of Workers Exposed, and OSHA Standards for Ten Carcinogenic Substances

Carcinogenic substance[a]	Annual U.S. production in millions of pounds	Estimated number workers exposed	OSHA standard (TWA in ppm)
EDC	10,000	33,000–2,000,000	50
EDB	330	9,000–660,000	20
VC	7,000	27,000–2,200,000	1
VDC	200	6,500–58,000	none
VB	>5[b]	360–26,000	none
TCE	301	282,000	100
PCE	700	500,000	100
ECH	470	85,000	5
$CHCl_3$	302	40,000–135,000	50
CCl_4	623	160,000–2,000,000	10

[a] Key for substances indicated in Fig. 1.
[b] Undisclosed.

Table 2
Industrial Uses and Potential Occupations with Exposure to Ten Carcinogenic Substances

Substance	Major uses	Major occupational exposure industries
EDC	chemical manufacturing, fumigant, fuel additive, organic solvent	petroleum products, chemical workers, fumigators
EDB	fumigant, fuel additive, organic solvent	chemical workers, petroleum products, exterminators and fumigators
VC	chemical manufacturing, plastics manufacturing	chemical workers, manufacturers
VDC	manufacture of copolymers and modacrylic fibers	chemical workers, plastics workers
VB	flame retardant, textiles, chemical manufacturing	chemical workers, textile workers
TCE	metal degreasing, organic solvent	chemical workers, metal workers, textile processing
PCE	dry cleaning and textiles, metal cleaning, chemical manufacturing	dry cleaners, chemical workers, textile workers, metal workers
ECH	manufacture of epoxy resins, surface active agents, and other chemicals	assemblers, machinists, painters, chemical workers
$CHCl_4$	chemical manufacturing, extraction solvent for drugs, dyes, and vitamins, anesthetics	chemical manufacturing, food and drug production
CCl_4	chemical manufacturing, grain fumigation, organic solvent	chemical workers, grain fumigators

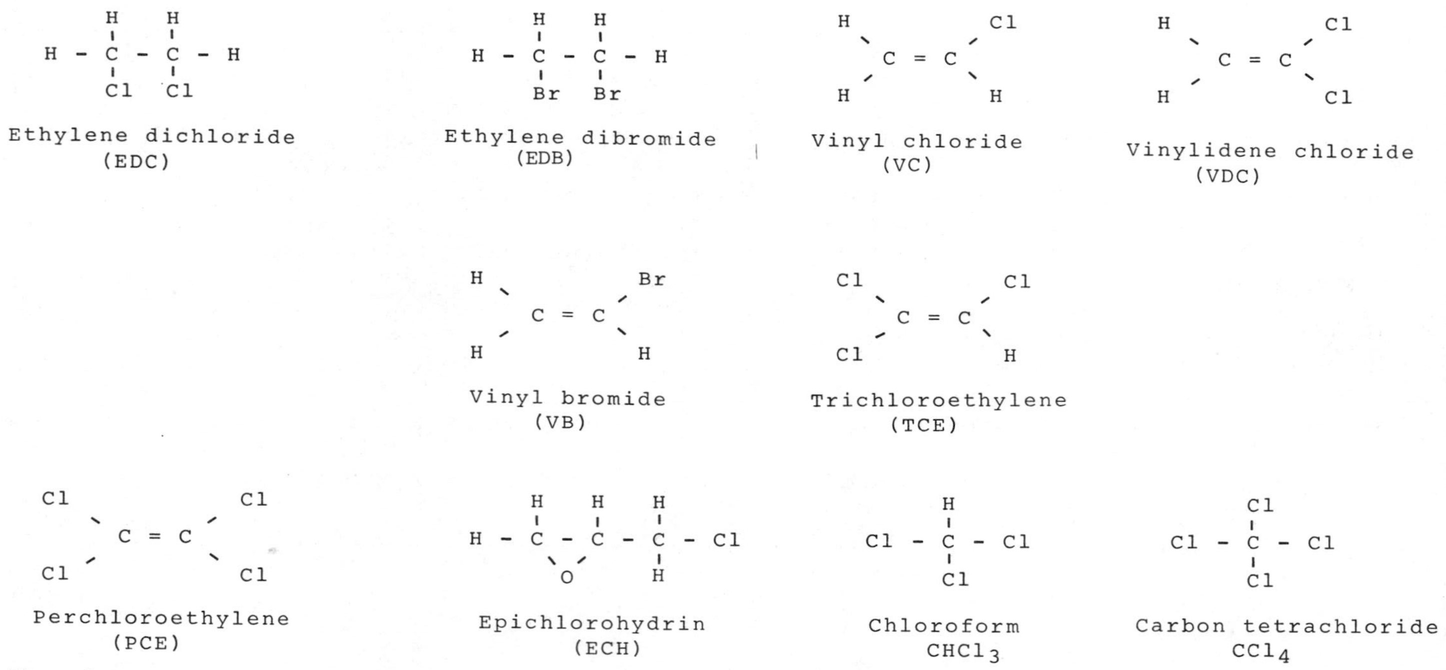

Ethylene dichloride (EDC)

Ethylene dibromide (EDB)

Vinyl chloride (VC)

Vinylidene chloride (VDC)

Vinyl bromide (VB)

Trichloroethylene (TCE)

Epichlorohydrin (ECH)

Perchloroethylene (PCE)

Chloroform CHCl₃

Carbon tetrachloride CCl₄

Figure 1
Structural formulae of ten halogenated hydrocarbons

Carbon atoms are most unique in their ability to link with themselves by single or double bonds to form stable chains or rings. The combination of hydrogen with the linked carbons affords an extraordinarily wide variety of compounds. Most hydrocarbons react readily with the halogen gases, chlorine or bromine, and the resulting products contain halogen atoms attached to carbon atoms by a covalent bond.

EXPERIMENTAL EVIDENCE

Carcinogenic Responses in Rats and Mice

Table 3 illustrates the overall carcinogenic responses found in rats and mice to ten halogenated hydrocarbons. All of these substances demonstrate the induction of cancer in experimental animals. Seven of the substances have induced cancer in both rats and mice, while two substances, TCE and PCE, have induced cancer in mice, but not in rats. One substance, VB, has not been tested in mice.

Target Organs in Rats and Mice

The broad spectrum of carcinogenic activity associated with these chemicals in rats and mice is shown in Table 4. Most of the substances have elicited a response in more than one organ per species. For two substances, EDC and EDB, there was a somewhat similar response not only in the same species, but also in the same strain. For example, both EDC and EDB induced cancers in the forestomach and spleen of the Osborne-Mendel rat (NCI 1978a; 1978b). The

Table 3
Carcinogenic Response in Rats and Mice to Ten Halogenated Hydrocarbons

Substance[a]	Response observed by species	
	rats	mice
EDC	+[b]	+
EDB	+	+
VC	+	+
VDC	+	+
VB	+	*[d]
TCE	−[c]	+
PCE	−	+
ECH	+	+
$CHCl_3$	+	+
CCl_4	+	+

[a] Key for substances indicated in Fig. 1.
[b] Carcinogenic.
[c] Not shown to be carcinogenic.
[d] No data available.

Table 4
Target Organs in Rats and Mice by Carcinogenic Substance

Substance[a]	liver	fore-stomach	spleen	lung	mammary gland	adrenals	kidney	nasal cavity	ear canal	skin	thyroid
EDC	M-B6	R-OM	R-OM	M-B6	R-OM M-B6					R-OM	
EDB	R-OM R-F3 R-SD	M-B6 R-OM	R-F3 R-OM R-SD	M-B6	M-B6 R-F3 R-SD	R-SD	R-SD	R-F3 M-B6		R-SD	R-F3 R-OM
VC[c]	R-CD M-Sw R-SD M-CD1			M-CD1 M-Sw	M-CD1 M-Sw		R-SD		R-SD	R-Wi M-Sw	
VDC	M-CD1			M-CD1 M-Sw	R-SD M-Sw		M-Sw				
VB	R-SD										
TCE	M-B6										
PCE	M-B6										
ECH								R-SD	R-SD	M-IC	
CHCl$_3$	M-B6						R-OM				R-OM
CCl$_4$	R-OM M-B6 R-Ja R-Wi		R-Ja R-OM			M-B6					R-Ja R-OM

[a] Key for substances is indicated in Fig. 1.
[b] Species, R = rat; strains: CD = Charles River, OM = Osborne-Mendel, SD = Sprague-Dawley, F3 = Fischer 344, Wi = Wistar, Ja = Japanese. Species, M = mouse; strains: B6 = B6C3F1, CD1 = Albino CD-1, Sw = Swiss; IC = ICR/Ha.
[c] VC also induced brain tumors.

primary target organs in the B6C3F1 mice, however, for both of these two-carbon, saturated halides were the lung and mammary glands (NCI 1978a; 1978b).

EDB has demonstrated a carcinogenic response in all except one of the target organs listed. It should also be noted that EDB has induced positive carcinogenic responses in both Sprague-Dawley and Fischer 344 strains of rat (Plotnick et al. 1979; NCI, unpubl. results).

Of the five two-carbon, unsaturated halides (or olefins) listed in Table 4, two of them, VC and VB, induced cancers of the ear canal and liver in Sprague-Dawley rats (Maltoni 1977; W. Busey, unpubl. results). In addition, VC and VDC have also shown positive carcinogenic responses in two strains of mice, the Swiss and the CD-1 (Maltoni 1977; Maltoni et al. 1977; Lee et al. 1978). The target organs of VC in both strains of mice were the liver, lungs, and mammary glands. On the other hand, for VDC, the liver was a target organ in the CD-1 mice, whereas the kidney was the target organ in the Swiss strain. VC has demonstrated carcinogenicity in two additional strains of rat, the Wistar and the CD (Viola et al. 1971; Lee et al. 1978). Although VC has been tested and found positive in more than one species, VB has been tested only in the Sprague-Dawley rat. The other double-bonded chlorinated hydrocarbons, TCE and PCE, were positive in the B6C3F1 mice, and the target organ for both substances was the liver (NCI 1976a; 1977; MCA 1978). Since these two substances have been found to elicit a positive carcinogenic response only in mice, they may prove to be examples of carcinogens that are species specific.

Only one three-carbon, fully saturated chlorinated hydrocarbon is shown in Table 4, namely epichlorohydrin (ECH; 1-chloro-2,3-epoxypropane). This substance has induced nasal cavity cancers in Sprague-Dawley rats and skin cancer in ICR/Ha mice (Laskin et al. 1980; Van Duuren et al. 1974).

Both chloroform ($CHCl_3$) and CCl_4 are well-known single-carbon, chlorinated hydrocarbons. (The close structural similarities would suggest that they might elicit a similar carcinogenic response.) As shown in Table 4, both of these substances induced cancer in Osborne-Mendel rats and B6C3F1 mice. The liver and the thyroid gland seemed to be the most commonly affected sites (NCI 1974; 1976b). The remaining target organs for these two substances were different in the two species. Generally, CCl_4 has induced cancers in more organs than chloroform, with the liver being the most common target organ in the various rat strains (Reuber and Glover 1970; Weisburger 1977).

Routes of Administration Versus Tumor Sites

As shown in Table 5, the liver is a target organ for nine of the substances, regardless of the route of administration. TCE induced hepatocellular carcinomas in mice when given either by gavage or by inhalation (NCI 1976a; MCA 1978).

Forestomach tumors developed after gavage with either EDC or EDB, however, this route of administration did not induce forestomach tumors with

Table 5
Site-Specific Tumor Induction by Substance and Route of Administration

Substance[a]	Route of administration			Tumor sites										
	gavage	inhalation	injection	liver	fore-stomach	spleen	lung	mammary gland	adrenal	kidney	nasal cavity	ear canal	skin	thyroid
EDC	X			X	X	X	X	X					X	
EDB	X			X	X	X	X							X
EDB		X		X					X	X	X		X	X
VC[b]	X	X		X				X		X		X	X	
VDC		X		X			X	X		X				
VB		X		X								X		
TCE	X	X		X										
PCE	X			X										
ECH		X									X			
ECH			X										X	
CHCl₃	X			X						X				
CCl₄	X			X		X			X					
CCl₄			X	X		X								X

[a] Key for substances is indicated in Fig. 1.
[b] VC also induced brain tumors.

five other substances shown in Table 5 (NCI 1978a; 1978b). The gavage of EDC and EDB also elicited carcinogenic responses in the liver, spleen, and lungs of the animals. Two inhalation studies using EDB also have demonstrated its carcinogenicity by this route of administration (Plotnick et al. 1979; NCI, unpubl. results).

In one inhalation study, Sprague-Dawley rats were subjected to a single concentration of 20 ppm EDB (Plotnick et al. 1979), resulting in tumors in the liver, spleen, mammary glands, adrenals, kidneys, and skin. The second inhalation study included two concentrations, 10 ppm and 40 ppm of EDB, and targeted some of the same organs as the preceding study, such as the liver, spleen, and mammary glands (NCI, unpubl. results). Most interestingly, this inhalation study induced cancers predominantly in the nasal cavity, a target organ not present in the preceding study.

When administered by inhalation, the three vinyl compounds, VC, VDC, and VB, all induced cancer of the liver (Viola et al. 1971; Maltoni 1977; Maltoni et al. 1977; Lee et al. 1978; W. Busey, unpubl. results), whereas two, VC and VB, induced cancer of the ear canal (Maltoni 1977; W. Busey, unpubl. results). Inhalation of VC also has induced cancer at multiple sites. Table 5 also shows that gavage with PCE or TCE resulted in the induction of hepatic tumors.

Only one substance, ECH, was administered both by inhalation and by s.c. injections. In the first study, the inhalation of ECH induced cancers of the nasal cavity in Sprague-Dawley rats, whereas in the second study, s.c. injections of the substance led to the induction of skin cancers (Laskin et al. 1980; Van Duuren et al. 1974).

Chloroform was administered by gavage to Osborne-Mendel rats and B6C3F1 mice (NCI 1976b). By this route, the liver, thyroid, and kidney were the target organs in the rat; the liver was the main target organ in the mice.

CCl_4 was administered by gavage and s.c. injections (Reuber and Glover 1970; NCI 1974). By gavage, the target organs in the B6C3F1 mice were the adrenal glands and the liver, and in the Osborne-Mendel rats the liver and spleen were the target organs (NCI 1974). Injections (s.c.) of CCl_4 were administered twice a week to three strains of rats—Osborne-Mendel, Japanese, and Wistar (Reuber and Glover 1970). Although not pronounced, carcinogenic responses did occur in the Osborne-Mendel rat in two of the same target organs that had responded to the gavage of CCl_4, i.e., the liver and spleen. As shown in Tables 4 and 5, the liver, thyroid, and to some extent the spleen, were also target organs in the Japanese strain.

Sites and Types of Tumors in Rats and Mice

As shown in Tables 5 and 6, all ten halogenated hydrocarbons induced carcinomas, a malignant type of tumor. It should be noted, however, that not all three tumor types (carcinomas, sarcomas, and adenomas) were induced in the same animal or found in the same target organ.

Table 6
Site and Types of Tumors by Substance

Substance[a]	Tumor sites											Types of tumors[b]		
	liver	fore-stomach	spleen	lung	mammary gland	adrenal	kidney	nasal cavity	ear canal	skin	thyroid	carcinoma	sarcoma	adenoma
EDC	X	X	X	X	X					X		X	X	X
EDB	X	X	X	X	X	X	X	X		X	X	X	X	X
VC[c]	X			X	X		X		X	X		X	X	X
VDC	X			X	X		X					X	X	X
VB	X								X			X	X	
TCE	X											X		
PCE	X											X		
ECH								X		X		X	X	
CHCl$_3$	X						X				X	X		X
CCl$_4$	X		X			X					X	X		X

[a] Key for substances indicated in Fig. 1.
[b] All tumor types not induced in same animal or found in same target organ.
[c] VC also induced brain tumors.

Carcinomas of the forestomach and liver and adenocarcinomas of the mammary gland were induced in both rats and mice by EDC, whereas angiosarcomas of the spleen were found only in the rats. Adenomas of the mammary gland and lung and fibromas of the skin also developed after EDC administration (NCI 1978a).

Of the ten halogenated hydrocarbons studied, EDB elicited carcinogenic responses in the greatest number of organs, and induced carcinomas, sarcomas, and adenomas. Similar to EDC, EDB induced carcinomas of the forestomach in both Osborne-Mendel rats and B6C3F1 mice (NCI 1978b). Hepatocellular carcinomas and splenic angiosarcomas developed in all three strains of rat tested with EDB, although the B6C3F1 mice seemed resistant to both tumor types (NCI 1978b; Plotnick et al. 1979; NCI, unpubl. results). EDB also induced adenomas of the thyroid gland in the Osborne-Mendel and in the Fischer 344 rats (NCI 1978b; and unpubl. results). There was, furthermore, extensive development of carcinomas of the nasal cavity in Fischer 344 rats after EDB inhalation. Nasal cavity cancers also developed to a lesser extent in the B6C3F1 mice. Alveolar-bronchiolar adenomas and carcinomas of the lung were the most predominant tumor type induced by EDB in this strain of mice.

Inhalation of VC induced carcinomas of the ear canal, angiosarcomas of the liver, neuroblastomas of the brain, and nephroblastomas in Sprague-Dawley rats (Maltoni 1977).

VC also induced mammary gland carcinomas, hepatic angiosarcomas, and lung adenomas in Swiss and Albino CD-1 strains of mice (Maltoni 1977; Lee et al. 1978). Moreover, Swiss mice developed a variety of skin tumors, ranging from squamous cell carcinomas to subcutaneous angiomas (Maltoni 1977).

In female Sprague-Dawley rats, VDC primarily induced carcinomas or fibroadenomas of the mammary gland (Maltoni et al. 1977). Two strains of mice, Swiss and Albino CD-1, also were sensitive to VDC (Maltoni et al. 1977; Lee et al. 1978). Although both strains of mice developed adenomas of the lung, only the Swiss mice had adenocarcinomas of the kidney (Maltoni et al. 1977) and only the CD-1 strain developed hepatomas or angiosarcomas of the liver (Lee et al. 1978). The brominated olefin, VB, was shown to induce carcinomas of the ear canal, as well as hemangiosarcomas and carcinomas of the liver in Sprague-Dawley rats (W. Busey, unpubl. results). As shown in Tables 4 and 6, the remaining two-carbon, unsaturated halides, TCE and PCE, induced the same tumor type, hepatocellular carcinomas, in B6C3F1 mice (NCI 1976a, 1977; MCA 1978).

The three-carbon halide, ECH, was responsible for the development of squamous cell carcinomas of the nasal cavity in Sprague-Dawley rats (Laskin et al. 1980) and injection-site adenocarcinomas or sarcomas of the skin in ICR/Ha mice (Van Duuren et al. 1974).

$CHCl_3$ induced carcinomas or adenomas of both the kidney and the thyroid gland in the Osborne-Mendel rat. The tumor type observed in the B6C3F1 mice was hepatocellular carcinoma (NCI 1976b).

Administration of CCl_4 elicited the development of hepatocellular carcinomas, as well as adenomas or pheochromocytomas in the adrenal glands of B6C3F1 mice (NCI 1974). Considering the effects observed in three strains of rats exposed to CCl_4, as shown in Tables 4 and 6, carcinomas of the thyroid and hemangiomas of the spleen developed in two of the strains, and carcinomas of the liver appeared in all three strains, i.e., Osborne-Mendel, Japanese, and Wistar (Reuber and Glover 1970).

EPIDEMIOLOGIC EVIDENCE

All available published and unpublished reports of epidemiologic studies of workers exposed to the selected halogenated hydrocarbons, excluding VC, were reviewed. Cohort studies have been reported for workers specifically exposed to ECH (Enterline 1978; Shellenberger et al. 1979), TCE (Axelson et al. 1978), VDC (Ott et al. 1976), and EDB (Ott et al. 1980). A study of laundry and dry cleaning workers, who had potential exposure to the dry cleaning solvents TCE, PCE, CCl_4, and benzene (Blair et al. 1979) was also considered.

A study of workers exposed to ECH was conducted by Enterline (1978) and further analyzed using the NIOSH modified life-table program (D. P. Brown and R. A. Rinsky, unpubl. results). Cohort members had to be employed for more than one quarter prior to 1966. Vital status was determined through December 1977. The number of observed deaths was contrasted with the expected based on rates from the standard U.S. white male population adjusted for age and calendar time period. As shown in Table 7, for those individuals with 15 or more years of latency 12 deaths from all cancers were observed versus 8.9 expected deaths. Seven lung cancer deaths (lung and bronchus considered as "lung" here) were observed versus 3.5 expected. Enterline has stated "the data should be viewed as highly suggestive of a carcinogenic risk of ECH for man." These data should be interpreted with caution because part of the cohort was exposed to a process known to be associated with an elevated cancer risk to the respiratory system, i.e., some cohort members had been employed in the manufacturing of isopropyl alcohol, which has been associated with the development of maxillary sinus cancer, but not lung cancer.

Information on smoking histories for the cohort was not available and this could have led to either an overestimation or underestimation of the lung cancer risk. The size of the cohort, 864 individuals, was small and the follow-up period was short. The average age of cohort members during the follow-up period was only 48 years. These latter factors particularly may have led to an underestimate of the risk of cancer in the study cohort. The data, however, are consistent with laboratory studies demonstrating respiratory cancer in experimental animals. A second epidemiologic study of workers exposed to ECH was conducted by Shellenberger et al. (1979). Two deaths from total malignancies were observed versus 3.5 expected. The sample size and latency consideration for this study are not sufficient at this time to merit further discussion of the study results.

Table 7
Observed and Expected Deaths from Studies of Workers Exposed to Selected Halogenated Hydrocarbons

Investigators	Compound	Total cancer deaths			Site-specific cancer deaths		
		observed	expected	RR[a]	observed	expected	RR
Shellenberger et al. (1979)	ECH	2	3.5	0.6		N.A.[b]	
Enterline (1978); D. P. Brown and R. A. Rinsky (unpubl. results) ≥15 yr latency	ECH	12	8.9	1.3	lung 7	3.5	2.0
Axelson et al. (1978); ≥10 yr latency, high exposure	TCE	3	1.8	1.7		N.A.	
Ott et al. (1976); ≥15 yr latency	VDC	1	0.5	2.0	lung 1	0.2	5.0
Ott et al. (1980); ≥15 yr latency	EDB	5	4.4	1.1		N.A.	
Blair et al. (1979)	TCE, PCE, CCl$_4$, benzene	87	67.9	1.3[c]	lung 17 liver 4 leukemia 5 cervix uteri 10	10.0 1.7 2.2 4.8	1.7[c] 2.4 2.3 2.1[c]

[a] Relative risk.
[b] Not analyzed.
[c] $p < 0.05$.

Axelson et al. (1978) reported on the mortality experience of 518 male workers exposed to TCE in the 1950s and 1960s and followed through 1975. Nine deaths from all types of cancer were observed compared to 9.5 expected. As shown in Table 7, for individuals with 10 or more years of latency there were three cancer deaths observed versus 1.8 expected. The data were too few to analyze by site-specific cancer risk. The authors stated the identification of one case of liver cancer would have resulted in a relative risk of 3.4 for the total cohort and 25.0 for the high-exposure cohort, and they made the reasonable conclusion that "the cancer risk to man from trichloroethylene can by no means be ruled out from this study, particularly with regard to uncommon malignancies such as liver cancer."

Ott et al. (1976) followed a cohort of 138 individuals employed sometime during 1942-1969 through 1973 to determine mortality experience. In 1973, 103 were still employed by the company, 3 were retired, 5 were deceased, and 27 who left the company were presumed living on the basis of Social Security follow-up. In this cohort, 40% had less than 15 years since initial exposure to VDC.

For all causes of death, 5 were observed versus 7.5 expected. As shown in Table 7, one death from respiratory cancer was observed in an individual with 15 plus years since initial exposure as compared to 0.2 expected (relative risk = 5.0). Because of the insufficient cohort size and latency period, it is not possible from this study to evaluate the carcinogenic risk of VDC to humans.

Ott et al. (1980) conducted a cohort mortality study of 161 individuals exposed to EDB sometime between 1940-1975 and followed through December 1975. As shown in Table 7, for individuals with 15 or more years of latency, 5 deaths from all cancers were observed versus 4.4 expected. The cohort was too small for analysis by specific sites of cancer. Again, the sample size, latency characteristics do not allow the development of valid inferences regarding the risk of cancer from individuals exposed to EDB.

Blair et al. (1979) conducted a proportionate mortality study of 330 deaths among laundry and dry cleaning workers that occurred between 1957-1977. These workers had potential exposure to several halogenated solvents—TCE, PCE, CCl_4, plus benzene. As shown in Table 7, a significant excess was observed for mortality from all types of cancer combined (87 versus 67.9 expected) and for lung and cervical cancer. Also noted by the authors was the more than twofold excess of liver cancer and leukemia. Liver cancer has been induced in experimental animals exposed to TCE, PCE, and CCl_4, whereas benzene has been used as a spot remover in dry cleaning.

SUMMARY AND CONCLUSIONS

Experimental studies have now demonstrated the carcinogenicity of nine VC-related compounds. Six studies have addressed the carcinogenicity among workers exposed to a limited number of these substances. One epidemiologic study suggests an elevated risk of lung cancer among workers exposed to ECH. Although a second study suggests a disproportionate number of dry cleaning workers dying from cancer of multiple sites, the study lacks specificity regarding the carcinogenic substances to which the workers were exposed. Four additional epidemiologic studies available for review do not allow for the assessment of carcinogenic risk among humans exposed to these substances because the number of workers available for study who had achieved an adequate latency period was small. Information presented indicates that all of the substances studied are in high-volume production with large estimated numbers of workers exposed; however, retention of personnel records containing information necessary for epidemiologic study of health hazards is not a requirement in the U.S. and adds to the problem of insufficient sample size.

On the basis of these observations, it is apparent that, in general, qualitative carcinogenic risk of a specific chemical substance to humans must be estimated through the conduct of experimental studies. Epidemiologic studies should be conducted when adequately sized cohorts of workers who have achieved sufficient latency are available, but should not be used as a basis for delay in public health decisions when adequate experimental data indicate cancer induction.

In light of the demonstrated carcinogenicity of these VC-related compounds, will we as a society continue subjecting workers to the current exposure levels until epidemiologic confirmation is achieved, as was the case with VC, or will a concerted effort be made to reduce human exposure to these carcinogenic substances beginning at the site of industrial origin? Rather than controlling carcinogens in the work environment, some segments of society have been preoccupied with concern for alleged government overregulation of toxic substances. Let us now hope that this same concern can be directed toward the protection of its workers.

REFERENCES

Axelson, D., K. Anderson, C. Hogstedt, B. Holmberg, G. Molina, and A. de Verdier. 1978. A cohort study on trichloroethylene exposure and cancer mortality. *J. Occup. Med.* **20**:194.

Blair, A., P. Decoufle, and D. Grauman. 1979. Causes of death among laundry and dry cleaning workers. *Am. J. Publ. Health* **69**:508.

Enterline, P. E. 1978. Updated mortality in workers exposed to epichlorohydrin. Submission to OSHA, Docket No. H-100, File 4.

Laskin, S., A. R. Sellakumar, M. Kuschner, N. Nelson, S. LaMendola, G. M. Rusch, G. V. Katz, N. C. Dulak, and R. E. Albert. 1980. *Inhalation carcinogenicity of epichlorohydrin.* Inhalation Center, Institute of Environmental Medicine, New York. (In press)

Lee, C. C., J. C. Bhandari, J. M. Winston, W. B. House, R. L. Dixon, and J. S. Woods. 1978. Carcinogenicity of vinyl chloride and vinylidene chloride. *J. Toxicol. Environ. Health* 4:15.

Maltoni, C. 1977. Vinyl chloride carcinogenicity: An experimental model for carcinogenesis studies. *Cold Spring Harbor Conf. Cell Proliferation* 4:119.

Maltoni, C., G. Cotti, L. Morisi, and P. Chieco. 1977. Carcinogenicity bioassays of vinylidene chloride. *Med. Lav.* 68:241.

MCA (Manufacturing Chemists Assoc.). 1978. Final report of audit findings, administered trichloroethylene (TCE) chronic inhalation study at Industrial Bio-Test Laboratory, Inc., Decatur, Illinois.

NCI (National Cancer Institute). 1974. *Report of carcinogenesis bioassay of carbon tetrachloride.* Division of Cancer Cause and Prevention, National Institutes of Health, Bethesda, Maryland.

―――――. 1976a. *Carcinogenesis bioassay of trichloroethylene.* Technical report series no. 2. DHEW publication number (NIH) 76-802, Government Printing Office, Washington, D.C.

―――――. 1976b. *Carcinogenesis bioassay of chloroform. Report by carcinogenesis program.* Division of Cancer Cause and Prevention, National Institutes of Health, Bethesda, Maryland.

―――――. 1977. *Carcinogenesis bioassay of tetrachloroethylene.* Technical report series no. 13. DHEW publication number (NIH) 77-813, Government Printing Office, Washington, D.C.

―――――. 1978a. *Carcinogenesis bioassay of 1,2-dichloroethane.* Technical report series no. 55. DHEW publication number (NIH) 78-1361. Government Printing Office, Washington, D.C.

―――――. 1978b. *Carcinogenesis bioassay of 1,2-dibromoethane.* Technical report series no. 86. DHEW publication number (NIH) 78-1336. Government Printing Office, Washington, D.C.

Ott, M. G., W. A. Fishbeck, J. C. Townsend, and E. J. Schneider. 1976. A health study of employees exposed to vinylidene chloride. *J. Occup. Med.* 18:735.

Ott, M. G., H. C. Scharnweber, and R. R. Langner. 1980. The mortality experience of 161 employees exposed to ethylene dibromide in two production units. *Br. J. Ind. Med.* (in press).

Patty, F. A., W. P. Yant, and C. P. Waite. 1930. Acute response of guinea pigs to vapors of some new commercial organic compounds. *Public Health Reports* 45:1963.

Plotnick, H. B., W. Weigel, D. Richards, K. Cheever, and C. Kommineni. 1979. *Dietary disulfiram enhancement of the toxicity of 1,2-dibromoethane.* Division of Biomedical and Behavioral Sciences, National Institute for Occupational Safety and Health.

Ramsey, J. C., C. N. Park, M. G. Ott, and P. J. Gehring. 1979. Carcinogenic risk assessment: Ethylene dibromide. *Toxicol. Appl. Pharm.* 47:411.

Reuber, M. D. and E. L. Glover. 1970. Cirrhosis and carcinoma of the liver in male rats given subcutaneous carbon tetrachloride. *J. Natl. Cancer Inst.* 44:419.

Shellenberger, R. J., C. D. McClimans, M. G. Ott, R. E. Flake, and R. L. Daniel. 1979. An evaluation of the mortality experience of employees with potential for exposure to epichlorohydrin. Submission to OSHA, Docket No. H-100, File 8.

Van Duuren, B. L., B. M. Goldschmidt, C. Katz, I. Seldman, and J. S. Paul. 1974. Carcinogenic activity of alkylating agents. *J. Natl. Cancer Inst.* 53:695.

Viola, P. L., A. Bigotti, and A. Caputo. 1971. Oncogenic response of rat skin, lungs, and bones to vinyl chloride. *Cancer Res.* 31:516.

Weisburger, E. K. 1977. Carcinogenicity studies on halogenated hydrocarbons. *Environ. Health Perspect.* 21:7.

COMMENTS

PLOTNICK: I've always had mixed feelings about such studies. When you are examining an occupational setting, do you use a technical grade of perchloroethylene (PCE) as a standard for exposure or do you use purified material? The real question is, what do we know about the materials that the workers are really exposed to?

BUSEY: Where did you get the VC-induced brain tumor?

MARLOW: From Maltoni's 1977 study. Dr. Busey, would you tell us about the TCE inhalation study in which hepatocellular carcinomas were induced?

BUSEY: An inhalation study was conducted for 2 years on B6 mice using TCE that contained EDC and all other stabilizers. There were several flaws in that study; in particular, the exact concentration to which the mice were exposed was not known and, a very important factor, the controls were probably from a different population, because they arrived at the laboratory somewhat later than the animals that were under exposure.

Anyone who has worked with B6 mice knows that the instance of liver tumors can vary upwards to 40% in males. The results of this study showed more liver tumors in the treated animals than there are in the controls.

In the NCI study (NCI 1974) there was an approximately 90% incidence of liver tumors, is that right?

HOOPER: The incidence of liver tumors was 65% in the high-dose male mice and 52% in the low-dose mice for the NCI study. What level of ECH and other impurities are there in the test TCE?

MARLOW: Please, this is an evaluation—that's why we're here.

HOOPER: I'd like to quote from the NCI report the stated impurities in the test sample of TCE. "The test trichloroethylene is greater than 99% pure for each of several batches." Earlier in the discussion it was suggested that the impurity of ECH was about 2%. It states here that it is less than one-twentieth of this. The other impurities are listed:

1,2-epoxybutane (0.19%),
ethyl acetate (0.4%),
N-methyl pyrrole (0.02%),
and diisobutylene (0.03%).

Thus, TCE appears in fact to be quite pure, with a very low level of ECH contaminant. I would also like to add that I don't see evidence that the animals at high doses are severely stressed, as was stated yesterday. In the male and female mice, there was no decreased survival or loss of weight among the high-dose animals, although these were the animals with significant incidence of hepatocellular carcinomas. Comparing incidence of liver carcinomas among controls (1/20) with low-dose male mice (26/50), we have a situation in which a dose level causing no shortening of life or lowering of body weight causes significant tumor incidence. What's causing these tumors? I doubt it is the ECH.

TER HAAR: I think the concentrations of ECH are typically 0.1 and 0.2%.

WARD: I'd like to add additional information that should be considered for the evidence of the carcinogenicity of TCE. In the NCI study, the results were dose-related in both sexes of mice. The mortality was dose-related, which was related also to the appearance of liver cancer in animals that died early.

Dr. Van Duuren (this volume) stated today that he believed that ECH given by gavage would not reach the liver, and so he would think that ECH administered by gavage would not induce liver tumors. If you calculate the dose of ECH that might be present after gavage of rats with TCE, you get an approximate maximum dose of something like 2.4 mg/kg ECH per day. If that dose induced liver tumors, then ECH would have to be a fairly strong carcinogen. From the results of the inhalation studies by Dr. Van Duuren (I'm not sure if it's been published yet on ECH) about 30% of the rats got nasal carcinomas after being given 100 ppm ECH by inhalation. From the evidence it appears that ECH is not a strong carcinogen, and would expect 2.4 mg/(kg · day) not to induce very many liver tumors.

HOOPER: What was the dose used in the ECH study that you mentioned?

MARLOW: In Dr. Van Duuren's study (Van Duuren, this volume), it was 100 ppm for male animals only.

WARD: Then the additional findings by Industrial Bio-Test Laboratory (IBT) would support our findings. Considering all those things, there was more evidence that the NCI study does show that most likely TCE, itself, is carcinogenic and not a contaminant. We can't, of course, eliminate any cocarcinogenic effect of TCE and impurities.

PLOTNICK: First about that statement on ECH. I thought it was a study on the tissue distribution of [^{14}C]ECH by oral gavage. The tissue distribution data, which appeared to have some work done by Dow on inhalation of [^{14}C]ECH was identical, at least, on a relative basis with respect to organ distribution. This would argue against destruction of significant amounts of ECH in the gastrointestinal tract before absorption.

If there's any question about the purity in the EDB used in the Midwest Research Institute study, I personally analyzed that and it was better than 99.9%.

MARLOW: Our listing of these compounds as a very rough comparison does show similarities, not only in dose levels but also in the ultimate target area.

TER HAAR: It shows similarities.

INFANTE: This is just information that has been published. For example, Dr. Maltoni hasn't published any of his recent data, and we don't have access to it.

TER HAAR: We started by looking at it this way too, but of course then you learn that other data were available. I was trying to figure out if there's anything else we ought to be doing to teach us if there are any differences among our work.

INFANTE: At Ethyl do you have any of Maltoni's more recent results?

BUSEY: Dr. Maltoni has positive results down as far as 5 ppm with VC. When we started this study we were convinced that VB would be negative at 10 ppm, but, obviously it is not. I think that VC was shown to be as positive at these low dose levels as VB was.

HOOPER: Gary [Ter Haar], to calculate properly the carcinogenic potency of VB from your inhalation study, we need the life-table data for liver angiosarcomas and carcinomas. Can we cooperate with you on this?

TER HAAR: Yes.

HOOPER: One point about VC—we wouldn't have identified it as a human carcinogen except that it caused a rare tumor. For what other sites do we have good epidemiological evidence that VC is a human carcinogen?

INFANTE: We've got brain cancer and lung cancer.

HOOPER: Yes, but these effects are detected because we are focusing on VC. We probably wouldn't have identified VC as a human carcinogen from these brain and lung tumors alone.

INFANTE: Probably not. However, even if the study were done, there's the problem of the high background level. VC was identified before the study was ever done because of the rare tumor.

The other problem is an inadequate period of follow-up. Some of the studies I've described show that the populations are not followed long enough to allow these cancers to become clinically manifest. Studies that follow workers for the first 10 or 15 years after initial exposure provide little data, really, to assess whether or not the substance is carcinogenic. In fact data for substances that produce cancer in working populations usually show a deficit of cancer for the first 10-15 years. This is expectable because you're comparing a healthy population of new employees to a group including the infirmed, the institutionalized, and people who already have cancer in the general population. Our reference population is usually the general population.

But there are many other problems in doing these studies, for example, lack of a lifetime study results. If we want to talk about quantitative risk assessment, we have to follow the individuals over a lifetime. We can't follow a few populations of VC polymerization workers and then try to extrapolate to all workers exposed to VC on the basis of a few small populations.

Underascertainment due to incorrect diagnosis, particularly for rare types of disease is another problem. I think VC and angiosarcomas are a very good example of this. Some of these workers, in fact, did not have cause of death listed on their death certificate as angiosarcoma. It was only later when we went back to look at that site that we found some of these cases were angiosarcoma deaths and not cirrhosis of the liver.

Information on confounding factors is another difficulty. Every time a study shows an excess of lung cancer, we tend to say it could be due to smoking. If we don't have adequate smoking histories and assuming there is only a 20-50% increase in the lung cancer risk, I think the only way to answer this question is to start getting information on smoking habits of workers so that, rather than sitting down and arguing about it, we would have the data for which we can make the judgment.

An additional problem is secular changes in nosology—the logic that you go through to indicate cause of death on the death certificate. This has changed over time. I think prior to 1949, for example, you couldn't separate out leukemia from some of the other blood diseases, or lymphomas. So, if you want to do a study of leukemia, for example, myelogenous leukemia versus other leukemias, you simply don't have a comparison population in the U.S. because it wasn't coded in that manner in the past. The same thing is true with mesothelioma. You can't generate an expected number of mesothelioma deaths because that code is mixed in with other causes of death.

Incorrect completion of death certificates is a problem. For example you may know that an individual has lung cancer, but the information is not on the death certificate, even as an underlying cause of death. Instead pneumonia is listed as the immediate cause of death. You cannot count that individual in your analysis. Finding an appropriate control population is also a problem. We usually use the U.S. white male population. But should we at times use national rates, state rates, county rates, or other industry controls? There really isn't an ideal population, so you have to use the best available.

I think the sum of all of these problems is that we can't really quantify the exposure and then look at the cancer risk because we lack complete information.

References

NCI (National Cancer Institute). 1974. *Report of carcinogenesis bioassay of carbon tetrachloride.* Division of Cancer Cause and Prevention, NIH, Bethesda, Maryland.

Evidence of the Mutagenicity of Ethylene Dichloride and Structurally Related Compounds

JILL D. FABRICANT
Department of Preventive Medicine and Community Health
Division of Environmental Toxicology
University of Texas Medical Branch
Galveston, Texas 77550

JOHN H. CHALMERS, JR.
Department of Biochemistry
Baylor College of Medicine
Houston, Texas 77030

A number of short-term methods have been developed in recent years to test chemical compounds and other agents for mutagenic activity. These tests involve a number of species, both mammalian and nonmammalian, and are conducted both in vivo and in vitro. A correlation has been shown between the mutagenicity and the carcinogenicity of a compound. It has been estimated that approximately 80% (B. N. Ames, pers. comm.; M. S. Legator, pers. comm.) of all mutagenic compounds are also carcinogenic, although not all carcinogens are mutagens. Therefore, the value of tests to determine the mutagenicity of a compound is important in that these tests may, at least in some cases, give some indication of the carcinogenicity of that chemical as well. The mutagenicity tests alone, however, cannot definitively identify a carcinogen. However, some of these tests can be used for monitoring at-risk population groups and to aid in the evaluation of acceptable exposure standards.

Mutational events may occur either at the gene level, where they result in mutant alleles, or at the chromosome level, where they cause chromosomal aberrations (usually the deletions or addition of some part of the chromosome arm). These may occur either during meiosis (in germ cells resulting in possible genetic transmission of that mutation) or during mitosis (occurring in somatic cells and not hereditary). Chromosomal aberrations may be either numerical or structural. Numerical aberrations affect the whole chromosome set by the addition or deletion of one or more chromosomes from the cell. Structural anomalies, however, are generally deletions, breaks, gaps, or translocations in chromosomes and do not affect the chromosome number. Although many of these mutational events may result in cellular death, many may be repaired.

In addition to the genetic end points of gene mutation or chromosomal aberration, primary DNA damage may result from chemical mutagenesis. There are a number of tests that evaluate damage-induced genetic recombination. These include mitotic crossing over, gene conversion, sister chromatid exchange, and unscheduled DNA synthesis, as well as others.

Table 1
Evidence for the Mutagenicity of EDC and Structurally Related Compounds

Short-term assays	Results
EDC	
Bacteria	+/−[a]
Drosophila	+
Plants	+/−
Cell culture	+
EDB	
Bacteria	+/−
Yeast, *Neurospora*	+
Plants	+/−
Drosophila	+
Cell culture	+/−
VDC	
Bacteria	+/−
Cell culture	+/−
TCE	
Bacteria	+/−
Yeast	+
Cell culture	+
PCE	
Bacteria	−
Yeast	+

[a] Cases in which different laboratories performed the same test with inconsistent results.

Table 1 shows the different short-term mutagenicity assays and results obtained for ethylene dichloride (EDC; 1,2-dichloroethane) and the structurally related compounds ethylene dibromide (EDB; 1,2-dibromoethane), vinylidene chloride (VDC; 1,1-dichloroethene), trichloroethylene (TCE; 1,1,2-trichloroethene), and perchloroethylene (PCE, tetrachloroethylene). In this table, the results column indicates the mutagenic capability of the given compound. A brief description of the short-term tests used in the testing of these chemicals is included in the following sections.

GENE MUTATIONS ASSAYS

The simplest and most widely used short-term test for mutagenicity is the *Salmonella*/mammalian microsome assay commonly called the Ames test (Ames 1973). This assay is based on the reversion of sensitive histidine auxotrophs with known types of genetic lesions (base-pair substitution, frameshifts, etc.) to prototrophy. For example, TA1535 is used to detect mutagens causing base-pair substitutions and TA98 is used for frameshifts. Mutagens are screened either by spotting small quantities of the substance onto seeded petri plates or quantitatively assayed by incorporating them into an agar overlay containing the test organism. By adding a rat liver microsomal preparation from animals pretreated with either Aroclor or phenobarbital (PB) to induce the drug catabolizing systems, the effects of in vivo metabolism are simulated. The presence or absence of the S-9 microsomal fraction with or without prior induction may account for much of the variation in mutagenicity reported by different laboratories, although strain differences between the animals may also be important. Other types of bacterial assays based upon the enhanced killing of repair-deficient mutants (Brem et al. 1974), forward and reverse mutation (Greim et al. 1975), and the induction of phage λ in lysogenic strains (Kristoffersson 1974) have been employed in some cases. Tests with eukaryotic microorganisms have also been developed using yeast (Murthy 1979) or other fungi, such as *Neurospora* (deSerres and Malling 1970).

Insects are also used in the detection of gene mutations. The most commonly used test of this type is the sex-linked recessive lethal test (Abrahamson and Lewis 1972) with the fruit fly, *Drosophila melanogaster*. This test has the advantage of rapidly screening a large number of genes. In this system, change in the X chromosome is evaluated in the second generation after treatment. However, certain translocations (II-III), chromosome loss (entire X, Y and partial Y), dominant lethals, nondisjunction as well as other end points, may also be identified with the use of certain strains (Wurgler et al. 1977).

Gene mutations can also be evaluated by using in vitro techniques, involving mammalian somatic cell lines or by in vivo studies involving mammals. Specific-locus effects may be detected after exposure to mutagenic chemicals. Direct-acting mutagens are generally detected by the induction of a forward mutation in the in vitro tests. The phenotypic markers may include drug resistance (e.g., 8-azaguanine or 6-thioguanine), serological (e.g., loss of an antigen), nutritional (e.g., amino acid auxotrophy), or temperature sensitivity. The most common cell lines are those derived from Chinese hamster ovary or embryo.

One of the most important assays presently available to evaluate the mutagenicity of a chemical in humans employs chromosomal analysis. Cytological techniques may include either in vivo or in vitro assays (Evans 1976). They are designed to determine genetic damage expressed in germ cells during early embryogenesis (transmitted and nontransmitted) as well as in somatic tissue (acute and chronic effects). For the germ cell studies, testes are generally used whereas bone marrow or peripheral lymphocytes are selected for adult somatic tissue cultures. Both chromatid and chromosome breaks are noted, and stable intra- and interchromatid and chromosome rearrangements are analyzed. Usually in vivo mammalian studies involve the analysis of bone marrow cells whereas circulating peripheral lymphocytes are the preferred in vitro system.

Chromosomal studies are also performed on plants and *Drosophila*. *Tradescantia* clone 02, for example, is a diploid plant that is very sensitive to mutation induction by radiation (Sparrow et al. 1974). In this system, mutagenic effects are scored as color changes in stamen hair cells, presumably because of chromosome aberrations. Other measurements of chromosome aberrations have also been obtained using onion (*Allium cepa*) and barley. In *Drosophila*, chromosomal changes can be determined simply by breeding experiments. Some mutagenicity experiments in *Drosophila* include meiotic and mitotic recombination, dominant lethality, and chromosomal rearrangements.

The yeast, *Saccharomyces cerevisiae*, has been used for mitotic gene conversion studies as well as mitotic crossing over and induced nondisjunction. For mitotic crossing over experiments, the reciprocal exchange of genetic information is determined.

The dominant lethal test assay involves the scoring of early fetal loss resulting from pre- and postimplantation mortality. The developmental anomalies leading to the embryonic mortality detected in this test are thought to result from chromosomal aberrations produced in sperm. In these tests, mice or rats are generally mated for 8–10 weeks. The pregnant mothers are usually sacrificed at midgestation and the number of live embryos, total implants, dead implants, and corpora lutea are recorded.

As can be seen from the data presented in Table 1, many chemicals are not found to be positive in all tests. Thus, one may consider that there is a continuous scale of mutagenic activity, which may be modified by environmental and genetic factors. In the case of EDC, for example, positive results were obtained with *Drosophila* tests, some *Salmonella*/mammalian microsome tests, and the *Escherichia coli* polA⁻ test as well as tests using barley and Chinese hamster cells. In tests using onions and some *Salmonella* assays, however, it was reported to be negative. Because of these differences, multiple test systems are necessary to evaluate fully the potential mutagenicity and carcinogenicity of chemicals to man.

EDC

Table 2 lists the tests performed using EDC on *Salmonella* and the reported mutagenic activity. Also listed are the concentrations tested (expressed as μmoles per plate), the inducers used (either PB or Aroclor, a mixture of polychlorinated biphenyls), the use of activation or not, and the bacterial strains used in testing. The most significant increase seen here is that observed in Rannug's study (Rannug et al. 1978 and this volume) in which a tenfold increase was reported in the S-9-activated TA1535 strain. When S-9 activation was not used, however, Rannug still reported an increase, although only twofold over that seen in the controls. The differences observed in mutagenic activity of EDC in *Salmonella* as seen in some of the other investigators' data (in which no effect is observed, for example) result from a variety of causes. Differences in the rat

Table 2
Mutagenic Effects of EDC on *Salmonella*

Reference	Concentration (μmole/plate)	Inducer	S-9	Strains rat	Strains bacterial	Effects
Rannug et al. (1978)	60, 40, 20	PB	+/−[c]	R strain	TA1535	10X/2X increase
McCann et al. (1975)	131	N.R.[b]	+/−	N.R.	TA100	+?[f]
Kanada and Uyeta (1978)	N.S.[a]	Aroclor	+/−	N.S.	complete set[e]	+[g]
King et al. (1979)	36	Aroclor	+/−	Sprague-Dawley	complete set	−
Brem et al. (1974)	10 (disk)	N.R.	−[d]	N.R.	TA1530 TA1535 TA1538	2X 2X −
Simmon et al. (1978)	N.S.	N.R.	−	N.R.	TA100	+

[a] Not specified.
[b] Not relevant.
[c] With and without S-9.
[d] Data in which S-9 was not used.
[e] TA1535, TA1538, TA98, TA100, TA1537.
[f] Slight increase noted.
[g] Data in which S-9 was used.

strains appear to affect the mutagenicity of the compound. R-strain rats were used in Rannug's study in which positive results were seen, whereas in King's study (1979), Sprague-Dawley rats were used and negative results were reported. Unfortunately, rat strain was not specified in Kanada and Uyeta (1978). Other differences (as in the case of Kanada and King) may be explained by the different inducers used. From this data, it appears that PB is a better inducer than Aroclor and that EDC is a weak direct mutagen since it appears that activation is required for any kind of mutagenic activity. Additional studies using *E. coli* (polA⁻ and K12) are negative in that significant cell kill was not reported (Brem et al. 1974; King et al. 1979). A final study (Kristoffersson 1974) using phage λ K39 involves a phage induction assay. Here, negative results were also reported.

The effect of glutathione (GSH) S-transferase on the mutagenicity of EDC on *Salmonella* strain TA1535 is seen in Table 3 (Rannug, this volume). Here, increased mutagenic activity is seen in *Salmonella* treated with EDC following the administration of GSH (120 vs 270 colonies).

The mutagenicity of EDC in *Drosophila* is seen in Table 4. In the three reports using the sex-linked recessive assay, increased mutation was observed (King et al. 1979; Rapoport, 1960; Shakarnis 1969). Although there are quantitative differences in these reports, they are all positive and clearly show mutagenic activity of EDC in *Drosophila*. The somatic mutation assay (Nylander et al. 1978) also showed positive results ranging from 0.045% (controls) to 7.21% mutations. The study involving nondisjunction of an X chromosome in *Drosophila*, however, was negative (Shakarnis 1969). In the micronucleus assay using mammalian cells, negative results were also reported (King et al. 1979).

From these data on mutagenicity studies of EDC, it appears that EDC is a rather weak direct mutagen that can be activated to a more effective species. The nature of the proximate mutagen is not known, however, although S-2-chloroethyl GSH has been suggested as an intermediate (Rannug, Anders, this volume). This compound itself may be mutagenic, or it may be converted to the very reactive episulfonium derivative (Anders, this volume). In any case, GSH and GSH S-transferases appear to be necessary for efficient activation (Rannug

Table 3
Effect of GSH on the Mutagenicity of EDC on *Salmonella* (TA1535)

Treatment	EDC concentration (μmole/plate)	
	0	60
S-9 + GSH	12	270
S-9	12	120
Heated S-9 + GSH	12	84
Autoclaved S-9 + GSH	12	15

R-strain rats, PB-induced, 1.6 mmole/ml GSH. Data from Rannug et al. (1978).

Table 4
Mutagenicity of EDC in *Drosophila*

	Strain	Mutation (%)	Males	Females	Comments	Reference
			Sex-linked recessive lethal			
Concentration						
0		0.21			spermatids	King et al. (1979)
50 mM		1.94			most sensitive	
0		0.21			added for 1 hr to test tube	Rapoport (1960)
					containing larvae	
35 mg		2.63				
Hours						
0		0.3			treated females 4.7 ppm	Shakarnis (1969)
4		3.22				
8		5.90				
			Somatic mutation assay			
Concentration						
0	stable	0.045 (2/4441)[a]				Nylander et al. (1978)
10	stable	4.20 (263/6260)[a]				
50 mM	stable	7.21 (44/610)[a]				
			Nondisjunction of X chromosomes			
Hours						
0			0	0		Shakarnis (1969)
4			0.3	0.01		
8			0.18	0.09		

[a]Sectored eyes.

et al. 1978). This conclusion is strengthened by the recent studies of Rannug on model compounds, such as S-2-chloroethylcysteine, which have been found to be much more mutagenic than EDC itself. Further reactions of the S-2-chloroethyl GSH conjugate may involve the sulfenyl chloride of GSH, a reactive compound of unknown mutagenic potential (Anders, this volume).

Metabolism of EDC by P450-mediated systems could lead to the highly reactive and mutagenic chloroacetaldehyde (McCann et al. 1975; Anders, this volume). This compound, after further transformation, could yield the chloroacetate metabolites reported by Yllner (1971) and others in the mouse. Consistent with this idea is the observation of Plotnick (this volume) that, in rats, the carcinogenicity of EDB is potentiated by disulfiram (Antabuse) which should cause haloacetaldehydes to accumulate. However, EDB may be metabolized by different routes than EDC due to the greater reactivity of the bromine as compared to the chlorine derivative. Thus, the question of the chemical nature of the mutagenic species remains unclear. It appears, however, that different mutagens may be generated in different strains or species depending on prevailing experimental conditions.

EDB

The mutagenicity of EDB is seen in Table 5. As can be seen here, EDB is much more mutagenic than EDC in the test systems studied. In all bacterial assays (Rannug et al. 1978; Brem et al. 1974), except for the phage λ test (Kristoffersson 1974), EDB was reported to be mutagenic. In the yeast gene conversion test (Murthy 1979) as well as in two of the plant systems (*Tradescantia* and barley), mutagenic activity was also found (Sparrow et al. 1974; Ehrenberg 1974). No effects, however, were seen using *Allium* (Kristoffersson 1974). Further positive results were seen in the sex-linked recessive studies reported by Vogel and Chandler (1974) in *Drosophila*.

Interesting studies of bull sperm morphology after the administration of EDB have shown an increased number of alterations in sperm head morphology that may correlate with the mutagenic potential of EDB (Amir et al. 1977). This data is discussed in more detail by Simmon (this volume). However, the results of the dominant lethal test, in which postimplantation mortality at midgestation in mice is ascertained, show no increase. The protocol for the dominant lethal assay involves the administration of the chemical to males 2-8 weeks before mating. Therefore, if spermatogenic aberrations occur that could result in abnormal embryos, then increased embryonic mortality should be observed. However, in the test using mice no differences were seen. This contrasts with the previous data in which an increase in morphological alterations of sperm heads was found in the bulls treated with EDB. The differences in results in these two reports can be ascribed to a number of causes, including different animals tested (bulls vs mice—bull sperm may be more sensitive), different concentrations of EDB (in the bull sperm study, Amir used 10 oral

Table 5
Mutagenicity of EDB

Test system	Mutagenic effects	References
Bacterial systems		
Salmonella TA1535	+	Rannug et al. (1978)
Salmonella TA1535	+	Brem et al. (1974)
Salmonella TA1530	+	Brem et al. (1974)
	+	Ames (1971)
E. coli pol A$^-$	+	Brem et al. (1974)
E. coli λ K39	−	Kristoffersson (1974)
Yeast gene conversion	+	Murthy et al. (1979)
Neurospora forward mutation	+	de Serres and Malling (1970)
Plant systems		
Tradescantia	+	Sparrow et al. (1974)
Barley	+	Ehrenberg (1974)
Allium	−	Kristoffersson (1974)
Drosophila sex-linked lethals	+	Vogel and Chandler (1974)
Cell systems		
Spermatozoa (bulls) morphology	+	Amir et al. (1977)
Dominant lethal (mice)	−	Epstein et al. (1972)
Lymphocyte cultures (human)	−	Kristoffersson (1974)

doses, each 4 mg/kg, and in the mouse dominant lethal study, Epstein used either one i.p. injection of EDB of 18 or 90 mg/kg or 5 oral doses, each of 2 or 40 mg/kg), or the fact that the dominant lethal test is an insensitive test for mutagenicity.

VDC

VDC is another halogenated hydrocarbon that has some mutagenic activity. Table 6 shows the results of a number of tests in which VDC has been tested and found to be mutagenic. In *Salmonella* strain TA100, using PB induction, positive results were reported (Bartsch et al. 1975, 1979). Greim et al. (1975) report a slight increase in mutagenic activity using *E. coli* K12.

    The dominant lethal test was performed in both rabbits and rats by chronic administration of the chemical in drinking water at 200 ppm or by inhalation at 150 ppm for 7 hours a day for a total of 9 days (Murray et al. 1979). In this report, species differences were noted as a slight increase in embryonic mortality in the rabbit embryos, whereas no differences were seen in the rat pups. These data imply the need for testing of a chemical in a number

Table 6
Mutagenicity of VDC

Test system	Mutagenic effects	References
Bacterial systems		
Salmonella TA100 plus S-9	+ PB induction and NADPH-dependent	Bartsch et al. (1975)
Salmonella TA100 minus S-9	+	Simmon et al. (1978)
Salmonella TA100 plus S-9	+	Bartsch et al. (1979)
E. coli K12	+?[a]	Greim et al. (1975)
Cell systems		
Dominant lethal		Murray et al. (1979)
Rabbits	+?	
Rats	−	
Cytogenetics		
Chinese hamsters	+?	Unpubl. results (cited in Maltoni et al. 1977)

[a] Slight increase.

of animal species before firm conclusions can be drawn. Cytogenetic studies in Chinese hamster cells in which 223 mg/kg of VDC was given in a single dose to the animals by oral gavage or 100 ppm by inhalation for a total of 6 weeks showed very minor changes (unpubl. results cited in Maltoni et al. 1977). These data, nonetheless, suggest mutagenic activity for VDC although the mechanism of its mutagenic action is not clearly understood. Data is lacking for this chemical in other mutagenicity test systems such as unscheduled DNA synthesis.

TCE

The mutagenicity of TCE is seen in Table 7. In this table, positive results were found in *Salmonella* (Simmon et al. 1977; Bartsch et al. 1979) as well as in *E. coli* K12 (Greim et al. 1975). In both the yeast gene conversion and the mitotic recombination assays, positive results were also reported (Shahin and von Borstel 1977; Bronzetti et al. 1978; Callen et al. 1980), as were those seen in the cytogenetic studies using human lymphocytes (Konietzko et al. 1978). In the latter report, 9 of the 28 people treated were found to be hypodiploid. In addition, one of the 28 people was found to have an extra small metacentric chromosome. Mutagenic activity was also reported in the mammalian spot test using mice (Fahrig 1977). In this test, an increase in brown spots was observed in the TCE-treated animals.

Table 7
Mutagenicity of TCE

Test system	Mutagenic effects	References
Bacterial systems		
Salmonella TA100	+?[a]	Bartsch et al. (1979)
Salmonella	+	Simmon et al. (1977)
E. coli K12	+?	Greim et al. (1975)
Yeast	+	Shahin and von Borstel (1977)
Yeast	+	Bronzetti et al. (1978)
Yeast	+	Callen et al. (1980)
Cell systems		
Human lymphocytes	+	Konietzko et al. (1978)
Mammalian spot test (mice)	+	Fahrig (1977)

[a] Slight increase.

PCE

PCE is a structurally-related compound that has been tested in both *Salmonella* (strain TA100, Bartsch et al. 1979) and in *E. coli* (K12, Greim et al. 1975) and negative results were reported in both systems. However, in yeast a positive effect was seen in both the gene conversion and the mitotic recombination assays using *S. cerevisiae* strain D7 (Callen et al. 1980).

EPILOGUE

Common metabolites have been found for four of the environmentally prevalent chloroethanes. Vinyl chloride, VDC, TCE, and EDC are apparently converted to monochloroacetic acid and via GSH conjugates to various cysteine derivatives and thioacids (Hathway 1977; E. S. Reynolds, pers. comm.). Although monochloroacetic acid was not mutagenic (McCann et al. 1975), the question of the mutagenicity of the other compounds is indeed a very important one that must be answered.

    It is important to consider the mutagenicity of EDC and structurally related compounds because of the widespread distribution of these compounds in our environment and because of the continual exposure of the population to them (previously reviewed by Fishbein 1976). Because of the positive results in the mutagenicity assays for EDC and some structurally related compounds, it is essential to reduce worker exposure to these chemicals to the extent feasible and to conduct periodic monitoring for genetic damage among exposed populations.

ACKNOWLEDGMENTS

Thanks are due to Robbie Allyn for his assistance in reviewing the literature.

REFERENCES

Abrahamson, S. and E. Lewis. 1972. The detection of mutation in *Drosophilia melanogaster*. In *Chemical mutagens: Principles and methods for their detection* (ed. A. Hollaender), vol. 2, p. 461. Plenum Press, New York.

Ames, B. N. and C. Yanofsky. 1971. The detection of chemical mutagens with enteric bacteria. In *Chemical mutagens: Principles and methods for their detection* (ed. A. Hollaender), vol. 1, p. 267. Plenum Press, New York.

Ames, B. N., F. D. Lee, and W. E. Durston. 1973. An improved bacterial test system for the detection and classification of mutagens and carcinogens. *Proc. Natl. Acad. Sci. USA* **70**:782.

Amir, D., C. Esnault, J. C. Nicolle, and M. Courot. 1977. DNA and protein changes in the spermatozoa of bulls treated orally with ethylene dibromide. *J. Reprod. Fertil.* **51**:453.

Bartsch, H., C. Malaveille, R. Montesano, and L. Tomatis. 1975. Tissue mediated mutagenicity of vinylidene chloride and 2′-chlorobutadene in *Salmonella typhimurium*. *Nature* **255**:541.

Bartsch, H., C. Malaveille, A. Barbin, and G. Planche. 1979. Mutagenic and alkylating metabolites of halo-ethylenes, chlorobutadienes and dichlorobutenes produced by rodent or human liver tissues. Evidence for oxirane formation by P450-linked microsomal monooxygenases. *Arch. Toxicol.* **41**:249.

Brem, H., A. B. Stein, and H. S. Rosenkranz. 1974. Mutagenicity and DNA modifying effect of halo-alkanes. *Cancer Res.* **34**:2576.

Bronzetti, G., E. Zeigler, and D. Frezza. 1978. Genetic activity of trichloroethylene in yeast. *J. Environ. Path. Toxicol.* **1**:411.

Callen, D. F., C. R. Wolf, and R. N. Philbot. 1980. Cytochrome P-450 mediated genetic activity and cytotoxicity of 7 halogenated aliphatic hydrocarbons in *Saccharomyces cerevisiae*. *Mutat. Res.* **77**:55.

deSerres, F. J. and H. V. Malling. 1970. Genetic analysis of *ad-3* mutants of *Neurospora crassa* induced by ethylene dibromide—a commonly used pesticide. *Environ. Mutat. Soc. Newsletter* **3**:36.

Ehrenberg, L., S. Osterman-Golkar, D. Singh, and V. Lundqvist. 1974. On the reaction kinetics and mutagenic activity of methylating and β-halogenoethylating gasoline additives. *Radiat. Bot.* **15**:185.

Epstein, S. S., E. Arnold, J. Andrea, W. Bass, and Y. Bishop. 1972. Detection of chemical mutagens by the dominant lethal assay in the mouse. *Toxicol. Appl. Pharmacol.* **23**:288.

Evans, H. J. 1976. Cytological methods for detecting chemical mutagens. *Chem. Mutagens* **4**:1.

Fahrig, R. 1977. The mammalian spot test (Fellfleckentest) with mice. *Arch. Toxicol.* **38**:87.

Fishbein, L. 1976. Industrial mutagens and potential mutagens. 1. Halogenated aliphatic derivatives. *Mutat. Res.* **32**:267.

Greim, H., G. Bonse, F. Radwan, D. Reichert, and D. Henschler. 1975. Mutagenicity *in vitro* and potential carcinogenicity of chlorinated ethylenes as a function of metabolic oxirane formation. *Biochem. Pharmacol.* **24**: 2013.
Hathway, D. E. 1977. Comparative mammalian metabolism of vinyl chloride and vinylidene chloride in relation to oncogenic potential. *Environ. Health Perspect.* **21**:55.
Kanada, T. and M. Uyeta. 1978. Mutagenicity screening of organic solvents in microbial systems. *Mutat. Res.* **54**:215.
King, M. T., H. Beikirch, K. Eckhardt, E. Gocke, and D. Wild. 1979. Mutagenicity studies with x-ray contrast media, analgesics, antipyretics, antirheumatics and some other pharmaceutical drugs in bacterial, *Drosophila* and mammalian test systems. *Mutat. Res.* **66**:33.
Konietzko, H. W. Haberlandt, H. Heilbronner, G. Keill, and H. Wechandt. 1978. Cytogenetische Untersuchungen an Trichloräthylen – Arbeitern. *Arch. Toxicol.* **40**:201.
Kristoffersson, Y. 1974. Genetic effects on some gasoline additives. *Hereditas* **78**:319.
Maltoni, C., G. Cotti, L. Morisi, and P. Chieco. 1977. Carcinogenicity bioassays of vinylidene chloride. *Med. Lavoro* **68**:241.
McCann, J., V. Simmon, D. Steitwieser, and B. N. Ames. 1975. Mutagenicity of chloroacetaldehyde, a possible metabolic product of 1,2-dichloroethane (ethylene dichloride), chloroethanol (ethylene chlorohydrin), vinyl chloride and cyclophosphamide. *Proc. Natl. Acad. Sci. USA* **72**:3190.
Murray, F. J., K. D. Nitschke, L. W. Rampy, and B. A. Schwetz. 1979. Embryotoxicity and feto-toxicity of inhaled or ingested vinylidene chloride in rats and rabbits. *Toxicol. Appl. Pharmacol.* **79**:189.
Murthy, M. S. S. 1979. Induction of gene conversion in diploid yeast by chemicals. Correlation with mutagenic action and its relevance in genotoxicity screening. *Mutat. Res.* **64**:1.
Nylander, P., H. Olofsson, B. Rasmusson, and H. Svahlin. 1978. Mutagenic effects of petrol in *Drosophila melanogaster*. 1. Effects of benzene and 1,2-dichloroethane. *Mutat. Res.* **57**:163.
Rannug, U., A. Sundvall, and C. Ramel. 1978. Mutagenic effect of 1,2-dichloroethane on *Salmonella typhimurium*. 1. Activation through conjugation with glutathione *in vitro. Chem. Biol. Interact.* **20**:1.
Rapoport, I. A. 1960. The reaction of genic proteins with 1,2-dichloroethane. *Dokl. Biol. Sci.* **134**:745.
Shahin, M. M. and R. C. von Borstel. 1977. Mutagenic and lethal effect of alpha benzene hexachloride dibutyl phthalate and trichloroethylene in *Saccharomyces cerevisiae. Mutat. Res.* **48**:173.
Shakarnis, V. F. 1969. Induction of X chromosome nondisjunction and recessive sex-linked lethal mutations in females of *Drosophila melanogaster* by 1,2-dichloroethane. *Sov. Genet.* **5**:1666.
Simmon, V. F., K. Kauhanen, and R. G. Tardiff. 1977. Mutagenic activity of chemicals identified in drinking water. In *Progress in genetic toxicology* (ed. D. Scott, B. A. Bridges, and F. H. Sobels), p. 249. Elsevier North-Holland, Inc.

Simmon, V. F., J. Kauhanen, K. Mortelmans, and R. G. Tardiff. 1978. Mutagenic activity of chemicals identified in drinking water. *Mutat. Res.* **53**:262.

Sparrow, A. H., L. A. Schairer, and R. Villalobos-Peitrini. 1974. Comparison of somatic mutation rates induced in *Tradescantia* by chemical and physical mutagens. *Mutat. Res.* **26**:265.

Vogel, E. and J. L. R. Chandler. 1974. Mutagenicity testing of cyclamates and some pesticides in *Drosophila melanogaster*. *Experientia* **30**:621.

Wurgler, F., F. Sobels, and E. Vogel. 1977. *Drosophila* as an assay system for detecting genetic change. In *Handbook of mutagenicity test procedures* (eds. B. J. Kiley, M. Legator, W. Nichols, and C. Ramel), p. 335. Elsevier Sci. Publ. Company, Amsterdam.

Yllner, S. 1971. Metabolism of 1,1,2-trichloroethane-1-2-^{14}C in the mouse. *Acta Pharmacol. Toxicol.* **30**:257.

COMMENTS

INFANTE: You have the positive results there for spermatozoa morphology in bulls and negative results for the dominant lethal. Is this a species difference, or is the dominant lethal test in the mouse insensitive, or is it more sensitive in the rat? What reliance could you have in such a test? If you have effects on the sperm, you don't see it on other . . .

FABRICANT: First of all, the dominant lethal test is not that sensitive a test system. However, I don't know of any other data in sperm with mice or rats. It very easily could be a species difference. Does anyone here know of any rodent data with sperm?

HINDERER: I think the question is not only directed towards the dominant lethal test, but is also directed towards whether or not sperm morphology has any meaning at all. A lot of assays in species models have a great deal of variability—this is a very gray area right now, so the answer could go either way.

FABRICANT: Well, the other issue was sperm morphology. In many cases there is only an increase of 1-3% sperm-head morphological aberrations, and that is considered positive, at least in the rodent system. We don't know what that means and those aberrations are heterogeneous in terms of shape. It's probably not only one type of single genetic damage.

BUSEY: Was that just straight semen evaluation on the bulls?

TER HAAR: It's reversible. In fact, in at least one of the studies the fertility wasn't quite that strong.

BUSEY: Did they do histopathology on the testes?

FABRICANT: No, I do not believe Amir did any histopathology studies, although both morphology and viability studies were conducted on ejaculation and epididymal sperm.

HOOPER: The dominant lethal test misses many compounds which are mutagens in the other short-term tests for mutagenicity, so a negative result in this test may not mean much.

FABRICANT: You miss an awful lot with this test.

HOOPER: You miss a lot of things that are mutagenic in other test systems.

HINDERER: None of these results are absolute, and that is the problem.

FABRICANT: What one could do, which would be more sensitive, is to look at cytogenetic abnormalities in embryos.

BUSEY: Well, semen evaluation doesn't necessarily correlate with dominant lethal testing.

FABRICANT: In certain cases it should. If there is a genetic aberration in the sperm, you could expect that sperm could fertilize the egg, and it could lead to mortality.

BUSEY: An abnormal spermatozoa does not necessarily mean a genetic aberration.

HINDERER: The correlation has never been shown.

BUSEY: There are an awful lot of things that cause abnormal spermatozoa in bulls. They're not necessarily a genetic problem absolutely.

FABRICANT: Yes. I think it can and can not be genetic. I think there are cases where this has been shown.

HOOPER: I'm not sure that such a study has ever been done. We are speaking of a correlation . . .

BUSEY: If you expose sperm to increased heat and you get abnormal sperm, that's not a genetic problem.

HOOPER: No, but there has to be a study done that looks at chemicals causing sperm abnormalities in mixed doses that cause dominant lethals. I don't think that study has been done. Sperm abnormalities are very poorly studied. Very few chemicals have been run.

FABRICANT: Wyrobek and Bruce did do a genetic study with sperm abnormalities, and found that in certain cases there was a genetic effect. That is, you could induce aberrations in mouse sperm in the F_1 generation, and that same aberration could be apparent in the F_2 generation. So there was a genetic change with certain chemicals.

HOOPER: That abnormality is transmitted.

FABRICANT: It can be transmitted with some chemicals. That was reported about 2 years ago.

Have either EDC or VDC been tested for effects on unscheduled DNA synthesis?

BUSEY: Have there been some subsequent tests with pure TCE where epichlorohydrin (ECH; 1-chloro-2,3-epoxypropane) was not used in it?

FABRICANT: Not that I know of.

WARD: The first two studies done in the mid-1960s, which Dr. Fishbein referred to, were positive. At that time they didn't know ECH was a contaminant. Subsequently there was a paper showing that pure TCE was not mutagenic.

FABRICANT: I am not aware of those studies.

REITZ: I have some additional data I would like to add to your excellent summary here regarding PCE and VDC. We've gone beyond the in vitro short-term tests. We made some in vivo measurements in rats and mice.

In the case of PCE and VDC, we have quantitated the alkylation of DNA that one observed in vivo as an indicator of mutagenicity, and this is orders of magnitude less than with other, genetically active chemicals. We have also measured unscheduled DNA synthesis in vivo with VDC and found that that is very low or absent. This may indicate that the chemicals don't have as much mutagenic potential in vivo as would be suggested by the in vitro tests.

FABRICANT: Thank you for that data. Is it published yet?

REITZ: No, it is in press in *Toxicology and Applied Pharmacology*.

KARY: Just one comment in regard to monitoring the worker population as a predictive tool. We need to have methods that are specific enough to suggest to us that the changes we are seeing are attributable to the compound.

FABRICANT: I agree, but I think that at this stage the only thing we can do is monitor the population with the tests that are available at this time. And in many cases, that's true, they may not be specific. We can, at least, monitor them and, where there are increases, make the appropriate changes—either by moving that person to an unexposed area, or perhaps encouraging them to work in another plant or in another facility. The strength of cytogenetic techniques is that we can monitor several concurrent chemical insults.

HOOPER: We need base-line data on cytogenetic effects, unscheduled DNA synthesis, sperm abnormalities, etc., to see what fluctuations there are each day, during a 1-month cycle, with smoking, etc. How difficult would it be to initiate a program to collect such data?

FABRICANT: It would be very easy. When a worker first comes to industry, that person could go through certain medical testing programs, including the cytogenetics study. The same cytogenetic studies could be repeated on that same person 6 months or 1 year later, with a continuing monitoring of the assays.

HOOPER: But if there are large daily fluctuations, such monitoring would be difficult. The sperm abnormality assay gives fluctuations with smoking, testes temperature, drug usage, diet, etc., and to get a significant result you may have to look at workers on a regular basis.

FABRICANT: I don't think that you would expect to have that much fluctuation. I think there are enough cytogenetic methods available now for the fluctuations that would occur. For example, there is a new technique to determine these fluctuations using gentian violet that increases the difference in structural chromosomal aberrations. Sister chromatid exchange (SCE), for example, is a very good indicator for some mutagens, and we can pick up very low level changes.

HOOPER: But on a day-to-day basis then, you don't see large changes. With sperm abnormality I think that you do see that. It's very sensitive to cigarette smoking, drugs, and diet, and to get a significant result, you have to look at people on a regular basis to fully document it.

FABRICANT: Well, I think with any kind of chromosome testing we really couldn't consider doing that so frequently. At this stage it would be financially difficult, too time consuming, and perhaps not even necessary if we were running concurrent controls.

HINDERER: Isn't it more appropriate to establish a model in an animal system? It would be more appropriate to try to establish that the event you are looking at has some risk significance in an individual animal. This would be a more useful tool.

FABRICANT: I agree with you completely. In fact, this has been accomplished for a number of years and we can now study these chemicals in both animal models and in man. A major disadvantage of the animal model, however, is that, in some cases, it may not be relevant to man.

RANNUG: One thing you can do is measure the degree of alkylation of hemoglobin in exposed people. Such a system has been developed by Ehrenberg and coworkers (Calleman et al. 1978) and gives a good estimation of the dose and thereby a possibility for a risk evaluation.

INFANTE: Are you saying (in response to the specificity question) that if you go in and characterize the environment and expose the workers to EDC that by measuring alkylation you have biological evidence that they are indeed exposed and you will then look at their chromosomes as well? How do you know, in terms of the specificity, that you have an elevated level of chromosomal breakage, and sometimes chromosomal aberration, and that then it is due to that agent?

Can't you do biological monitoring for EDC and see if the individuals with the elevated aberrations have elevated body burdens? Or perhaps you can do alkylation studies?

KARY: We can monitor the environment of the worker while he is on the job, but we cannot monitor the environment of the worker for that two-thirds of his life when he is not in the workplace environment. Recognizing that this is 1979 and that individuals are exposed continuously to a number of different agents, the need for specificity is all the greater, because two-thirds of the worker's life is off the job. What we really need is the specificity to link the changes with the chemical, and right now from what I see, we don't have that degree of specificity.

HOOPER: For cancer, with a long lag-time between the exposure and the cancer, we need an early-warning system. It may be true that the worker is on the job only one-third of his or her day, and the cancers that arise may result from exposures from the other two-thirds of the day. When there are exposures to carcinogens in the workplace, then we need methods of monitoring the worker population for DNA damage.

FABRICANT: At this stage, that would be the best thing to do.

JOHNSON: I think it would be very difficult to make changes in employment of individuals on that basis. You would have to take a group of 500 people and do simple straightforward chromosome studies. It would cost approximately $250 per person if done commercially. If it's done in-house one technician can do 100 employees a year.

FABRICANT: I agree with you that it's expensive. A first step would be to test 50 people exposed to EDC and 50 nonexposed controls.

JOHNSON: It's much more practical to study this group of workers and a control group. If this group is high then we'll maybe confirm that, and if it's persistent we'll say something is happening to this group and, engineering-wise, change their exposure.

FABRICANT: I think there's less variability doing an analysis of chromosome abnormalities than there is with the sperm morphology test—at least at this stage.

INFANTE: I don't think anyone proposed doing a general surveillance of the entire industry. Perhaps an ad hoc study should be done with a particular group that you're concerned about—take those individuals and see if there is an increase in some type of a precursor. What does it really mean in terms of a precursor state where you're going to prevent mortality?

FABRICANT: It certainly would be good to do that kind of study on an industry-wide basis. The best control, however, is still a concurrent control. This together with a previous study on the same individual will considerably strengthen any conclusions and would tend to minimize any false positive.

TER HAAR: What is the effect of alcohol in those systems? That's the toughest aspect for us to control for when it comes to VC. You tell somebody that they can't drink before they have this test . . .

HOOPER: Or cigarette smoking. Drinking or cigarette smoking may give changes in the number of chromosome aberrations.

GOLD: There is suggestive evidence from two studies. Lambert et al. in Sweden found significantly higher SCE frequencies among cigarette smokers than nonsmokers. Obe and Hera in Germany found that people smoking more than 40 cigarettes a day have several times more exchange-type aberrations in their peripheral lymphocytes than controls do.

INFANTE: I think this is a study design problem. How many people had colds, smoked, and drank alcohol—these categories occur in any appropriate control group. We always have to do this epidemiologically since we don't have a laboratory environment.

TER HAAR: I bring up that example because with VC we've got some real feeling about enzyme changes in the liver in relation to alcohol content. I believe most of us don't know what those things mean and yet we think that there's a relationship. We just can't make anything out of it.

HOOPER: How do we determine the health risk from a detected biochemical change in the liver that is related to exposure?

TER HAAR: We don't know what to do with these.

PLOTNICK: Are you familiar with the studies that showed an increase in hemangiosarcomas and earlier onset with VC exposure and ethanol in the drinking water?

TER HAAR: No.

PLOTNICK: You're talking about effects of alcohol and it apparently does increase the percentage of incidence.

References

Calleman, C. J., L. Ehrenberg, B. Jansson, S. Osterman-Golkar, D. Segerbäck, K. Svensson, and C. A. Wachtmeister. 1978. Monitoring and risk assessment by means of alkyl groups in hemoglobin in persons occupationally exposed to ethylene oxide. *J. Environ. Pathol. and Toxicol.* 2:427.

Metabolism of 1,2-Dihaloethanes

MARION W. ANDERS AND JOHN C. LIVESEY
Department of Pharmacology
University of Minnesota
Minneapolis, Minnesota 55455

Ethylene dichloride (EDC; 1,2-dichloroethane) and ethylene dibromide (EDB; 1,2-dibromoethane) are used in a large number of agricultural, industrial, and commercial processes (Fishbein, this volume). The use of EDC as an intermediate in the production of vinyl chloride accounts for its large annual production in the United States, which amounts to approximately 10 billion pounds (Infante, this volume). In addition to the large number of workers occupationally exposed to EDC, there is widespread exposure of the general population to EDC, since it is a contaminant of drinking water (Deinzer et al. 1978).

Both EDC and EDB are mutagenic in a variety of test systems (Fabricant; Hooper; Simmon; all this volume). Glutathione (GSH) and hepatic cytosol fractions are needed to convert EDC to a mutagen and the bile of rats given EDC, or EDB is mutagenic; these results suggest that a GSH conjugate of EDC or EDB is involved in the formation of a mutagen (Rannug et al. 1978; Rannug and Beije 1979; Rannug, this volume). Finally, EDC and EDB are suspected carcinogens (Olson et al. 1973; Ward, this volume), although evidence for a lack of carcinogenicity of EDC has been presented (Maltoni, this volume).

To better understand the relationship between the metabolism and toxicity of 1,2-dihaloethanes (1,2-DHE) (Hill et al. 1978; Rannug et al. 1978; Rannug and Beije 1979; Rannug, this volume), we have investigated their biotransformation to volatile metabolites. A summary of our earlier studies on the metabolism of 1,2-DHE to ethylene (Livesey and Anders 1979) and recent results on reaction mechanism studies of this biotransformation are presented in this paper. Finally, the overall metabolism of 1,2-DHE will be reviewed and related to the toxicity of these compounds.

MATERIALS AND METHODS

Chemicals

EDB (99%) and EDC (99+%) were purchased from Aldrich Chemical Company, Milwaukee. Deuterated (d_4)-EDB was obtained from Merck and Company/Isotopes, Los Angeles, and contained 99 atom percent deuterium.

S-(2-Chloroethyl)-DL-cysteine · HCl (CEC) was prepared according to the method of Carson and Wong (1964). cis- and trans-2-Butene (95%) were obtained from Matheson, Inc., Joliet, Illinois. meso- and racemic-2,3-Dibromobutane (DBB) were prepared according to the method of Young et al. (1929), and threo- and erythro-2-bromo-3-chlorobutane (BCB) were prepared according to the method of Hageman and Havinga (1966). The isomeric dihalobutanes were fractionally distilled at 50 torr, yielding materials having purities >98%, as determined by gas chromatography. Each synthetic product had a mass spectrum consistent with the structure. All other chemicals were of reagent grade and were obtained from commercial sources.

Enzyme Preparation and Incubations

Male Sprague-Dawley-derived rats (EDC or EDB experiments) or male Long-Evans-derived rats (dihalobutane [DHB] experiments) were used. Subcellular fractions of rat tissues were prepared by differential centrifugation (Livesey and Anders 1979) and, unless otherwise stated, hepatic cytosol was dialyzed overnight against 0.1 M sodium phosphate buffer (pH 7.4) at 4°C. Unless otherwise stated, incubation mixtures contained 50 μmole phosphate buffer (pH 7.4), 30 μmole GSH, 225 μmole substrate, and 6 mg cytosol protein in a total volume of 3 ml. When the pH dependency of the reaction was studied, 50 μmole of phosphate buffer was used from pH 6.5–8.0 and 50 μmole of Tris-HCl buffer was used from pH 8.0–9.0. Incubations were carried out in flasks closed with sleeve-type stoppers; shaking was done at 37°C for 30 minutes in an atmosphere of air. The reaction was stopped by placing the flasks on ice.

Analytical Methods

Ethylene was quantified via gas chromatography (GC) by injecting a 1-ml sample of the flask headspace gas through a gas-sampling valve and into a Varian series 1400 gas chromatograph, equipped with a flame ionization detector. Ethylene was separated from other components of the headspace gas on a 2 mm i.d. × 1.8 m glass column packed with Carbosieve B (Supelco, Bellefonte, Pennsylvania), 60/80 mesh, and maintained at 140°C. The injector and detector temperatures were 193°C and 220°C, respectively. A 100-ppm calibration standard of ethylene in nitrogen eluted in 1.5 minutes. cis- and trans-2-Butene were determined using a 3.2 mm i.d. × 1.8 m stainless steel column packed with 0.19% picric acid on 80/100 mesh Carbopack B (Supelco). The column and injector were operated at approximately 23°C; the detector was maintained at 150°C. Authentic cis- and trans-2-butene eluted in 9.5 and 10.0 minutes, respectively. Protein was measured according to the method of Lowry et al. (1951), with bovine serum albumin as the reference standard.

Gas Chromatography-Mass Spectrometry

A Finnigan model 3200E gas chromatograph-mass spectrometer (GC-MS) equipped with a 2 mm i.d. × 1.5 m glass column packed with Carbosieve B, 120/140 mesh, was used. The column was operated at 180°C. A Finnigan PROMIM multiple-ion monitoring device wa used to monitor the intensities of ions at m/e 26 and 30.

RESULTS

Metabolism of 1,2-DHE to Ethylene

The following is a summary of work published previously (Livesey and Anders 1979). The enzymic metabolism of EDC to ethylene was studied using dialyzed rat liver cytosol as the enzyme source. Ethylene formation was linear with time for at least 1 hour and with protein concentrations up to at least 8 mg of protein/ml. The reaction showed a temperature maximum at about 55°C. However, no distinct pH optimum was observed for the conversion of EDC to ethylene; instead the reaction rate increased as the pH was increased from 6.5 to 9.0. On the basis of these findings, the following conditions were employed for subsequent experiments: a 30-minute incubation at 37°C with 2 mg of protein/ml at pH 7.4.

EDC metabolism by rat liver cytosol was found to be independent of the presence of NADPH but highly dependent on the presence of reduced GSH; other thiols supported only low rates of ethylene production from EDC. The enzymes catalyzing the formation of ethylene from EDC were found to be present predominantly in the cytosolic fraction of hepatic tissue; nuclear, mitochondrial, and microsomal fractions exhibited only low rates of ethylene formation from EDC. Undialyzed cytosol prepared from rat kidney catalyzed the conversion of EDC to ethylene at a rate of about 50% of that observed in hepatic cytosol; cytosol prepared from lung, brain, or muscle showed little activity (<10% of that found in hepatic cytosol). The metabolism of EDC to ethylene was inhibited only by those reagents that react with sulfhydryl groups or are substrates for GSH S-transferases; p-chloromercuribenzoate and methyl iodide produced a marked inhibition, whereas metabolic inhibitors, such as cyanide or fluoride, showed no significant inhibition.

The substrate specificity was investigated using several vic-substituted ethanes (Table 1). The methanesulfonates showed only low reactivity and bromethylacetate produced negligible amounts of ethylene. Among the vic-dihalides, the halide order was observed and rates of ethylene production followed the order EDB > 1-bromo-2-chloroethane > EDC.

d_4-EDB was used to identify the source of the ethylene formed in the reaction mixtures. Figure 1 shows the mass fragmentograms of headspace gas samples in which EDB or d_4-EDB was incubated with rat liver cytosol. In the left panel, the peak at m/e 26, with a retention time of 3.6 minutes, corresponds

Table 1
The Metabolism of *vic*-substituted Ethanes to Ethylene by Rat Liver Cytosol

Substrate	Ethylene (pmole/min)	
	nonenzymic	enzymic
Ethylene-*bis*(methanesulfonate)	11.9 ± 1.0	8.3 ± 1.8
2-Chloroethyl methanesulfonate	6.2 ± 1.1	13.6 ± 4.7
Bromoethylacetate	0.4 ± 0.2	0
Bromoethylacetate + paraoxon	0.2 ± 0.1	0.4 ± 0.3
EDC	0.6 ± 0.2	26.8 ± 8.5
1-Bromo-2-chloroethane	5.9 ± 4.9	135.7 ± 55.6
EDB	323.5 ± 161.2	305.1 ± 87.6

Reprinted, with permission, from Livesey and Anders (1979).

Substrates and GSH were added to final concentrations of 25 and 10 mM, respectively. Paraoxon was added to a final concentration of 10^{-6} M. Nonenzymic reaction was determined by performing incubations of substrate with buffer in place of cytosol. Enzymic reaction is the difference between total reaction and nonenzymic reaction. Data represent the mean ± S.D. of three experiments.

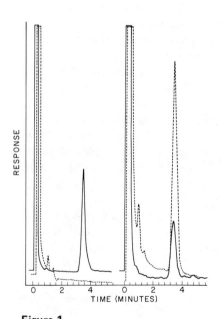

Figure 1
GC-MS analysis of ethylene produced by the metabolism of EDB (*left*) and d_4-EDB (*right*). (– – – –) m/e 30 ($C_2D_3^+$); (———) m/e 26 ($C_2H_2^+$ or C_2D^+). (Reprinted, with permission, from Livesey and Anders 1979.)

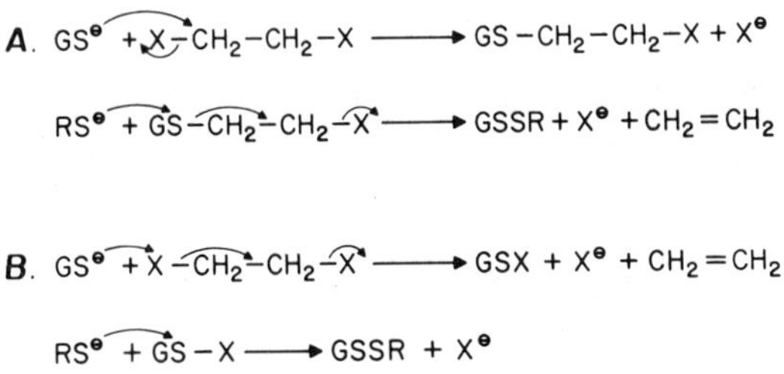

A. $GS^\ominus + X-CH_2-CH_2-X \longrightarrow GS-CH_2-CH_2-X + X^\ominus$

$RS^\ominus + GS-CH_2-CH_2-X \longrightarrow GSSR + X^\ominus + CH_2=CH_2$

B. $GS^\ominus + X-CH_2-CH_2-X \longrightarrow GSX + X^\ominus + CH_2=CH_2$

$RS^\ominus + GS-X \longrightarrow GSSR + X^\ominus$

Figure 2
Possible reaction mechanisms for the conversion of 1,2-DHE to ethylene. (Reprinted, with permission, from Livesey and Anders 1979.)

to the fragment $C_2H_2^+$ and is identical to that produced by authentic ethylene. In the right panel, the peaks at m/e 30 and 26 correspond to fragments $C_2D_3^+$ and C_2D^+, respectively. The ratio of these fragment peaks is in close agreement with the abundances reported for authentic d_2-ethylene.

Of the several reaction mechanisms that may be written for the metabolism of EDC to ethylene, that shown in Figure 2A involves the intermediate formation of S-(2-chloroethyl)-GSH. An analog of this compound, CEC, was tested for its ability to form ethylene. CEC conversion to ethylene was found to be independent of the presence of rat liver cytosol but was highly dependent on the presence of sulfhydryl compounds (GSH, cysteine, D-(-)-penicillamine or cysteamine).

Metabolism of 2,3-DHBs to 2-Butenes

The metabolism of *meso*- and *racemic*-DBB to *cis*- and *trans*-2-butene was also studied (Table 2). *meso*-DBB was almost exclusively converted to *trans*-2-butene; *racemic*-DBB gave rise to the *cis* isomer. Similar experiments were conducted using *erythro*- or *threo*-BCB as substrates. *erythro*-BCB was converted

Table 2
Metabolism of 2,3-DHBs to *cis*- and *trans*-2-Butene

Substrate	2-Butene		
	% *cis*	% *trans*	total (nmole/30 min)
meso-DBB	0.9	99.1	7.05
racemic-DBB	98.9	1.1	5.92
erythro-BCB	31.0	69.0	0.30
threo-BCB	89.0	11.1	0.25

Substrate and GSH concentrations were 30 and 10 mM, respectively. Incubation mixtures contained 6 mg of cytosolic protein/3 ml. Data represent the average of 2 experiments and are corrected for nonenzymic reaction.

to a mixture of *cis*- and *trans*-2-butene with the *trans* isomer predominating. *threo*-BCB was converted to a mixture of *cis*- and *trans*-2-butene, with the *cis* isomer being formed in greater amounts.

DISCUSSION

These results show that 1,2-DHEs are metabolized to ethylene by GSH-dependent cytosolic enzymes. GC-MS studies proved that the ethylene was derived from the 1,2-DHE. The subcellular location, cofactor requirements, and effects of inhibitors on the enzymes metabolizing 1,2-DHEs to ethylene are similar to those involved in the metabolism of halogenated methanes (Johnson 1966; Ahmed and Anders 1976) and suggest the participation of a GSH *S*-transferase (Jakoby et al. 1976). The involvement of a GSH *S*-transferase in the conversion of EDC to a mutagen has been demonstrated (Rannug et al. 1978; Rannug, this volume). The results also suggest that the first step in the reaction is the enzymatic formation of *S*-(2-haloethyl)-GSH, which undergoes a subsequent nonenzymatic conversion to ethylene. This conclusion is based on the observation that the analog, *S*-(2-chloroethyl)-D L-cysteine, was nonenzymatically converted to ethylene by a variety of sulfhydryl compounds.

At least two reaction mechanisms may be suggested to explain the biotransformation of 1,2-DHE to ethylene (Fig. 2). The first, shown in Figure 2A, involves the initial enzyme-catalyzed nucleophilic ($S_N 2$) attack of GSH on the electron-deficient carbon of the 1,2-DHE. The second step in the reaction involves the attack of GSH on the sulfur atom of *S*-(2-haloethyl)-GSH followed by β-elimination of a halide. The attack of sulfur on sulfur has been described in chemical and enzymatic studies (Oki et al. 1971; Hutson et al. 1976; Kosower and Kanety-Londner 1976). Alternatively, the conversion of 1,2-DHE to ethylene may be explained by an E2 elimination (Fig. 2B). Weygand and Zumach (1962) have suggested that this mechanism is involved in the oxidation of thiols by 1,2-diiodoethane.

To better understand the reaction mechanisms involved in the biotransformation of *vic*-dihaloalkanes to olefins, the metabolism of 2,3-DHB to *cis*- or *trans*-2-butene was investigated. These compounds have been widely used in chemical investigations to establish the mechanism of elimination reactions (Hine 1962; Saunders and Cockerill 1973). It was observed that *meso*- and *racemic*-DBB give rise to exclusively *trans*- and *cis*-2-butene, respectively. Furthermore, *erythro*- and *threo*-BCB were metabolized predominately to *trans*- and *cis*-2-butene, respectively. The *meso* and *erythro* isomers are stereochemically equivalent. The results obtained with the DBBs are similar to those obtained in the chemical systems using iodide as the nucleophile (Hine 1962; Saunders and Cockerill 1973) and suggest that an E2 elimination is involved (Fig. 3). In contrast, Sonnett and Oliver (1976) found that the iodide-catalyzed conversion of *erythro* and *threo* isomers of several *vic*-bromochlorides to olefins yielded *cis*- and *trans*-olefins, respectively, suggesting a $S_N 2$-E2 reaction

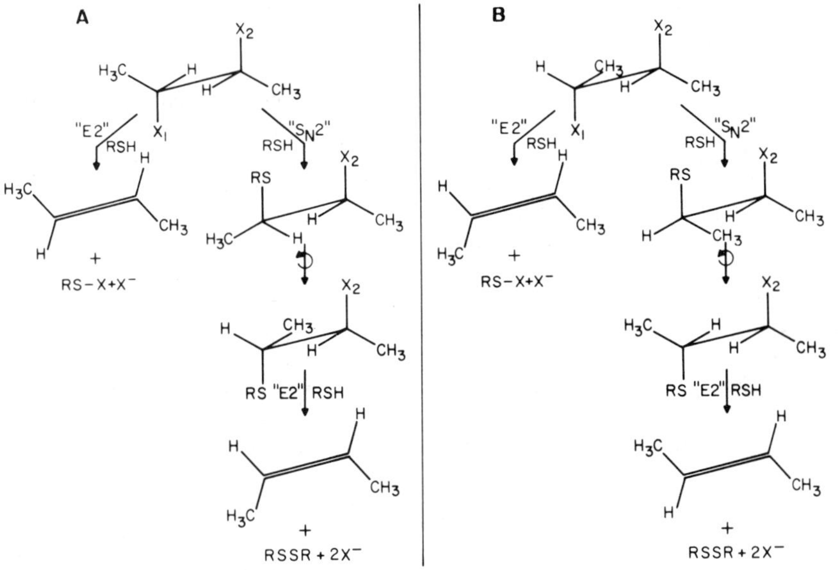

Figure 3
Possible reaction mechanisms for the metabolism of *meso-* or *erythro-*2,3-DHB (*A*) and *racemic-* or *threo-* (*B*) 2,3-DHB to *cis-* or *trans-*2-butene. For 2,3-DBB, $X_1 = X_2 = $ Br; for 2-bromo-3-chlorobutane, $X_1 = $ Br, $X_2 = $ Cl; *RSH* = glutathione.

sequence (Fig. 3). However, the enzymatic reaction shows a preference for the E2 elimination mechanism, since both *meso-*DBB and *erythro-*BCB are metabolized to mostly *trans-*2-butene.

It has also been reported that *meso-*1,2-dibromo-1,2-dideuteroethane is converted to *cis-*1,2-dideuteroethylene with iodide as the nucleophile (Schubert et al. 1955), suggesting a mechanism analogous to that shown in Figure 2A. If the results obtained with *vic-*dibromides in the chemical systems predict the mechanism of the enzymatic reaction, it would be expected that 1,2-DHE would be metabolized via a S_N2-E2 sequence (Fig. 2A). This would be consistent with the formation of the highly reactive episulfonium ion, which may be the reactive intermediate involved in the toxicity of 1,2-DHEs.

The metabolism of 1,2-DHEs has been well studied, and the metabolites identified in both in vivo and in vitro studies are listed in Table 3. Pathways to account for the formation of these metabolites are shown in Figure 4. The formation of inorganic halide (Heppel and Porterfield 1948) could occur as a consequence of attack by GSH or oxidative metabolism. In the former case, the expected intermediate would be *S-*(2-haloethyl)-GSH, which may be converted to ethylene (Livesey and Anders 1979) or, via the episulfonium ion (Fig. 4, shown in brackets), to *S,S'-*ethylene-bis(glutathione) or *S-*(2-hydroxyethyl)-GSH (Nachtomi 1970), formed by attack of GSH or water on the episulfonium

Table 3
Metabolites of 1,2-DHEs

Metabolite	Reference
Inorganic halide	Heppel and Porterfield 1948; Nachtomi 1970
S-(2-Hydroxyethyl)-cysteine	Nachtomi et al. 1966; Edwards et al. 1970
N-Acetyl-S-(2-hydroxyethyl)-cysteine	Edwards et al. 1970
N-Acetyl-S-(2-hydroxyethyl)-cysteine-S-oxide	Edwards et al. 1970
S-Carboxymethylcysteine	Yllner 1971
Thiodiglycolic acid	Yllner 1971
S-(2-Hydroxyethyl)-GSH	Nachtomi 1970
S-(2-Hydroxyethyl)-GSH-S-oxide	Nachtomi 1970
S,S'-Ethylene-bis(glutathione)	Nachtomi 1970
Chloroacetic acid	Yllner 1971
Bromoacetaldehyde	Hill et al. 1978
Ethylene	Livesey and Anders 1979

ion, respectively. S-(2-Hydroxyethyl)-GSH may undergo sulfoxidation to yield S-(2-hydroxyethyl)-GSH-S-oxide (Nachtomi 1970) or be further metabolized to S-(2-hydroxyethyl)-cysteine (Nachtomi et al. 1966; Edwards et al. 1970), which in turn may undergo sulfoxidation to yield N-acetyl-S-(2-hydroxyethyl)-cysteine-S-oxide (Edwards et al. 1970).

The oxidative metabolism of 1,2-DHE by cytochrome P450-dependent mixed function oxidases would be expected to yield a 2-haloacetaldehyde (Hill et al. 1978) as the initial product. This may be converted by dehydrogenases to a 2-haloacetic acid (Yllner 1971) or undergo attack by GSH and subsequent dehydrogenase activity to give rise to S-carboxymethylglutathione; metabolism of this intermediate by peptidases would yield S-carboxymethylcysteine (Yllner 1971). S-Carboxymethylcysteine may be further metabolized to thiodiglycolic acid (Yllner 1971).

There are at least two possible reactive intermediates formed during the metabolism of 1,2-DHEs that may be involved in the production of tissue damage. The episulfonium ion, which would be formed by GSH-dependent metabolism, would be highly reactive and may play a role in the mutagenic action of DHEs (Rannug et al. 1978; Rannug and Beije 1979; Rannug, this volume). Alternatively, the 2-haloacetaldehyde intermediate, formed during oxidative metabolism, would also be very reactive and it is thought to be involved in the covalent binding of 1,2-DHE metabolites to tissue macromolecules (Hill et al. 1978). Furthermore, chloroacetaldehyde has been reported to be mutagenic (McCann et al. 1975). Further studies are needed to elucidate the relative roles of these intermediates in the toxicity of 1,2-DHEs.

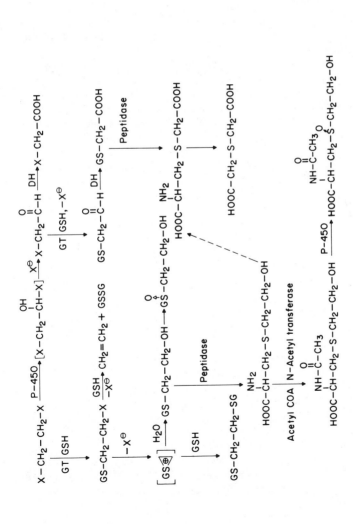

Figure 4
Metabolism of 1,2-DHEs. (GT) GSH transferase; (DH) dehydrogenase; (P-450) cytochrome P450.

ACKNOWLEDGMENT

This research was supported by National Institutes of Health grants ES 01082 and GM 07397.

REFERENCES

Ahmed, A. E. and M. W. Anders. 1976. Metabolism of dihalomethanes to formaldehyde and inorganic halide I. In vitro studies. *Drug Metab. Dispos.* **4**:357.

Carson, J. F. and F. F. Wong. 1964. The synthesis of L-1,4-thiazane-3-carboxylic acid 1-oxide. *J. Org. Chem.* **29**:2203.

Deinzer, M., F. Schaumburg, and E. Klein. 1978. Environmental health sciences center task force review on halogenated organics in drinking water. *Environ. Health Perspect.* **24**:209.

Edwards, K., H. Jackson, and A. R. Jones. 1970. Studies with alkylating esters II. A chemical interpretation through metabolic studies of the antifertility effects of ethylene dimethanesulphonate and ethylene dibromide. *Biochem. Pharmacol.* **19**:1783.

Hageman, J. J. and E. Havinga. 1966. The addition of bromine chloride to some cyclohexene derivatives. *Recl. Trav. Chim. Pays-Bas Belg.* **85**:1141.

Heppel, L. A. and V. T. Porterfield. 1948. Enzymatic dehalogenation of certain brominated and chlorinated compounds. *J. Biol. Chem.* **176**:763.

Hill, D. L., T.-W. Shih, T. P. Johnston, and R. F. Struck. 1978. Macromolecular binding and metabolism of the carcinogen 1,2-dibromoethane. *Cancer Res.* **38**:2438.

Hine, J. 1962. *Physical organic chemistry*, p. 208. McGraw-Hill, New York.

Hutson, H. D., D. S. Holmes, and M. J. Crawford. 1976. The involvement of glutathione in the reductive dechlorination of phenacyl halide. *Chemosphere* **5**:79.

Jakoby, W. B., W. H. Habig, J. H. Keen, J. N. Ketley, and M. J. Pabst. 1976. Glutathione S-transferases: Catalytic aspects. In *Glutathione: Metabolism and function* (ed. I. M. Arias and W. B. Jakoby), p. 189. Raven Press, New York.

Johnson, M. K. 1966. Studies on glutathione-S-alkyl transferase of the rat. *Biochem. J.* **98**:44.

Kosower, E. M. and H. Kanety-Londner. 1976. Glutathione. 13. Mechanism of thiol oxidation by diazenedicarboxylic acid derivatives. *J. Amer. Chem. Soc.* **98**:3001.

Livesey, J. C. and M. W. Anders. 1979. In vitro metabolism of 1,2-dihaloethanes to ethylene. *Drug Metab. Dispos.* **7**:199.

Lowry, O. H., N. J. Rosebrough, A. L. Farr, and R. J. Randall. 1951. Protein measurement with the folin phenol reagent. *J. Biol. Chem.* **193**:265.

McCann, J., V. Simmon, D. Streitwieser, and B. N. Ames. 1975. Mutagenicity of chloroacetaldehyde, a possible metabolic product of 1,2-dichloroethane (ethylene dichloride), chloroethanol (ethylene chlorohydrin), vinyl chloride, and cyclophosphamide. *Proc. Natl. Acad. Sci. USA* **72**:3190.

Nachtomi, E. 1970. The metabolism of ethylene dibromide in the rat. The enzymic reaction with glutathione *in vitro* and *in vivo*. *Biochem. Pharmacol.* **19**:2853.

Nachtomi, E., E. Alumot, and A. Bondi. 1966. The metabolism of ethylene dibromide in the rat. I. Identification of detoxification products in urine. *Isr. J. Chem.* **4**:239.

Oki, M., W. Funakoshi, and A. Nakamura. 1971. The reaction of α-carbonyl sulfides with bases. I. The reaction between α-carbonyl sulfides with thiolates. *Bull. Chem. Soc. Jpn.* **44**:828.

Olson, W. A., R. T. Habermann, E. K. Weisburger, J. M. Ward, and J. H. Weisburger. 1973. Induction of stomach cancer in rats and mice by halogenated aliphatic fumigants. *J. Natl. Cancer Inst.* **51**:1993.

Rannug, U. and B. Beije. 1979. The mutagenic effect of 1,2-dichloroethane on *Salmonella typhimurium*. II. Activation by the isolated perfused rat liver. *Chem.-Biol. Interact* **24**:265.

Rannug, U., A. Sundvall, and C. Ramel. 1978. The mutagenic effect of 1,2-dichloroethane on *Salmonella typhimurium*. I. Activation through conjugation with glutathione *in vitro*. *Chem.-Biol. Interact.* **20**:1.

Saunders, W. H., Jr. and A. F. Cockerill. 1973. *Mechansims of elimination reactions*. John Wiley & Sons, Inc., New York, p. 332.

Schubert, W. M., H. Steadly, and B. S. Rabinovitch. 1955. The stereochemistry of the debromination of *meso*-1,2-dibromo-1,2-dideuteroethane by iodide ion. *J. Am. Chem. Soc.* **77**:5755.

Sonnet, P. E. and J. E. Oliver. 1976. Olefin inversion. 2. Sodium iodide reductions of *vic*-bromochlorides and *vic*-dichlorides. *J. Org. Chem.* **41**:3284.

Weygand, V. F. and G. Zumach. 1962. Cystinpeptide durch Dehydrierung mit vicinalen dihalogeniden. *Z. Naturforsch.* **17B**:807.

Yllner, S. 1971. Metabolism of 1,2-dichloroethane-^{14}C in the mouse. *Acta Pharmacol. Toxicol.* **30**:257.

Young, W. G., R. T. Dillon, and H. J. Lucas. 1929. The synthesis of the isomeric 2-butenes. *J. Am. Chem. Soc.* **51**:2528.

Summary

ROBERT K. HINDERER
B. F. Goodrich Chemical Company
Cleveland, Ohio 44131

In our two days of discourse on ethylene dichloride (EDC; 1,2-dichloroethane), we have heard and discussed a wide range of information. This exchange has emphasized the economic importance of EDC and has underlined the basis for concern about its potential health risks.

The reasons for our interest in and concern about EDC are quite clear. Annual production in the United States is the range of 10 billion pounds per year. Furthermore, millions of people may potentially be exposed to EDC as the result of emissions during manufacture, processing, and gasoline distribution and as the result of other sources. Although studies of the health affects of EDC are conflicting, they do support the general concern about potential health risks. Present studies do not provide any evidence that EDC causes birth defects or reproductive effects. However, other studies suggest that EDC may pose a carcinogenic and a mutagenic hazard.

Drs. Maltoni and Ward reported on two separate studies of the effects of long-term exposure of rats and mice via inhalation and gavage, respectively. Although no carcinogenic effects were evident in the inhalation study, EDC was carcinogenic when force-fed. The numerous flaws in the latter study, including the lack of definitive information on sample purity, significantly clouded the interpretation and meaning of the positive response and prompted Dr. Maltoni to request the assistance of the National Cancer Institute (NCI) in repeating the gavage study in his laboratory. However, the results of this blemished study and the mutagenic responses noted in some short-term screens were suggestive evidence of a carcinogenic potential.

It is noteworthy that several other halogenated hydrocarbons such as vinyl bromide, vinyl chloride, ethylene dibromide (EDB; 1,2-dibromoethane), were described as being carcinogenic. The carcinogenic response of EDB in both long-term inhalation and gavage studies is particularly interesting in light of the different responses in the EDC studies mentioned above. Although the flaws in the long-term forced-feeding study may be important factors in the response, it is very likely that the difference in responses between the two studies are the result of species differences. Such species differences, in the case of trichloroethylene, may also account for the lack of a carcinogenic response reported by both Dr. Maltoni and Dr. Van Duuren, and the positive carcinogenic response observed in long-term inhalation and gavage studies sponsored by EDC manufacturers under the auspices of the Chemical Manufacturers Association and NCI, respectively. Dr. Maltoni has offered to repeat the NCI gavage study of EDC, but the question of species differences might be better addressed by a long-term inhalation study using the same species studied by NCI.

Another possibility that deserves mention is the effect of route of exposure. One might speculate that the differences noted are the result of different routes of exposure and ultimately different risks. However, Dr. Reitz reports that they have not yet found much difference in distribution and elimination of EDC between the oral and inhalation routes. Further investigations in this area are essential to our understanding of risk.

Although our understanding of the metabolic-pharmacokinetic handling of EDC is limited, several of our participants confirm that EDC is metabolized via a glutathione (GSH)-dependent system and that it is cleared rapidly from the body. Boyd et al. (1979) point out that high concentrations of GSH in the glandular stomach and liver of rodents may play an important role in gastric cancer induction.

Furthermore, both the biological status of the animal and certain chemicas are known to affect GSH levels. Such effects are particularly pertinent to EDC and may provide some insight into differences in carcinogenic responses.

Because of potential species differences, it is important to only conduct comparisons on studies which have used the same species. Furthermore, it is prudent to only compare data from the same laboratory if possible. In NCI cancer bioassays using the B6C3F1 mouse and the Osborne-Mendel rat, we see that EDC is less potent than EDB. Other studies which have been reported by Dr. Maltoni show that vinyl chloride (VC) produces cancer in the Sprague-Dawley rats while EDC does not. If we assume that different routes of exposure only affect risk and not its ability or propensity to cause cancer, then one can conclude that EDC is also less potent than VC.

Well, how does all this relate to human risk? In order to answer this question, we must consider both potency and potential exposure. Data presented by Robert Kellam of the Environmental Protection Agency indicates that 99+% of the estimated population exposures (expressed as an annual average) to atmospheric EDC are in the 10 to 990 parts per *trillion* range. When we compare these EDC exposures with one part per *million* (eight hour TWA) OSHA standard, the 10 part per *million* EPA emission standard, and the 10 part per *billion* California ambient standard for VCM, this suggests that the relative human risk is low.

REFERENCES

Boyd, S. C., H. A. Sasame, and M. R. Boyd. 1979. High concentrations of glutathione in glandular stomach: Possible implications for carcinogenesis *Science* **205**:1010.

Appendix

REITZ ANALYSIS

Analysis of NCI Retainer Sample of EDC

The National Cancer Institute (NCI) was requested to supply a sample of ethylene dichloride (EDC; 1,2-dichloroethane) employed in their bioassay so that it could be analyzed. However, although they agreed to do this, a sample had not been received in time to complete an analysis before publication deadlines. Consequently, a sample of EDC was obtained from Kim Hooper, University of California, Berkeley Campus. This material had been obtained by Dr. Hooper at an earlier date from NCI, and is reported to be from the lot of EDC used in the NCI bioassay. Great care was taken in the handling of this sample. It was only exposed to glass or Teflon® during transfer and analysis. Therefore, it is felt that the results reported here accurately reflect the composition of the sample obtained by Dr. Hooper from the NCI.

Experimental

EDC sample CR 58-17-2 #9300 was analyzed qualitatively by gas chromatography (GC) with flame ionization detection (HP 5700), gas chromatography-mass spectrometry (GC-MS) (LKB 90005, 70 eV) and gas chromatography-infrared spectroscopy (GC-IR) (Digilab FTS-10). Quantitative analyses were performed using GC with a flame ionization detector. EDC was assayed by comparison to a standard known to be 99.95% pure. EDC (10.0 μl) was added to 10.0 ml pentane (containing 5.0 μl undecane as an internal standard) through a Mininert® cap. Two standards and two samples were prepared and each was analyzed twice.

Impurities in the EDC were determined by external standard methods. Microliter amounts of known impurities were added through a Mininert® cap to 40.0 ml EDC standard.

A 2m × 2mm i.d. glass column packed with 10% SP-1000 on 100/120 CWHP was used for the quantitative separations. Helium was used as the carrier gas at 18 ml/minute. When assaying the EDC, the column temperature was maintained at 80°C isothermal. For determination of the minor components, the column temperature was held at 60°C for 8 minutes before programming at 8°C/minutes to 140°C, with a 4-minute hold at the upper limit.

Other column packings evaluated for this separation were 10% OV-101 on 100/120 CWHP, 10% Carbowax 20M on 100/170 CWHP, and 2.5% Oronite NIW on 60/80 Carbopack B.

Summary of Results

The EDC sample CR 58-17-2 #9300 was assayed by internal standard procedures to be 99.3% ± 0.23% (S.D., n = 4) pure. The impurities identified by GC-MS and quantified by GC are listed in the attached table. An authentic standard of cis-1,2-dichloroethylene was not available for direct comparison. Vinylidene chloride (VDC; 1,1-dichloroethene) and epichlorohydrin (ECH;

Table 1
Impurities in EDC CR 58-17-2 #9300

Compound	Retention time (min) (SP-1000)	Concentration (μg/ml)
Vinyl chloride (VC)	0.7	13
Unknown 1[a]	0.8	2
VDC	1.1	n.d.[d] (5)
Acetone	1.8	350
trans-1,2-Dichloroethylene	2.2	93
1,1-Dichloroethane	2.6	57
Unknown 2[b]	3.3	4
cis-1,2-Dichloroethylene[c]	4.9	69
Chloroform	5.9	2200
Unknown 3[c]	12.2	18
Unknown 4[c]	12.6	79
ECH	13.2	n.d.[d] (10)
1,1,2-Trichloroethane	15.0	77
Unknown 5[c]	17.3	10
1,2-Dibromo-3-chloropropane (DBCP)	25.9	40

Relative S.D. ± 10% for impurities. Assay by difference assumes all components were measured.
[a] Calculated using response for vinyl chloride.
[b] Calculated using response factor for 1,1-dichloroethane.
[c] Calculated using response factor for 1,1,2-trichloroethane.
[d] Not detected; detection limit in parentheses.

1-chloro-2,3-epoxypropane) were sought but not detected by GC or GC-MS at the indicated detection limits. A total of 0.3% of the sample was accounted for as chromatographically distinct impurities.

The presence of the two largest impurities, acetone and chloroform, as well as the identity of EDC were confirmed by GC-IR. Furthermore, no significant impurities were observed to coelute with EDC. GC separations using the other columns produced no unexplained peaks.

It should be kept in mind that this sample (~0.5 ml) had undergone at least three transfers prior to being received at this laboratory. Thus some losses or contamination are conceivable.

PLOTNICK ANALYSIS

An analysis of sample 9300 CR 5B-17-2, EDC utilized in the NCI bioassay of the chemical was completed. The procedures utilized and the results obtained are presented below.

GC Analysis

A 1.0 μl aliquot of the neat liquid received was injected into a Perkin-Elmer Model 3920B GC equipped with a 20 ft × 1/8 inch (o.d.) stainless-steel column packed with 10% SP-1000 on 80/100 mesh Chromosorb W, AW/DMCS. The injector and detector were each maintained at 200°C. The carrier gas was nitrogen at a flow rate of 25.0 ml/min. The column oven temperature was programmed from 80°C to 180°C at a rate of 2°/minute following an 8-minute delay at the lower limit. A flame ionization detector was employed to detect the materials emerging from the column. Under these chromatographic conditions, EDC exhibited a retention time of 16.15 minutes. Area analysis indicated a purity of greater than 98%. Thirteen other peaks were detected. The largest of these contaminant peaks had a retention time of 12.92 minutes.

GC-MS Analysis

1.0–2.0-μl aliquots of the neat liquid received were injected into a Hewlett-Packard model 5711 gas chromatograph interfaced with a Hewlett-Packard Model 5982 mass spectrometer to obtain the mass spectra of the principal component and impurities present. The GS column utilized was packed with the same material utilized in the GS studies presented above. The mass spectrum of the principal component was compared with the mass spectrum of an authentic sample of EDC obtained (MC/B Manufacturing Chemists, Norwood, Ohio). The two spectra obtained were indistinguishable, indicating that the principal component of the sample received is EDC. The largest of the contaminant peaks was tentatively identified as chloroform from the mass spectrum obtained. When an authentic sample of chloroform (MC/B Manufacturing

Chemists) was injected, the chloroform exhibited the same retention time and virtually the same mass spectrum as the principal contaminant.

Summary

The sample received has been identified as EDC with a purity greater than 98%. Thirteen contaminant peaks were detected by GC analysis utilizing a flame ionization detector. The principal contaminant was found to be chloroform. The other contaminants were not identified due to a lack of sensitivity of the mass spectrometer.

619250

3 1378 00619 2507

THE LIBRARY
UNIVERSITY OF CALIFORNIA
San Francisco
666-2334

THIS BOOK IS DUE ON THE LAST DATE STAMPER BELOW

Books not returned on time are subject to fines according to the Library Lending Code. A renewal may be made on certain materials. For details consult Lending Code.

14 DAY	14 DAY	14 DAY
JUN 30 1981 RETURNED JUN 30 1981	DEC -4 1983	NOV -7 1994 RETURNED NOV -9 1994
14 DAY NOV 28 1982 RETURNED DEC -2 1982	14 DAY DEC 18 1983 RETURNED DEC 13 1983	
14 DAY MAY 30 1983 RETURNED JUN -7 1983	14 DAY AUG 6 1987 RETURNED AUG 7 1987	

Series 4128